Contested Knowledges

Contested Knowledges: Water Conflicts on Large Dams and Mega-Hydraulic Development

Special Issue Editors

Esha Shah
Rutgerd Boelens
Bert Bruins

MDPI • Basel • Beijing • Wuhan • Barcelona • Belgrade

MDPI

Special Issue Editors
Esha Shah
Wageningen University
The Netherlands

Rutgerd Boelens
Wageningen University
The Netherlands

Bert Bruins
Wageningen University
The Netherlands

Editorial Office
MDPI
St. Alban-Anlage 66
4052 Basel, Switzerland

This is a reprint of articles from the Special Issue published online in the open access journal *Water* (ISSN 2073-4441) in 2019 (available at: https://www.mdpi.com/journal/water/special_issues/ Water-Conflicts)

For citation purposes, cite each article independently as indicated on the article page online and as indicated below:

LastName, A.A.; LastName, B.B.; LastName, C.C. Article Title. *Journal Name* **Year**, *Article Number, Page Range.*

ISBN 978-3-03897-810-7 (Pbk)
ISBN 978-3-03897-811-4 (PDF)

Cover image courtesy of Juan Pablo Hidalgo Bastidas, CEDLA, University of Amsterdam.

Contents

About the Special Issue Editors

Esha Shah is an Assistant Professor with the Water Resource Management Group at Wageningen University. She has held research and teaching positions at the Institute of Social and Economic Change (ISEC) in Bangalore, Institute of Development Studies (IDS) at University of Sussex, UK, and Faculty of Arts and Social Sciences at Maastricht University, the Netherlands. Between 2013 and 2015, she was a fellow at the Indian Institute of Advanced Study in Shimla. For 9 months, between October 2017 and October 2018, she was a fellow with the Nantes Institute of Advanced Study in Nantes, France. Her research interests include history and anthropology of science and technology, including water control infrastructure, debates on modernity and agrarian development in India, history and genealogy of development co-operation, debates on genetically modified crop biotechnology and, more recently, the role of subjectivity in shaping objective knowledge. https://www.wur.nl/en/Persons/Esha-E-Esha-Shah-PhD.htm https://www.researchgate.net/profile/Esha_Shah6.

Rutgerd Boelens is a Professor of 'Water Governance and Social Justice' at Wageningen University, a Professor of 'Political Ecology of Water in Latin America' with CEDLA, University of Amsterdam, and a Visiting Professor at the Catholic University of Peru and the Central University of Ecuador. He directs the international Justicia Hídrica/Water Justice alliance (www.justiciahidrica.org). His research focuses on political ecology, water rights, legal pluralism, cultural politics, governmentality and social mobilisation. Among his latest books are "Water Justice" (with Perreault & Vos, Cambridge University Press, 2018), "Water, Power and Identity. The Cultural Politics of Water in the Andes" (Routledge, 2015) and "Out of the Mainstream: Water Rights, Politics and Identity" (with Getches & Guevara-Gil, Earthscan, 2010). https://www.researchgate.net/profile/Rutgerd_Boelens/publications and http://www.cedla.uva.nl/20_research/researchers/pub_list/Rutgerd_Boelens_pub.html.

Bert Bruins is is a lecturer at the Water Resources Management group of Wageningen University. In addition to teaching, he was involved in a research program entitled 'Hydropower development in the context of climate change—Exploring conflicts and fostering cooperation across scales and boundaries in the Eastern Himalayas' and in various higher education capacity building projects (in Bangladesh, Benin, Ethiopia and Nepal). He is also the Programme Director of the BSc and MSc programmes on International Land and Water Management at Wageningen University.

Preface to "Contested Knowledges: Water Conflicts on Large Dams and Mega- Hydraulic Development"

Since the early 1980s, large dams have been at the centre of intensely contentious debates regarding their profound social and environmental impacts. All over the world, the design, construction and operation of large dams are amongst the most prestigious, but also the most sensitive and contested, developmental issues. As a result of the long-fraught conflict and controversy regarding large dams, which has spanned over two decades, the World Commission on Dams (WCD) was constituted. In 2000, it published its renowned report *Dams and Development: A New Framework for Decision-Making*. The WCD is an independent, international commission comprised of leaders from all sides of the debate; it has proposed guidelines for improving dam performance and governance by incorporating principles of participation, equity and transparency. The WCD has, so far, conducted the most rigorous evaluation of the role and impact of large dams on society and the environment, and its formation was accompanied by hopes that broad-based agreements would be forged to control the adverse effects of large dams on society and the environment. Despite the WCD process, the issue of large dams continues to remain highly contentious with respect to the conflict over providing hydropower, irrigation and flood control to limited segments of society while devastating the basic rights and livelihoods of many others and damaging shared rivers and ecosystems. More recently, the debates on climate change, increasing energy demand and challenges to the use of fossil fuels have prompted a renewed interest in hydropower, which is increasingly promoted as a source of clean energy. In addition, traditional development banks and developers have been steadily challenged by competitors from the private sector and the Capital invested by rising and powerful economies, such as China, for positions as primary investors in large dams in Africa and Latin America. These private actors barely follow the WCD guidelines and have adopted aggressively competitive approaches to investment in large dams, often in combination with the goal of strengthening their interests in extractive industries, such as large-scale mining and agribusiness. In many ways, these new forms of water grabbing and hydro-territorial transformation are far more destructive than the traditional funders' intervention practices, such as those of national governments and the World Bank, which at least provisionally adopted the WCD guidelines. As a result of this changing scenario, an upsurge of dam projects has been witnessed all over the globe and, in particular, in the global South. At the same time, the broader constellation of opponents of large dams have become more sophisticated in their understanding of the issues and modes of action. On the other side of the spectrum, a strong coalition of local and transnational environmental and social movements against large dams have emerged both in the Global North and South, bringing political awareness to even the most marginalised communities and empowering them to find new ways of engaging in and challenging decisions regarding dams. The nature of the debate now is more than the simplistic 'small versus big' binary of the past. In this edited volume, we argue that the debates around large dams increasingly involve disputes among different philosophical views. These encounters encompass fierce and complex interactions among diverse trans-boundary actors in dynamic, power-laden arenas involving, for example, multinational donor agencies, private consultancy firms, hydraulic construction companies, state bureaucracies, civil society movements, indigenous groups, environmental NGOs and different natural, as well as social, science activist-scholars. This edited volume addresses these contested philosophies in relation to a variety of large dams and mega-hydraulic projects from various parts of the world, from both the Global North and South. A diverse set of contributions in this edited volume discuss the mega-hydraulic dynamics of techno-scientific choices, discursive constellations and material transformations, as well as the diverse ways in which they interact with societal responses and alternative epistemologies and worldviews in specific ecological contexts. Last, but not the least, we would like to thank all the authors for their enthusiastic participation and thought-provoking contributions to the edited volume. They have helped us scrutinise the issues and reach a nuanced understanding of the socio-material and political-discursive networks and negotiations as they become manifest in the conceptualisation, design, development, implementation and governance of large dams and mega-hydraulic infrastructure projects. We also appreciate the editorial support of the Water journal.

Esha Shah, Rutgerd Boelens, Bert Bruins
Special Issue Editors

water MDPI

Editorial

Contested Knowledges: Large Dams and Mega-Hydraulic Development

Rutgerd Boelens [1,2,3,4], **Esha Shah** [1,*] **and Bert Bruins** [1]

[1] Water Resources Management Group, Department of Environmental Sciences, Wageningen University, P.O. Box 47, 6700 AA Wageningen, The Netherlands; rutgerd.boelens@wur.nl (R.B.); bert.bruins@wur.nl (B.B.)

[2] Centre for Latin American Research and Documentation (CEDLA), University of Amsterdam, Roetersstraat 33, 1018 WB Amsterdam, The Netherlands

[3] Faculty of Agricultural Sciences, Universidad Central del Ecuador, Ciudadela Universitaria, Quito 170129, Ecuador

[4] Department of Social Sciences, Catholic University Peru, Avenida Universitaria 1801, Lima 32, Peru

[*] Correspondence: esha.shah@wur.nl; Tel.: +31-317-484190

Received: 18 September 2018; Accepted: 3 January 2019; Published: 26 February 2019

Abstract: Locally and globally, mega-hydraulic projects have become deeply controversial. Recently, despite widespread critique, they have regained a new impetus worldwide. The development and operation of large dams and mega-hydraulic infrastructure projects are manifestations of contested knowledge regimes. In this special issue we present, analyze and critically engage with situations where multiple knowledge regimes interact and conflict with each other, and where different grounds for claiming the truth are used to construct hydrosocial realities. In this introductory paper, we outline the conceptual groundwork. We discuss 'the dark legend of UnGovernance' as an epistemological mainstay underlying the mega-hydraulic knowledge regimes, involving a deep, often subconscious, neglect of the multiplicity of hydrosocial territories and water cultures. Accordingly, modernist epistemic regimes tend to subjugate other knowledge systems and dichotomize 'civilized Self' versus 'backward Other'; they depend upon depersonalized planning models that manufacture ignorance. Romanticizing and reifying the 'othered' hydrosocial territories and vernacular/indigenous knowledge, however, may pose a serious danger to dam-affected communities. Instead, we show how multiple forms of power challenge mega-hydraulic rationality thereby repoliticizing large dam regimes. This happens often through complex, multi-actor, multi-scalar coalitions that make that knowledge is co-created in informal arenas and battlefields.

Keywords: mega-hydraulic projects; modernist traditions; knowledge arenas; manufactured ignorance; depoliticization; UnGovernance; dehumanizing rationality; multi-actor multi-scalar alliances; co-creation; power

---

1. Mega-Hydraulic Dams, Socioenvironmental Impacts and Knowledge Contestations

In 1984, Charles Perrow's book, *Normal Accidents,* was published, which has since become recognized as a classic. After its publication a series of major technological disasters occurred—the industrial chemical leak in Bhopal in India in 1984, the explosion of the US Challenger space shuttle and the Chernobyl nuclear accident in Russia in 1986. The book makes a persuasive case that we should view technological failures as a product of complex interacting systems. Perrow argued that multiple and unexpected failures are an inherent feature of complex technological systems; that these accidents are unavoidable, and even 'optimal technological designs' are unable to avoid them: they happen because of human error and organizational failure. For Perrow it is inevitable that such high-risk technological systems will eventually suffer what he called the 'normal accidents'.

This is not the place to rehearse the arguments of the book and its critiques; we evoke the book to provide a counterpoint to this special volume, which takes the opposite view to Perrow. Instead of examining the failure of complex technological systems as a matter of inevitability, and hence the end-product, we want to examine the front-end contestations of one such complex technological system that can be found all around the world: large dams and mega-hydraulic projects. Perrow takes the designs of high-risk technological systems as given. We argue that the back-end results, such as accidents, are not inevitable, and their conceptualization and production needs to be examined prior to the design process. In contradistinction to Perrow's project, we will examine how the conceptualization, design, production, implementation and operation of mega-hydraulic systems are focal points of epistemological contestation; how mega-technological systems are not a given but are socially produced and politically constructed; how mega-dams and the ways in which they are designed are not neutral, objective or 'optimal' but a biased and contextual response to particular problems that are framed and invented by a particular, and dominant, water culture. We will argue that large dam and mega-hydraulic knowledge schemes form the core of crucial power and knowledge encounters because they represent 'universalized' solutions that sideline many alternative arrangements. This special issue examines various facets of these knowledge contestations.

We propose that, explicitly or not, the design and planning of all mega hydro-technological systems involves conflicts between social groups and disputes among different knowledge regimes. In the history of debates on water governance in general, and large dams and mega-hydraulic systems in particular, such conflicts and the attendant contestation of knowledge have most conventionally been played out between water user groups (and/or dam-affected communities) and engineers or bureaucrats, for example, about a design of a dam, its location, or the allocation and distribution of water and electricity. These days, however, these conflicts around knowledge involve increasing numbers of arenas and transboundary actors, for example, multinational donor agencies, global policy institutes, international human rights courts, local and international civil society movements, indigenous groups, environmental Non-Governmental Organizations, media, government and bureaucratic agencies, experts and engineering schools, and independent scholars and activists. This makes the process far more complex. A recent example may provide an illustration:

Oroville—On 12 February 2017, Californian authorities ordered 180,000 people to evacuate their homes as the spillway of the Oroville dam was at risk of collapsing, threatening to release a deluge of water over large areas of land and thousands of houses. Built between 1962 and 1968, the 230 m high Oroville dam is the tallest in the United States. The structure is meant to protect downstream areas from floods and includes a hydropower plant.

How could this masterpiece of water engineering, that was meant to protect people from too much water, become the source of a looming disaster? The newspapers of February 2017 report that, in the first instance, it was due to extreme weather conditions—an 'atmospheric river'—that unleashed quantities of rainfall far beyond the storage capacity of the dam. As per the dam's design, the excess water was flowing over the dam's spillway. Many scientists warned that such extreme weather conditions could be expected to occur more often as a result of climate change.

While it was true that more water than usual was being discharged, the situation became critically dangerous because of a massive hole in the primary spillway and erosion in the emergency spillway. If not repaired, these could undermine the integrity of the dam. The structural state of the dam was, therefore, another cause of the threat of an emergency. Knowing that hydraulic structures like the Oroville dam have a design life of about 50 years, there are voices warning that many of America's great dams are gradually becoming obsolete and dangerous. These warnings have come from different directions. The American Society of Civil Engineers included Oroville as one of the 'high hazard' dams. As early as 2005 various environmental groups pleaded for reinforcements of the emergency spillway, but the government insisted that the dam was safe. The authorities averted the risk of dam failure by filling the hole and the emergency spillway with rocks, but it is clear that this was just a temporary measure.

Was building the Oroville dam justified in terms of historical, current and future socioenvironmental costs, risks and benefits? As all other large dams, Oroville opens our eyes to many thorny questions. How can we define and compare the, often noncommensurate, burdens and benefits? Is large dam building still a solution to claimed water scarcities, flood risks or energy shortages? Are large dams a green and safe response to multiple societal needs? Or, are they a manifestation of powerful economic, political, professional and institutional groups pursuing their interests? In such scenarios, which knowledge and truth claims are more legitimate, valid, appropriate and fair? Whose knowledge will count in the final analysis? These are not questions that should be left to engineers and politicians alone. A host of actors, including (to name just a few) environmental groups, the Native Americans who lived in the area before the dam was constructed, voters, taxpayers, local dwellers who build their homes and livelihoods near the dam site, international human and environmental rights courts—interact in this knowledge battlefield that will codetermine the future of Oroville.

The Oroville event is an illustration of the challenges and limits facing mega-dams. This special issue includes articles on the politics of contested knowledge as they become manifest in the development and operation of mega-hydraulic infrastructure projects in all continents. We witness situations where multiple knowledge regimes interact and conflict with each other, and hydrosocial realities are constructed using different grounds for claiming the truth. The concept of knowledge encounters is fundamental here. In this respect, as Norman Long argued, knowledge is a cognitive and socio-political construction that "results from and is constantly shaped by the experiences, encounters and discontinuities that emerge at the points of intersection between different actors' life-worlds ... It is entangled with power relations and the distribution of resources ... Hence, knowledge emerges as a product of interaction, dialogue, reflexivity, and contests of meaning, and involves aspects of authority, control and power" [1] (p. 30). Dam development interventions are focal points of intense knowledge battlefields, where entirely different, and often divergent, epistemologies interplay with, and confront each other: dominant 'expert' and multiple 'lay' forms of knowledge, beliefs and values, producing fierce struggles over their legitimation, validation, exclusion, and commensuration. In this sense, knowledge encounters around large dams combine, as objects of struggle, both the socioenvironmental issues at stake and the powerful contestation of knowledge regimes (see, e.g., [2,3]). We make particular use of Foucault's concept of power/knowledge not only to show how mega-hydraulic projects are planned, designed and constructed and what the socioenvironmental impacts of large dams are, but also to show how, and by whom, the very problems and their solutions are defined, and who has the power and/or knowledge to shape dam related techno-political realities (see also [4–6]).

While these knowledge confrontations and epistemological encounters materialize around the construction of dams they also go far beyond the geographies of the actual dam sites: the object of these encounters and conflicts is not just the site-particular, hydrotechnological artefacts or system. As Sanjeev Khagram argues in his work 'Dams and Development: Transnational struggles for water and power' [7], behind the intense struggles over large dams, mega-hydraulics and hydropower projects there are fierce struggles among competing visions of the past, the present and models of future societal development. "A range of powerful, transnationally allied groups and organizations have historically promoted the construction of these projects: politicians, bureaucrats, landed classes, and industrialists, multinational corporations, the World Bank, and other international organizations, as well as transnational professional associations of engineers and scientists" [7] (p. 3), which he calls an informal international "big dam regime". Underlying this regime are the deeply rooted values, norms and principles that, together, have promoted a development vision that was conceptualized nearly a century ago and which has been unleashed since the 1950s and 1960s (see, in this issue [8–17]). This vision equated development as the largescale, top-down, techno-centric pursuit of economic growth through the intensive exploitation of natural resources, that commonly disregards alternative knowledge systems, development trajectories and human suffering. Nehru's dams, 'temples of modernization' were presented as bringing development for all. "Hence critics of large dam building have had to challenge not only powerful interests and dominant institutions, but also hegemonic ideas about development in their struggles against big dam projects" [7] (p. 4).

For this reason, the contributions to this special issue highlight the multidimensionality of large dam projects; they include the power/knowledge dynamics of projects that extend far beyond 'just' water governance or water system development. In this issue, we are especially interested in addressing a number of questions. Firstly, we want to understand which (dominant and non-dominant) knowledge regimes are encountered, shaped, and validated in mega-hydraulic development. How do these knowledge regimes claim legitimacy and authority in concrete contexts? And, whose knowledge counts and whose knowledge is sidelined in particular conflict situations? We also want to examine how different "epistemic communities" are formed in relation to design, planning and construction. And, how race, class, caste, ethnicity and gender as well as professional identity interplay and influence the formation of epistemic communities around dams and mega-hydraulics. We also want to address the historical and contemporary processes of how both dominant and contested knowledge are formed, how they (interactively) shape norms, rules, beliefs and values about water problems and solutions, and how they become embedded in material artefacts. This includes the ways in which societal values (for instance, notions of justice, citizenship, progress, and democracy) are deployed and coproduced in the contested epistemologies on large dams and hydraulic infrastructural development projects.

In the following section we first discuss how mega-hydraulic projects, despite international critique and a temporary withdrawal of multilateral funding agencies, have recuperated their prominence worldwide. Here we also discuss how 'the dark legend of UnGovernance', as the epistemological backbone of mega-hydraulic regimes, makes invisible the world's diverse water cultures, imaginaries and hydrosocial territories. In Section 3 we show how this results in modernist regime practices that subjugate other knowledge systems and dichotomize civilization and backwardness. Sections 4 and 5 discuss how mega-hydraulic epistemic systems and methods depend upon depersonalized water planning models, whereby manufactured ignorance (including manufactured risks) is inherent to designing and implementing large dam projects. While showing the intricacies of how dominant mega-hydraulic rationality shapes values and artefacts, in Section 6 we also warn against romanticizing and idealizing local and indigenous knowledge, which paradoxically can seriously threaten dam-affected communities. Sections 7 and 8 detail how dam-related knowledge is typically produced in arenas that become battlefields; we examine how multiple forms of power are involved in the efforts to challenge and (re-)politicize large dam regimes and projects, increasingly manifested in and through complex, multi-actor and multi-scalar coalitions.

2. The Return of Mega-Hydraulics: Modernity and Control over Nature

Hydropower and other mega-hydraulic projects have long been a deeply controversial issue, generating intense local, national and transnational disputes among proponents and opponents. Large-scale water infrastructure development has been shown to generate profound social and environmental impacts, the more so since the burdens and benefits are unevenly distributed among population groups and locations [8–17]. Commonly, mega-hydraulic projects aim to supply water and/or energy to industrial growth sectors, large-scale capitalist export agriculture, and the growing thirst of mega-cities and urban zones [18,19]. As Duarte-Abadía et al. [20] (p. 244) note, mega-dams grossly change hydrological regimes and tend to irreversibly alter the ways in which local communities are able to make their livelihoods. Fainguelernt explains how, characteristically, the Belo Monte hydropower dam in the Brazilian Amazon, the third largest in the world and an icon of the modernist-developmentalist model, " … disrespects Brazil's environmental legislation and the rights of indigenous populations, who are considered 'hindrances' to economic development" [21] (pp. 257–258). In many places, people affected by a hydropower project, through dispossession, expropriation or resettlement—have been left on their own, bearing the burdens but receiving no benefits ([22–26]).

The grassroots, activist and academic worlds have not remained silent. Over the past few decades, mega-hydraulic interventions in local territories and hydrological regimes have received growing media attention, scholarly scrutiny and public critique, all denouncing the multiple ways in which they generate environmental damage and human suffering. They have also been characterized as icons of civil engineering, top-down and supply-side water resource development [27–31]. Mehta [32,33], Swyngedouw [34], Hommes and Boelens [35] and Lynch [36], among others, explain how this dominant techno-economic regime's narrow epistemic focus on a few issues such as 'solving water scarcity', actively legitimizes capital-intensive supply-side investments (the dams). Paradoxically, rather than solving water scarcities the regime often, not just discursively but also materially, actually produces them, by fiercely promoting the expansion of multiple, high-consumptive water use sectors (see also [37–41]). As Mehta argues: "The extra-basin transfer of water from large dams is often considered the only way to mitigate the problem of water scarcity in semi-arid and arid areas, despite the counter-evidence discrediting them ... Such narratives tend to serve certain socio-political agendas and/or reflect the worldviews of their advocates instead of being rooted in local realities ... Scarcity is essentialized and naturalized" [32] (pp. 2037–2038).

Despite several decades in which mega dams were subject to widespread critique on social and environmental grounds, which led for a while to the withdrawal of international policy institutes and funding agencies from backing such projects, hydropower and large dam development have recently made their worldwide comeback [19,42,43]. Amidst rising awareness about climate-change, hydropower dams, in particular, are presented as key solutions that support the transition towards new 'green economy' and 'clean development' [25,35,42,44,45]. Menga and Swyngedouw [5] show how the hydropower industry, which in the last two decades of the 20th century had become highly controversial, has re-established its dominance as the world's main renewable energy source since the early 21st century. This trend is most notable in emerging markets of the Global South. They cite the World Energy Council figures for 2015, which state that 76% of all renewable electricity comes from hydropower plants [5]. In a similar vein, Barbara Lynch [15,36] argues that " ... criticized in the 1990s for its initial investment in the Chixoy, Narmada, and other ill-conceived dam and hydro-electric projects, the World Bank moved away from lending for big infrastructure projects. But by 2003, it had returned to its eagerness to fund big dam and water transfer project developments" [36] (p. 74). As Hommes et al. [43] argue, dam development has been reinvented and reframed in a renewed, strongly depoliticized language of overall progress, sustainable, clean development and efficient, rational water management. "This disregards competing claims and conflicts over water, landscape and hydropower development and assorted interrelated struggles over socio-cultural issues, problem definitions, knowledge frameworks, ontological meanings, decision-making and preferred solutions" [43] (p. 11). One commonly finds that this powerful new mega-works discourse ignores the lessons of the past decades and neglects large dams' negative contribution to climate change [25,46]. At the same time, grassroots organizations, displaced communities and indigenous protestors are increasingly criminalized and violently oppressed for defending their land and water rights (see, e.g., [15,42,47–49]).

These dams are promoted by 'hydrosocial networks' consisting of State bureaucrats and politicians, engineering schools, transnational consultancy firms, construction companies, investors and funding agencies, who decisively prioritize mega-hydraulic development over interactively designed, context-specific and less expensive alternatives. They discredit proposals that facilitate local control of water resources, embedded cultural practices and sustain existing livelihoods [50–52]. They are part of, and contribute to, capitalist/market based economic growth and support the process of resource accumulation by elites [6,25,52].

The big dam regime builds on a modernist epistemological discourse, which is importantly founded on what Boelens [53] called the "Dark Legend of UnGovernance" (or "UnGov Legend"), involving a deep, often subconscious, neglect of the existence of multiple hydrosocial territories and diverse water cultures and societies. This untold legend claims that local water territories are basically unruly, consisting of " ... disorganized humans, irrational values, unproductive ecologies, inefficient resource use, and continual water conflicts. The UnGov Legend disfigures water societies by overlooking water users, meanings, values, identities, and rights systems on the ground. It then constructs its own water users, with identities that conveniently fit the models, with needs and rationales matching the imaginations of those in power, shored up in their science, technology and policy towers ... " [53] (pp. 7–8). The UnGov Legend presents mega-hydraulic projects as being entirely coherent, benevolent, shedding light in the darkness and bringing rational order to the water regime. This means that modernist water governance models, as national and international mega-hydraulic policies and plans and State- and market-based water laws, do not need to adapt to the realities and practices of local populations. On the contrary, it is these local populations who need to adapt, not the plans. These modernist models seek to create their own, utopian, water world [39].

In a similar vein, Worster [31], Blackbourn [54], Kaika [29], Swyngedouw [55], and Boelens and Post Uiterweer [50] relate how, following 'the dark ages' of feudalism, colonialism and other forms of dystopian, structural, chaos, large dam building projects are presented as 'utopian techno-political arms' that will bring order to a disordered society and nature, and help to build new nations. Nixon states: "Megadams served as highly visible, spectacular statements that new nations were literally soaring toward development, by mastering rivers and reaching for the sky. Constructions on such a scale rendered material the trope of nation building: to erect a megadam was literally to concretize the postcolonial nation's modernity, prosperity, and autonomy. Each dam was simultaneously an act of national self-assertion—independence writ large across the landscape—and an act of natural conquest" [56] (pp. 65–66). Nixon [57] and Watts [58] also refer to the dominant, dark legend that legitimizes the epistemologically imagining of communities and the physical transformation of territories to fit the new extractivist, order that supports the construction of mega-dams. The flipside of this is that it also involves the active "un-imagination" and erasure of communities, knowledges, places and livelihoods that do not fit the model. "The idea of the modern nation-state is sustained by the production of imagined communities but also by the active production of unimagined communities", that is, "communities whose vigorously unimagined condition become indispensable to the maintenance of a highly selective discourse of national development" [56] (p. 62). The Dark Legend of Ungovernance [53] converts inhabitants of so-called 'hydrological zones' into "uninhabitants [...], the convergent unruliness of 'irrational' river people and an 'irrational' river must be straightened out and channeled into a national culture of rational development. We thereby witness a combined assault on an 'unregulated' river and purportedly 'lawless' people ... " [56] (p. 74).

Other mainstays of the big dam regime's modernity project also merit scrutiny. There are, obviously, "multiple modernities" [59] and the dam regime, its imaginaries, designs and practices are not monolithic: large-scale dam development's techno-political-economic power grid requires, builds on, and fosters 'thinking globally' but in order to be successful its need to mediate and engage with diverse localities, producing complex and sometimes divergent capitalisms. This means that modernization pushes in particular directions that steer hydro-territorialization in response to locally prevailing structural forces, contingencies, multi-scalar relationships, multiple knowledgeable social actors [1,60] and their complex interaction with socio-natural actants [61].

Nevertheless, we can find commonalities and characteristic features within modernist large dam development. The neglect or dismissal of the old, the past and of existing cultural and ecological diversity is one key element in constructing the modernist discourse. The emphasis on man's agency and ability to actively shape the physical and social environment and ability to construct their life and water worlds is another fundamental aspect. This reflects (and advances) the passage from one development stage to the next improved one, as epitomized in in 1960 in Rostow's classic (but widely discredited) 'stages of development' [62], where 'natural states of underdevelopment' simply require a big push to fast track onto a linear trajectory of evolutionary modernist development. The realization of large dam projects heralds a step towards civilization, which radically both changes the physical environment and reconfigures 'primitive' social structures. This view not only sees traditional societies as in need of being modernized, but sees nature as the Other, non-human, disordered, and savage: something that needs to be conquered, colonized and subjected to humanity's will and benefit [29,31,54,63,64]. One implication of this is that the mega-hydraulic modernist discourse inherently entails an epistemological and ontological divide between society and nature [61,65].

Other intrinsically related features of the modernist paradigm in large dam development include:

- 'De-rooting' the past, and ahistorical views that stress 'making a break' and discontinuity (in order to achieve development);
- The deep-grounded notion of the plannability of socio-natural futures;
- The need and possibility of reducing diverse cultural meanings, values, language and knowledges to a single rubric, in order to arrive at one common metric ('commensuration');
- The objectification of social values and relationships and the calculability of societal choices and preferences in order to derive socially engineered optimal outcomes;
- The deployment of instrumentalist rationalities that enable a universalist water governance culture, and;
- The commodification of nature and society in order to justify large-scale hydro-territorial development.

In addition, as Polanyi [66] argued, the modernist paradigm also assumes that the formal economy exists separately and independently of the societies and cultures in which it is embedded, while it can and should be based on the principles of extractivism, appropriation and productivism [64,67,68].

In line with Scott's notion of 'high modernism', the modernist large dam development paradigm entails a local-global project founded on "supreme self-confidence about continued linear progress, the development of scientific and technical knowledge, the expansion of production, the rational design of social order, the growing satisfaction of human needs, and, not least, an increasing control over nature" [69] (p. 89). This rational design of social, political and cultural order, commensurate with the laws of natural science, entails standardizing the subjects of development and eliminating attributes that are considered "situated", "deviant", and "contextual" [70]. Modernist water governance and dam development, thereby, entail power relationships and strategies that seek to produce a particular hydrosocial order.

Therefore, hydromodernity is an inherently socio-territorial project [44,52,71]. As Hommes et al. [43] argue, mega-hydaulic projects " . . . entail discourses implying concealed efforts to reconfigure existing socio-natural relationships and implant new meanings, values, distribution patterns and frames of rule-making and alignment; they aim to build profoundly new 'territory' matching powerful ruling group interests to self-governing citizens" [43] (p. 18). Large dam schemes connect the social and the natural in specific ways, and they construct precise patterns as to how water should be stored and distributed and how humans and non-humans need to be ordered in socio-technical hierarchies. They legitimize this through moral and symbolic orders that strengthen the status quo [52,72]. In other words, mega-hydraulic projects fundamentally (re-)configure hydrosocial territories (Boelens and colleagues [73] (p. 2) conceptualize the hydro-social territory notion as "the contested imaginary and socioenvironmental materialization of a spatially bound multi-scalar network in which humans, water flows, ecological relations, hydraulic infrastructure, financial means, legal and administrative arrangements and cultural institutions and practices are interactively defined, aligned and mobilized through epistemological belief systems, political hierarchies and naturalizing discourses"). This is not just a 'social affair': mega-dam based reterritorialization projects entail efforts to embed these new knowledge contents, principles, social–political norms, morals and hydro-cultural relations in material infrastructure, artefacts and technological network relationships. That means that these projects involve the 'moralization' of (hydro-)territorial infrastructures. As 'governmentality' endeavors, " . . . dominant hydrosocial configurations commonly curtail local sovereignty and create a political order that makes these local spaces comprehensible, exploitable and controllable" [43] (p. 11).

In fact, large dam knowledge encounters powerfully produce new (and always contested) social and symbolic materialities (cf. [74–79]).

3. The Civilized Self and the Backward Other: Battlefields of Modernist and Subjugated Knowledges

Throughout the world, the conquering of nature and the ordering of humans and ecology through large dams, reservoirs, tunnels and canals, to take water from the 'backward rural areas' to 'modern urban metropoles', has invoked utopian-inspired exaltations of modernity, civilization and progress [31,45,80]. Several papers in this special issue [9,11–17] detail how mega-hydraulic projects shape and materialize visions about technological and economic modernization, the mastery of territories and natural resources, and, as Teräväinen frames it, enhance the nation-state's depoliticized, techno-scientific knowledge base.

Lena Hommes describes how in Peru, for decades, water transfers from the remote highlands to the desert city of Lima have been framed in terms of enlightened utopianism, combining hydraulics, a civilizing mission and the disciplining of nature and humans (cf. [81–83]; see also: [31,44,50,84–86]). One of the national newspapers glorified Lima's water works in the following way: "Men's labour has dominated the landscape and regulated raging torrents. Works of civilization in its most exact sense: dominance and utilization, so the true conquest for the community's benefit. [. . .] Great victory for men, their science and determination!" (all quotes from [35,45]). The hydropower company adds that: " . . . transforming the dramatic topography of the Andes—a hostile barrier to Peruvian man and his life needs—is an idealistic goal and driving force of progress for a beautiful metropolis". In the same way, the Swiss co-financer praises the "technical means that liberate men . . . giving them a better opportunity to develop their personality and soul, to become a higher class of human being, more intelligent and technically better equipped to live peacefully in the future". Mega-hydraulic designs *'liberate by dominating'*—Peru's water engineering echoes George Orwell's linguistic inversions in 1984 Newspeak—they domesticate landscapes and water, but also Andean highland villagers. As Hommes' studies show, to justify transforming the highlands of the Andes, affected communities are presented as in need of civilizing. Hydropower's mission is to tame savage waters (the enemy) and save rural people from the dark, as an admirer of the water transfers and hydropower development expresses " . . . to enrich Lima's watershed from the virgin highland sources, full of wild waters and lagoons. A tremendous, frightening battle against an enemy that wouldn't bow down: the water behaved in outright confrontation with those water seekers—a heroic deed by the technicians" (from Buse's 1965 book «*Huinco 200,000 KW*», [35] (p. 71)).

Modernist firm belief in dichotomies, which also finds expression in opposing the highlands' abundance to the civilized megacity's lack of water, justifies largescale water extraction. Obsolete, leaking water infrastructure inside the city itself and huge water abundance, squandering and over-allocation to Lima's elite sectors—with filled swimming pools and intensively irrigated parks—is discursively obscured. Water scarcity, politically constructed in poor neighborhoods, is presented as nature's fault and a result of the surrounding desert, climate change and weather conditions, all of which legitimize mega-hydraulic water transfers that would supposedly 'benefit all'. And while massive investments of cash and engineering skills are made in channeling water to Lima, benefits for local highland communities remain limited [35,45].

Large dam regime's modernist dichotomies not only divide nature and society by portraying nature as the savage Other; in order to justify water extraction and territorial transformation, there is also a need to dichotomize rural backwardness and city life's civilization. As Hommes describes, the hydropower company's brochure contrasts a power line, symbolizing progress, with a llama, symbolizing Andean communities' backwardness [45]. This follows the historical discourse in which Indians are represented as 'talking llamas' [4]. A book published by Lima's drinking water utility, entitled *Land of the Lagoons*, reflects this ingrained paternalism: "Living in a natural paradise, these communities are at a distance from our country's reality and necessities. Accordingly, they showed indifference towards the great project that will benefit the regions of Lima and Callao with new water sources [...] Nevertheless, explaining the project's kind-heartedness conquered their resistance" (quoted by Hommes and Boelens [35] (p. 77)).

More generally, since the mid-20th century, development institutions and policies have emphasized this dichotomy between subject and object to justify 'superior' modernist knowledge/technology interventions; for this, they even created the subject as an object of intervention. As Michael Kearny argues: "With the disappearance of 'the primitive', 'the peasant' increasingly came to typify the generalized Other, but an Other seen not as primitive nor primordial but as 'underdeveloped'. This 'underdeveloped peasantry' thus became an inversion of 'the modern,' a new objectified and contrasting Other . . . " [87] (p. 35). In a similar vein, Ivan Illich [88], in his famously provocative style, once typified the stages in which the Other's values, needs and (non)knowledge have been constructed historically by the West, not with the intention to reflect reality but to foster the latter's own political projects. "Each time the West put a new mask on the alien, the old one was discarded because it was now recognized as a caricature of an abandoned self-image. The pagan with his naturally Christian soul had to give way to the stubborn infidel to allow Christendom to launch the Crusades. The wild man became necessary to justify the need for secular humanist education. The native was the crucial concept to promote self-righteous colonial rule. But by the time of the Marshall Plan, when multinational conglomerates were expanding and the ambitions of transnational pedagogues, therapists and planners knew no bounds, the natives' limited needs for goods and services thwarted growth and progress. They had to metamorphose into underdeveloped people" [88] (pp. 94–95) (see also [89]).

Indeed, throughout history, similar cultural-political constructs have been purposely invented for subordinating (or, on the contrary, reifying) the Others' identities, property relationships and knowledge systems. This prepared a valuable way for experimenting with how to organize people and property: presenting a civilized Self versus a barbarian Other. The West was unaware of non-Western governance and knowledge frames and so invented myths about them, to justify invasion and to introduce order to the ungoverned. The actual people and their forms of natural resource governance were conveniently ignored.

Ignorance of the diversity of governance and knowledge forms involves erasing localities' place-making, place-experience and meaning-giving—"local knowledge is a mode of place-based consciousness, a place-specific (even if not place-bound or place-determined) way of endowing the world with meaning" [90] (p. 153). Or, as Illich tellingly explained in *II₂O and the Waters of Forgetfulness* [91], the vernacular understandings and meanings of 'water' are always related to 'dwelling' and (highly dynamic) place-based experiencing. Similarly, Tim Ingold talks about local knowledges as "dwelling perspectives" [92] (pp. 153–154). By contrast, as we have outlined above, modernist mega-hydraulic discourses depict locally existing systems as representing disorder, ignorance and lack of governance. Characteristically they deny existing places (or, as Lefebvre [93] and Bauman [64] argue, they aim to replace them by an empty or fluid 'space'). As Barbara Lynch commented on the Chixoy Dam Project in Guatemala, " . . . in the process of (the) production of space, place was erased, first conceptually and then literally. Components of place that were obliterated in the transformation of the project area included its connection to ancestors, sacred elements in the landscape, the knowledge that resides in (the) landscape and its features, relations and networks of economic interaction, and knowledge about safety and danger" [70] (p. 15).

But beyond just aiming to obliterate such vernacular, locally embedded knowledge, modernist governance ideologies and strategies are Janus-faced: in addition to stressing the "radical differences" with the communities that need to be displaced and unimagined, they strategically and simultaneously adopt a liberal discourse of integration and participation, which *will include* peasants, indigenous and lay cultures and knowledge as "potentially equals" [4,53]. They are potentially equal, have the right to be equal, and should be equal. This complementary face of modernist water policies is based on 'equalizing expansionism', not on violent conquest but on universal water rationality.

Modernist water science and policy projects churn out recipes by 'certifying good water use': new water knowledge, rules and identities for becoming equal. Water cultures are judged (and made to self-evaluate) how well they meet these standards. Failure to meet these 'self-evident' principles is presented as a lack of capacity for reason and unwillingness to progress. Modern water resource policies promise to accelerate 'progress' through planned development and guarantee control over nature through advanced science; material wealth through superior water technology; and effective, good governance through the rational organization of water users. The idea is that local imperfections and inefficiencies, just like cultural differences, will disappear as people realize the effectiveness of rational, modern experts' capacity to meet water development needs.

In this Foucauldian game, mega-hydraulic interventions and water training projects exercise power that constantly generates new water knowledge. And, in turn, official water knowledge continually reinforces powerful hydro-political configurations. As Foucault stated, power and knowledge depend on each other: power cannot be exercised without knowledge, and knowledge necessarily engenders power [94] (p. 52). In this politics-of-truth, modernist water science, water governance and mega-hydraulic policy-making produce permanent, clear results, separating legitimate forms of water knowledge and rights from illegitimate ones. Power, thus, makes claims on reality, knowledge and truth, and even determines the ways in which 'truth is made true'. Foucault: "Truth is to be understood as a system of ordered procedures for the production, regulation, distribution, circulation and operation of statements. Truth is linked in a circular relation with systems of power which produce and sustain it, and to (the) effects of power which it induces and which extend it. A regime of truth" [94] (p. 133). Thus, through endless 'degrees of validity', valid water knowledge—although profoundly normative—is objectified and judged according to its deviation from the (hidden) norm or standard: efficient water use, effective infrastructure, productive irrigation systems, rational water rights, equitable water allocation, best watering practices, democratic water governance, sustainable water development, modern water users, and so on [4]. This entirely depoliticizes 'truthful knowledge', and the agents and relations that set the standards. The language of modernist mega-hydraulics actively subordinates the variety of existing knowledge claims about, and practices related to, water control and territorial rights to its functionalist, universalist, epistemology. Foucault [94] (p. 82) would frame the latter as sets of "subjugated knowledges" that are disqualified, hierarchically seen as inferior, and noncompliant with the required levels of cognition or scientificity. They are to be invaded and re-arranged by expert-thought and scientific/formalist thinking.

Foucault [95] identified four interrelated ways through which dominant knowledge projects discipline and control existing knowledge regimes and eradicate "false or non-knowledges" (that is, the non-compatible and non-commensurate knowledges): (1) selection; (2) normalization; (3) hierarchization; and (4) centralization. These processes are highly applicable to mega-hydraulic interventions in local, complex hydrosocial territories. Large-scale hydraulic projects that entail knowledge encounters conform perfectly with Foucault's description [95] (p. 179) of how "bigger, more general, or more industrialized knowledges, or knowledges that circulated more easily . . . annex, confiscate, and take over smaller, more particular, more local, and more artisanal knowledges". In a first stage, this is done "by eliminating or disqualifying what might be termed useless and irreducible little knowledges". Second, "by normalizing these knowledges, this makes it possible to fit them together, to make them communicate with another, to break down the barriers of secrecy and technological and geographical boundaries". Importantly, "this makes not only knowledges, but also who possess them, interchangeable". Thirdly, "the hierarchical classification of these knowledges allows them to become, so to speak, interlocking". Next, finally, dominant epistemology builds a "pyramidal centralization that allows these knowledges to be controlled" (all quotes from Foucault [95] (p. 180)). Foucault is describing the battle for truth, the struggle over "the ensemble of rules according to which the true and the false are separated. . . . It is not a matter of a battle 'on behalf' of the truth but of a battle about the status of truth and the economic and political role it plays" [94] (p. 132).

4. 'Dehumanizing' Rationality and Manufactured Ignorance

The modernist-scientific project needs to be 'objective', to keep a distance, and to avoid emotional contacts with the common people. This keeps most hydro-technological scientific research, hydraulic development, water governance and policy formulation from *feeling* what is actually happening, or from imagining what *could* happen in the hydro-territorial realities in which they are intervening. In this way, most dam-development expert institutions and funding agencies can more easily make far-reaching decisions about other people's lives. As Gunter Anders remarked, "the larger the distance, the proportionally smaller our capacity to imagine, and the less our actions are restrained" [96] (p. 15). In modernism's enlightened science and policy-making, knowledge, empirical perception and intellectual understanding are separated from the ability to creatively *imagine* human and non-human consequences (Anders argues that 'knowing' is not sufficient and that it "is the weakest existing form of involvement" [96] (p. 138)). In this respect, mega-hydraulic epistemic communities use 'puppet-based' depersonalized water planning models that, in fact, dehumanize water development and, as a result, avoid addressing the political roots of the problems of water scarcity and overabundance. As Boelens observes, "Water science and policy model-making ivory towers largely combat the generalized Water Crisis by inventing a 'hydro-political dream scheme'—an idealized socio-technical order aligning humans and non-humans—obscuring (the) day-to-day consequences of these policy models for real flesh-and-blood men and women" ([4] (p. 197). See also [97]). In this vein, the international consultants and academic directors of the new hi-tech Yachay University in Ecuador (ironically, *yachay* means 'knowledge' in the indigenous Kishwa language) explain that technology/knowledge development does not need adaptation to local society but that society must fit to new, external highly modernist knowledge: "If we can transform their lives by lightening their imagination (then) we will have engendered a new society based on science and technology ... We endeavor to make Yachay change the 'chip' of Ecuadorian mentality, the way in which Ecuadorians see life" (Vistazo, 2014:13,14,16, in [98]). Such epistemology absolutely ignores human diversity, and local water identities and the complexities of territorial water rights systems. Standardization and universalization equalize human actors and relationships, taking the dominant (mostly white, male, occidental, privileged class, and/or non-indigenous) as the referent. "Overlooking differences among actually existing water users and rights systems, generates biased user and rights representation, active commensuration, and thereby indifferences regarding real-life users and rights. Seen from high above, from the towers of indifference, everybody is equal and made equal" [4] (p. 197).

The modernist water policy modeling and mega-hydraulic engineering activities, which follow universalistic guidelines for building hydro-political dream schemes, lack the empathy to understand the very real concerns about their socioenvironmental impacts. They stimulate ignorance and an incapacity to think about the motives and effects of (technical and social) engineering decisions, which leads to indifference, and a neglect of actual consequences of their actions. The evaluation and reporting of the results of dam projects correspond to theoretical disciplinary assumptions and ignore multi-dimensional realities. The institutional and economic incentive structures (and even scientific credit systems) that support large dam projects carry no obligation to focus on the actual impacts in the hydrosocial territories, inhabited by real people. 'Success' is separated from real improvement as judged by local villagers and water users.

The power of ignorance, both conscious and unconscious, and its impacts, are fundamental to large dam building. Nikhil Anand [99], studying water supply in Mumbai, argued that, 'ignorance' plays as important a role as 'knowledge' in hydraulic engineers' claims of good water governance. 'Active ignorance' or 'consciously not knowing' goes beyond the arguments about the inevitability of failure embodied in mega-projects advanced by Perrow [100] (and discussed earlier in this article). It is better captured in the thinking of Nobel Prize-winning economist Albert Hirschman, progressive intellectual, activist, policy thinker, and World Bank consultant. In the 1960s, Hirschman was appointed to evaluate projects funded by the World Bank in many countries around the world. Barbara Lynch [15], in this volume, discusses Hirschman's argument about what he called the *"hiding hand"*—how the unknowns, hidden uncertainties and unpredictability in the planning of mega-projects, such as large dams, fosters decisiveness and creativity by deluding the project planners about the potential difficulties that the project will inherently encounter. Along much the same lines as Perrow, Hirschman argued that the planning of complex technological systems inherently, and inevitably, contains a high level of uncertainty. Whereas Perrow argued that this unpredictability would inevitably result in accidents (and hence was inherently dangerous) Hirschman argued the need to embrace this unpredictability, since knowing or imagining all the negative future scenarios beforehand would hinder the execution of all large-scale project and so, 'development' and 'progress'. Lynch directs a powerful question, and implicit critique to Hirschman, which should also be directed to Perrow. Should one close one's eyes to potential human and environmental suffering? And is ignorance (i.e., uncertainties and unpredictability—as in Perrow) about such difficulties merely a chance or coincidence, or is it systematically produced? Lynch discusses two projects, the San Lorenzo dam in in northern Peru (which was Hirschman's original case study) and the controversial Guatemalan Chixoy Dam and argues that the hidden costs and suffering were not inadvert, but came about as a result of a systematic production of ignorance. Manufactured ignorance in the planning process enabled international donors and other actors to disregard genocide, the systemic state-sponsored military campaigns against the indigenous Maya populations (see also [22,47,70]. Lynch makes a compelling case that it was not uncertainties and unpredictability, but the pattern of manufactured ignorance that guided the development of the projects and that this pattern was supported by the international mega-hydraulics culture of donor agencies, contractors, and experts. Lynch also argues that the Chixoy Dam project was not an isolated instance of development miscarried or a case of unintended side effects, but came about through the deliberate exclusion of the voices of the people affected by the project. This ignorance was followed by the exclusion of contextual evidences and local knowledge, which ultimately shaped how 'valid knowledge' was defined and accepted among these development actors. Some other scholars have also argued, contrary to Perrow, that the damage caused by dam disasters is often not unavoidable or unforeseen but instead allowed to happen [14,49].

The production of ignorance is an issue that is fundamentally neglected in studies of mega-hydraulic projects and large-scale territorial transformations (see also [101]. Proctor and Schiebinger [102] argue that a great deal of attention has been given to epistemology—the study of how we know. Sismondo [103] (p. 169) goes further, arguing that the entire discipline of science and technology studies is focused on knowing how we know and neglects the equally important question of what, how and why we don't know things. Ignorance is commonly seen as an absence of knowledge, something that needs correcting. It may even be interpreted as uncertainty or unpredictability (as Hirschman and Perrow do, in their own distinct ways). But as large dam developments have deep cultural, socioeconomic, psychological, environmental and political impacts, we also need to think about the conscious, unconscious and structural production of ignorance [102]. Lynch's contribution to this volume [15] powerfully shows how ignorance is a resource, a selective choice, a strategic play and an *active construct*.

Fox and Sneddon [11], in this volume, illustrate this point by providing glaring examples of such ignorance being strategically deployed for political ends—for instance, Cambodian Prime Minister Hun Sen, after visiting the highly contested Don Sahong dam (then being constructed in the Mekong river basin) said: "I visited the dam and it does not have any impacts". They discuss in detail the extensive body of evidence showing that this dam will have significant negative impacts, including the displacement of communities and threatening fish production and food security in the region. This systematic production of ignorance is not only promoted by government officials but also supported by the extensive involvement of experts. The Laos Ministry of Energy and Mines, involved in the project, promotes a narrative of hydropower as embodying sustainable development, verified through "consultation with experts". Fox and Sneddon asked why experts have such a prominent role in the face of the mounting conflicting evidence that has been painstakingly collected by local residents, NGOs and independent scientists. This leads us to argue the need to rethink the role of experts in the building of large dams since, most often, their technical and moral decision-making capacity 'is not relevant to the locally specific contexts in which they intervene': contexts that they may know nothing about and to which they may choose to turn a blind eye. In Gunther Anders' words, their schemes are based on profound 'subject- and fantasy-loss' [96].

5. Manufacturing Risks, the Commensuration of Values and Calculating Compensation

Given the colossal (foreseen or hidden) impacts of dams on nature and society, the calculation of risks and damage and the provision of adequate compensation for the later are crucial, controversial, yet also highly neglected issues. We want to highlight how manufactured ignorance is connected with manufacturing risk. Huber [14], in this volume, shows how willful ignorance regarding hydropower risks—a result of institutional complacency, technological hubris and manufactured uncertainty—contributes to the unequal production of risk, and the associated processes of marginalization and facilitation. Huber's paper links the issues of inequality and risk. She narrates the everyday experience of "ecological precarity" caused by the construction of hydropower infrastructure in fragile ecological settings, such as the Himalayas. She found that it is the poor people, with limited means, who suffer from the sudden appearance of cracks in their houses, the occurrence of landslides, or the sudden drying up of springs or declining soil moisture in agricultural land. It illustrates how environmental governance in the Indian state of Sikkim—an interplay of institutional mechanisms, policy lacunae and complacency by government departments and corporate leaders—has sidelined the prevention and mitigation of environmental risks and impacts associated with hydropower development in a context of heightened hazard potential. Huber illustrates how depoliticization of technological risks often works through particular, techno-scientific framings and erasure, i.e., by excluding risk from the terms of the debate.

Central to any analysis of this is the question of how knowledge about risks and vulnerabilities is produced, negotiated and contested. Ulrich Beck and Anthony Giddens argue that the discourse on risk adopted by regulatory authorities is inadequate and offer alternative ways of understanding and positioning risks. Beck [104] (p. 21) defines risk as "a systematic way of dealing with hazards of insecurities induced and introduced by modernization itself". Giddens [60] distinguishes between external risk and manufactured uncertainty, a point developed by Levitas [105] (p. 201): "If risks are perceived as external, only (the) consequences are addressed and if they are considered manufactured the causes will be called into question". For Beck, risks are "manufactured" in the production of scientific and technological knowledge. In his seminal work 'Risk Society', he argues that late modernity has involved a shift from a class society to a risk society—a shift from the questions of the production and distribution of wealth to the production, definition, and distribution of risks. He distinguishes between early modernity in which conflict over wealth (goods) production dominated and later modernity where conflict is focused on the production and mitigation (or not) of hazards ('bads'). Beck's claims (though criticized for being too linear, see, for example, Scott [106]), hold that the central political issue in late modernity concerns the reduction and legitimation of risks, rather than reducing or legitimating inequality. Underlying this there is also a repositioning of the concept of risk. Risks are not 'out there', as external phenomena that need to be deciphered and controlled. They are manufactured, not only through the application of technologies but also through the production and management of knowledge. If risks are manufactured then risk assessment must also address the causes of risks, rather than confining its mandate to assessing risk as a consequence. Second, all interpretations and knowledge of risks, including empirically driven knowledge, are inherently a matter of perception and hence subjective and political. Third, within risk assessment, the politics of risk definition become extremely important, and involve claims about the legitimacy (as opposed to merely the reliability) of particular forms of knowledge. In line with Foucault's 'battle for truth', this revised concept of risk highlights the contested nature of who defines what a risk is, and how. This contestation also calls for a re-examination of the role of expertise, not only in assessing, but also in manufacturing, risks. As Huber [14] explains, political and economic elites are able to capitalize on risky dam projects, and further marginalize weaker social groups. She argues that influencing the production of knowledge about risk is a fundamental way to challenge the top-down imposition of hydropower and the uneven risks it entails.

Given the importance of understanding and politically discussing such key issues as 'mega-hydraulic risks', 'valid (as well as ignored) dam knowledge and norms', or 'rightful compensations' for affected families, it is fundamental to scrutinize and bring to light the *'commensuration processes'* in norm and knowledge building around dam development projects. It refers to the ways in which experts' (explicit but, especially, implicit) norms, definitions, and values become the equalizing metric: in fact, the politics of how 'cultural particulars' are made universal. In general, the effectiveness and efficiency of large dam developments—including the way in which compensation measures are envisioned and negotiated, directly depend on objectifying and quantifying water resources: as H$_2$O without any of its cultural values and meanings, on de-personalizing and/or commodifying land, water rights, territories and natural resources, and access thereto within universalistic frameworks and forms of governance. Commensuration is central in that it seeks to standardize entirely different water governance rationalities and hydro-territorial contexts into one, single, common metric (see [107–109]). Hoogendam and Boelens [13], in this volume, elaborate on how the Misicuni Dam and Tunnel project, in Bolivia, was designed to transfer water to the Cochabamba Valley, in particular to the city, and had severe impacts on the rural communities affected by it. Even though the project aimed to compensate those families deprived of their agricultural land, houses and livelihoods, it proved hugely complex to find shared values and expressions of what the losses meant for the different actors involved. The skewed balance of power between the state agencies and indigenous peasant communities biased the terms of commensuration and subsequent compensation that was made.

Commensuration denies circumstantial power relationships, the relevance of water governance contexts, and their embeddedness in particular cultures, territorialities and histories. This also poses a dilemma for NGOs and grassroots movements, which when claiming 'rights', often have to choose how far they can, and wish, to frame their interests as 'commensurables'. This choice is also directly related to how they 'upscale', professionalize, universalize their worldviews and valuations and allow them to be converted into a technological fix (see [23,110–114]).

Espeland and Stevens [107] illustrate this point by using the example of indigenous Yavapai ancestral lands in Arizona, USA, that were threatened by a proposed dam. For the Yavapai (as in numerous cases elsewhere) the land was a 'constitutively incommensurable', intrinsically connected to their ways of life and being and their identification with their ancestral territory. "The Yavapai understood themselves in relation to this specific land. Valuing (the) land as an incommensurable was closely tied to what it means to be Yavapai. The rational decision models used by bureaucrats to evaluate the proposed dam required that the various components of the decision be made commensurate, including the cost and consequences associated with the forced resettlement of the Yavapai community" [107] (p. 327). The modernist commensuration required by the dam project officials involved 'objectively compensating' the affected communities according to 'rational standards' and the 'natural laws of economics' which were incompatible and deeply contradictory to the Yavapai's values, interests and knowledge frames.

6. From Ignoring to Reifying Local and Indigenous Knowledges?

Several papers in this volume discuss the way in which local territorial realities and knowledges are ignored by the big dam regime—the UnGov Legend is very much alive. Bakker and Hendriks [8], in this volume, examine the construction of the Site C Hydroelectric Project on the Peace River in British Columbia, Canada. They elaborate on how, from the outset, the government aimed to bypass regulations that would allow for an in-depth analysis of the socioeconomic, cultural and environmental impacts of the dam. In the process, local indigenous territorial livelihoods and worldviews were largely disregarded, and the treaty rights of indigenous peoples were conveniently overlooked. The article by Terävainen [16] in this volume studies divergent and competing expectations about water and technological development through the prism of the Coca Codo Sinclair Hydropower mega-project in Ecuador. During the design and implementation of the project the local peasantry and indigenous people were sidelined and their territorially rooted knowledge regimes ignored. She also elaborates how NGOs draw from techno-scientific approaches and environmental expertise and professionalize political activism in order to gain a political voice, a strategy that may ultimately reinforce the rationality of the techno-economic dam regime. Similar narratives can be found in Dukpa et al. [10,115], Hidalgo-Bastidas and Boelens [12], Huber [14], and Lynch [15], all this volume. Indeed, as Section 3 of this article shows, the dam planning process subjugates local and indigenous knowledges.

Yet, this should not lead us to see Western and indigenous knowledges as dichotomous. As we argue in the next section, 'Western' knowledge is necessarily mediated, hybrid, coproduced. And local and indigenous knowledge is never autarchic, it does not develop in isolation. Dichotomization and reification (and hierarchical 'recognition' of the one by the other) tend to lead to mistaken conclusions and often-dangerous practices, in particular for the marginalized communities themselves. As Cruishank argues, local knowledge " . . . is a concept (that is) often used selectively and in ways that reveal more about histories of Western ideology than about ways of apprehending the world. Its late 20-century incarnation as 'indigenous' or as 'ecological' knowledge continues to present local knowledge as an object for science" [116] (p. 358).

The legacies of primordialism (global), indigenism/*indigenismo* (e.g., Latin America), regenerationism (Spain), recent TEK 'traditional ecological knowledge' approaches (global), neoliberal multiculturalism (global), and similar ideologies, show the profound paradoxes (and complex politics) of modernist 'recognition' of local and indigenous knowledge. At the one hand, day-to-day norms, people's customs and wisdoms, applied to concrete situations, are central in these approaches. They argue that official laws and formally accredited knowledge systems must respect, support and recognize people's legitimate, everyday forms of customs and knowledge systems that have been molded by centuries of practical experience. But, at the other hand, to enforce these local norms and take them beyond solely 'internal use' these approaches commonly set out to systematize and codify them. In order to examine, select and codify customary law and local/indigenous knowledge, and to separate 'good' customs from 'bad' ones, they argue that we need positivist, universalistic science. Experts are then hired to evaluate 'good' local water knowledge and norms. These practices, currently framed as 'best practices', or local forms of 'transparent, good governance' characteristically aim to facilitate 'self-governance under oversight and tutelage' and favor 'compatible knowledges' (see also [111,114,117,118]). The paradox here is clear: while claiming to value customary knowledge and norms they are also considered "to be unsystematic and disorganized, [and] must be submitted to the universalistic rules of professionals, experts, scientists, to select and discipline. Expert intermediaries judge and promote 'universal truthfulness', as if there were one single truth about 'effective' water rights, norms of 'good governance' or 'optimal' agricultural practices" [50] (p. 56).

Ironically, this scientific examination and codification redefines, assimilates and marginalizes local, vernacular, water knowledge frameworks. Only the knowledge and principles that fit into official schemes and policies are approved, thereby muzzling the complex variety of 'unruly rules' and 'disorganized wisdoms'. These modernist recognition approaches ignore the fact that professionals and scientific experts are not disinterested agents but embedded in cultural and power relations. They seek to purify and universalize 'best practices', thereby denying the ability of local water users to actively recreate and regenerate their own water management practices.

At the core, indigenist and TEK approaches, which assume the existence of 'best practices' (that means, 'best knowledge'), implicitly assume that indigenous knowledges are static, non-relational, a-historical, and transferable to other contexts. The only knowledges and practices that are selected are those that are compatible (or can be made commensurate) with dominant modern (scientific, legal, and management) knowledge frameworks. It is not just the 'local actors' who are at risk of being subjugated, but also their modes of knowing and their epistemological worlds: they are " . . . subsumed within universalizing hierarchies; what is included and what is left out is not random" [116] (p. 371). By making everyday water control, social relations, territories and knowledge forms 'graspable' and 'controllable' such approaches subtly install the dominant culture's knowledge and frames of reference.

In the same vein, Gaventa and Cornwall [119] argue that by reifying local knowledge—by the dominant or the marginalized—and treating it as singular, " ... the possibility is rarely acknowledged that what is expressed as 'their knowledge' may simply replicate dominant discourses, rather than challenging them" [119] (p. 75). Sandy Marie Anglás Grande adds to this by explaining how " ... the age-old typification of the 'ecologically noble savage' is being resurrected and employed by certain factions of environmentalists, ecophilosophers, and ecofeminists alike ... these environmentalists simply add an academic riff to the pop construction ... as primitive savior, EcoGuru, keeper of mystical wisdom and romantic vision" [120] (p. 312). In the same vein as our observation about the modernist construct of 'equals' through the dehumanizing rationality of mega-hydraulic projects (Section 4), Grande expresses the worry that this essentialization may provide these self-styled 'ecosaviors' with an "oversimplified, and thus, dehumanized identity" [120] (p. 313). Far beyond understanding and supporting the claims of people affected by the loss of their territories and livelihoods, " ... the noble savage stereotype, often used to promote the environmentalist agenda, is nonetheless immersed in the political and ideological parameters of the modern project" [120] (p. 307).

The idealization of indigenous knowledges can therefore pose a real danger for communities affected by dam projects, or factions within them. For instance, as Paredes Penafiel and Li [121] (p. 16) state: "Writing about the existence of multiple ontologies or worlds risks romanticizing ways of being that do not conform to 'Western' modes of existence, as well as idealizing resistance to extractive industries. People's relationships with nonhuman beings are often translated into the language of the 'sacred' or interpreted as reverence for the natural world, and these ideas are adopted by environmentalist campaigns in ways that distort people's lived realities. This, in turn, essentializes cultural identity and makes differences seem fixed and incommensurable. This is not only theoretically problematic but may also undermine the political goals of those responding to mining activity". Horowitz elaborates on how the discourses that idealize indigenous people's 'natural' ecological rationality and wisdom often stand in sharp contrast to lived realities and may seriously hamper their economic, political interests: "When indigenous people do not attain 'the impossible standards of ecological nobility' set for them, they are judged as inauthentic and their concerns may be ignored [...]. Such discourses privilege Western environmentalist values, which can easily be turned into justifications for restrictions on local people's behaviour" [122] (p. 1383). There is a particular danger when activists and environmentalists deny internal divisions and ignore political-hierarchical realities within the communities they are 'defending'—presenting strategic simplifications to governments and donors that portray communities as uniform and united; "[these] simplistic portrayals can play into the hands of corporations who are able to coopt self-styled community representatives and completely overlook less powerful sub-groups, such as women and young people" [122] (p. 1383). In a related vein, Dukpa et al. [10], in this volume, elaborate on how unity among community members in Sikkim, India, beyond a presumed 'indigenous characteristic', was internally forced by community authorities who imposed the *Chya*, (a ritual cursing bringing death by supernatural forces to individuals and their descendants) on those villagers who supported hydropower development against the community's wish.

As we have made clear in the previous sections, we do not aim to deny the importance of the ecological, cultural and technological knowledge frameworks and practices that 'local' and indigenous people possess and deploy, particularly when they are confronted with the rationality of mega-hydraulic interventions and offer an alternative logic of how the territory can support their livelihoods. It is crucially important to understand their voices and the logic of 'other' modes of living. As Grande states, " ... the reason is not because we possess any kind of magical, mystical power to fix the devastating effects of generations of abuse and neglect, but because we stand as living critiques of the dominant culture" [120] (p. 320). In this respect we need to acknowledge that local, peasant, and indigenous socioenvironmental knowledge and practices are not exempt from, but are pervaded by power, internal and external divisions and hybridity. Cruishank [116] (p. 358), writing about the Canadian context, says that " ... local knowledge is not something waiting to be 'discovered' but, rather, is continuously made in situations of human encounter: between coastal and interior neighbors, between colonial visitors and residents, and among contemporary scientists, managers, environmentalists and First Nations". This is not simply an encounter between 'cultural' local/indigenous knowledge forms and 'technical' expert/Western knowledge systems, since " ... *both* local knowledge and expert/scientific knowledge are cultural, social, and political" [123] (p. 235). Given these dynamic hybrids, " ... the politics of knowledge production is not simply the challenge to official positivist accounts by local ones—rather it involves conflict within local and official communities themselves, who settle upon definitive accounts in a process of environmental struggle" [123] (p. 235). (Important to note that, in this struggle, responding to racist and modernist regimes, grassroots/indigenous movements also (re-)essentialize. De-constructivist schools often criticize this counter-representation (merely on 'scientific grounds,' applying the same objectivist perspective they claim they challenge) but neglect the political properties in and of these counter-discourses. Grassroots' essentialistic 'counter-images' require critical, contextualized examination as part of concrete struggles, including supralocal alliance strategies and subtle forms of self-representation [4,72]).

In the next section we therefore elaborate upon the relational coproduction of knowledge. Indigenous knowledges are not closed systems neither are 'Western modes of modernity', and, moreover, the latter have no monopoly on modernity. As Eisenstadt [59] (p. 24) argued, "within all societies, new questionings and reinterpretations of different dimensions of modernity are emerging ... While the common starting point was once the cultural program of modernity as it developed in the West, more recent developments have seen a multiplicity of cultural and social formations going far beyond the very homogenizing aspects of the original version". New, hybrid and alternative epistemological and technological trajectories challenge classical notions of Western (mega-hydraulic) modernity (see also [124]).

7. Sub-Politics, the Dimensions of Power and the (Relational) CoProduction of Knowledge

Complex socio-natural, techno-political constructs, such as large dams, can be viewed in different ways by different interest groups. These entities, and how they are known, are not fixed but sprout (socially and materially) from encounters and interactions between different, often divergent and sometimes incompatible knowledge systems (see [110,125,126]). A large dam can represent a modern feat of technical engineering that controls nature and brings progress. It can also represent the taming and, even death, of a living, animated river that can no longer flow freely, or be a symbol of greedy capitalism destroying ecology and creating inequality; the drowning of a territory that contains and supports human and natural communities; a source of new flora and fauna triggering tourism, or the key to future urbanization, and many other constructs. The mega-dam—as a contested hydrosocial territory—is the focus of conflicting knowledge and value systems and these overarching or hegemonic representations (which may be combined) express the prevailing power relationships at the local, regional, national and international level.

Consequently, the production of dam-related knowledge is an arena [1] or even battlefield, in which the values, understandings and interests of different actors are brought into confrontation with each other. The development of knowledge is a process of hybridizing as Norman Long says. It involves 'multiple social realities', unequal powers, and diverse ways " ... of construing and ordering the world, [and] not as a simple accumulation of facts or as being unified by some underlying cultural logic, hegemonic order or system of classification. Knowledge emerges out of a complex interplay of social, cognitive, cultural, institutional and situational elements. It is therefore always essentially provisional, partial, and contextual in nature, and people work with a multiplicity of understandings, beliefs and commitments ... " [1] (p. 15). This relational co-creation of knowledges also leads to the multiple modernities [59] we have mentioned in Sections 2 and 6. Or as Baud, Boelens and Damonte [127] elaborate in a recent special issue, under a seemingly uniform process of (market and/or state controlled) capitalist expansion, we can witness a multitude of 'capitalisms' and 'developmentalisms' evolving. New hybrid discourses, knowledge regimes and epistemological contestations are being developed, partly forced by public debates and social movements, but also resulting from how large corporate enterprises and the state have learned to adapt their discourses, ideologies and practices (with capitalisms expressing themselves increasingly as 'benevolent', 'green' and 'sustainable', 'pro-poor' or even 'pro-indigenous'). In all instances, even though dominant techno-political engineering knowledge paradigms are hugely powerful in mega-hydraulic projects, knowledge, like power, is a relational process and product, not an asset. Consequently, it cannot be possessed as a property or depleted in a zero-sum game but emerges from interactions between subjects [1,94].

That said, in these political battlefields of knowledge, actors often concentrate upon and reify particular forms of modernist hydraulic/hydrological knowledge: as Long says, as if this knowledge were an absolute property that can be accumulated. "[K]nowledge encounters involve the struggle between actors who aim to enroll others in their 'projects', getting them to accept particular frames of meaning, winning them over to their points of view. If they succeed, then other parties 'delegate' power to them" [128] (p. 27). In these encounters and contestations among 'expert' and 'non-expert' knowledge frames, fundamental political issues are transformed into issues of expertise, debated and fought over.

This mostly occurs outside the realm of the institutional politics of rule-directed struggle. One consequence is that, in the case of large, capital-intensive, public-private interventions, such as mega-hydraulic projects, which are dominated by expert knowledges and financial interests, democratic scrutiny over the planning, design, construction and implementation of mega-hydraulic schemes is very partial. Ulrich Beck [129] referred to such processes as 'sub-politicization': decisions that may affect large numbers of people are debated and settled outside of state-related political institutions and arenas. Here, politics is 'dispersed' or 'displaced'. In his book, *Risk Society*, Beck argues that real politics is made in the various realms of sub-politics such as the firm, the laboratory or the gas station. Collins and Evans [130] raise the fundamental question of whether the political legitimacy of technical decisions in the public domain should be maximized by referring them to the widest democratic processes, or whether such decisions should be based on the best available expert advice. They propose a "normative theory of expertise" that draws the line between appropriate and inappropriate inclusiveness in technical debates conducted in public domains. Sheila Jasanoff [131] responds to this question and argues that such a dilemma is a false one, since we need *both strong democracy and good expertise* for making important public domain decisions, and that the real question is how to integrate them to achieve an appropriate balance between power and knowledge. Wider participation of the lay public, and affected people in expert decision-making is needed in order to test and contextualize the framing of the issues that experts are asked to resolve.

Aside from the question of how dominant mega-hydraulic knowledge subjugates diverse vernacular knowledge (see Section 3), it is also useful to examine the different dimensions of power that influence the production of knowledge and how they are mediated and contested by grassroots and civil society institutions that aim, or claim, to democratize them. Following Lukes [132,133], Gaventa [134], Gaventa and Cornwall [119], Gramsci [135] and Foucault [94,136,137], in the battlefield of 'contested knowledge' the manifestations and modes of power that determine the role of dam-development will vary according to the specific context and situation. While in practice these different conceptual modes and empirical dimensions combine and become hybrids, each power dimension tends to lead to a different approach or strategy of contestation. We examine some of these below:

A first dimension of power relates to knowledge as a 'resource' that is deployed to inform decision-making on mega-hydraulic design and development. Here, power is 'visible' in Weberian terms and is expressed in vested rules, hierarchies, institutes and institutions. Typically, this expert and formal knowledge is *contested by counter-expertise*. Advocacy groups counter it by presenting 'more objective' and 'better-grounded' (mostly positivist) knowledge that seeks to (de)legitimize the knowledge claims of the dominant stakeholders.

A second dimension and working mechanism of power relates to how knowledge is purposely biased and how power colludes in "keep[ing] some issues and actors from getting to the table" [119] (p. 71): here, we are talking about 'hidden power' [134]. Dominant mega-hydraulic development players, who have control over the production of expert knowledge and may know about particular (e.g., negative) effects of dam schemes, may hide such information in order to promote their schemes. In this way, they aim to skew the agenda, leave out the thorny issues, exaggerate the positive impacts, and include or exclude particular views and knowledge agents. In response, grassroots alliances will engage in the knowledge battlefield by *involving and empowering sidelined actors and alternative issues* when presenting their knowledge and interests in the techno-political debate. Unequal societal structures (e.g., class, gender, ethnic relationships) that produce the knowledge that drives mega-hydraulic projects and excludes grassroots epistemologies and ontologies are actively challenged, bringing the grassroots actors and wisdoms to the knowledge-negotiation table.

Lukes [132] identified a third dimension of power, whereby consciously "winning the hearts and minds" [133] of those subjugated is crucial. Beyond the open or hidden use of power and knowledge, and the generation of conflicts over knowledge, the avoidance of conflict, the subtle silencing of counter-voices and influencing the consciousness of the dam-affected populations are key elements in this. This is done through controlling the production and dissemination of knowledge, the intentional construction of pro-dam narratives and discourses; education and socialization programs and the manipulation and steering of mass media; actively mobilizing ideology in order to shape consciousness, thereby fortifying a Gramscian-style hegemony ([135]; see also [119]). In response to this, grassroots and anti-dam advocates will strive to *curb the process of knowledge production itself*, actively seeking to shape consciousness, and the self-awareness of affected populations and their capacity for knowledge creation and dissemination.

As we outline in Section 3, in the fourth, 'Foulcauldian approach', knowledge and power are not 'possessed' by dominant groups, or strategically deployed to oppress, but are relational and productive forces. Discourses, institutions and practices link knowledge, power and truth in triangular relationships, whereby the normality and morality of large dam projects and acceptance of the mega hydraulic order are internalized. New, truthful and legitimate knowledge is produced—by both dominant and subjugated actors, who are 'subjectified', and this relational power web steers them towards 'correct behavior and thinking' and even to self-correction. This may lead grassroots and advocacy alliances to engage in the complex task of *questioning the normality of mega-hydraulic development, commensurate modernity, discourses of interconnected national and individual progress and collective harmony*. This is the complex struggle against the boundaries that contain and constrain dam-affected *and* nonaffected societies in a dominant, normalizing, hydro-political discursive network.

Beyond the conflict over the material means of production and the socio-political/hydro-technological re-patterning of humans and non-humans in dam-affected territories, these workings of power also involve a struggle for control over the means of knowledge production, and over who has the societal power to determine what counts as 'normal', legitimate and valid knowledge. In these knowledge battlefields it is not just hydro-political institutions and hydraulic-territorial *objects* and their claims to the truth that are contested, it is also that new hydro-territorial *subjects* are actively shaped, through a process of *co-constitution* (New subjectivities and identities are constructed; human agents and non-human actants are 'subjectified'. In this context the notion of 'subject' simultaneously refers to an actor capable of initiating action and to a being subjected by normalizing power [138,139]). Two articles in this special issue explore this theme. Duarte-Abadía et al. [9] and Hidalgo-Bastidas and Boelens [12] explore the examples of the Rio Grande Dam project in Málaga, Spain, and the Baba Dam project in coastal Ecuador, respectively. They explain how, in this 'subjectification game', the mega-hydraulic development regime deployed discourses that sought to establish what is 'true' and socio-naturally 'coherent' and how, through these means, they institutionally, materially, and symbolically sought to *shape the mega-hydraulic reality*. In both cases, the networks of dam opponents engaged in fierce knowledge battles through strategic interactions, and 'counter-conducts' that made strategic use of all the four dimensions of power outlined above. They consciously and unconsciously entwined positivist engineering, activist, grassroots and other knowledge systems—with new social relationships and material manifestations as outcomes.

8. Multi-Actor, Multi-Scalar Battlefields

Is it possible that a new and common societal project can democratize mega-hydraulic knowledge building and implementation? And, if so, how? In their contribution to this issue, Fox and Sneddon [11] ask if we can see evidence of new spaces for knowledge production emerging out of contestations over dams that shift epistemological boundaries? Certainly, large dam developments are increasingly contested and subject to alternative proposals from civil society which sometimes lead to their redesign. Examples abound, as in the UK where: " ... Swansea Corporation announced plans to build a reservoir in the Gwendraeth Fach Valley in Carmarthenshire, flooding parts of the village of Llangyndeyrn. This plan was met with resistance from the population of this area, who built barricades to stop construction workers entering the region [...]. This opposition was ultimately successful, with the Swansea Corporation finding an alternative site on unpopulated land and building the Llyn Brianne reservoir near Llandovery" [140] (p. 3). Duarte-Abadía et al. [9], in this volume, describe, how in the Rio Grande sub basin, near Malaga, residents mobilized on a mass scale in a creative network, integrating multiple forms of knowledge, stakeholders and scales to stop the damming of their river, while defending their collective water management practices and creatively constructing an anti-hegemonic alternative for water control. They autonomously constructed a deep-rooted hydrosocial territory that ensured the permanence of their water management legacy whilst, at the same time, renewing their water-cultural practices. Hidalgo-Bastidas and Boelens [12], also in this volume, explain how, in Ecuador's Baba Dam project, indigenous communities, peasant federations, environmental NGOs, critical scholars, water professionals and urban leaders influenced the very structures of knowledge and materials underpinning this mega-hydraulic scheme, and how their socioenvironmental demands helped shape the design of the project.

Sanjeev Khagram opens his earlier mentioned book, *Dams and Development*, by quoting the former president of the International Commission on Large Dams (ICOLD), who argued that " ... hydropower is the cheapest and cleanest form of energy, but environmentalists don't appreciate that. Certainly large dam projects create local resettlement problems, but this should be a matter of local, not international concern" (cited in [7] (p. 1). It is precisely this local/international interplay that creates, not just disturbances and threats, but also strong movements that challenge the dynamics of the modernist temple of mega-hydraulic progress. Keck and Sikkink [141] argue that such transnational grassroots networks are able to amplify local voices, which in turn provides a platform for those contestations and popular claims that are actively oppressed or fall upon deaf ears at home (see also [111,114,142]). When grassroots networks engage in scalar politics [142–144] they are able to link and pool particular domains and bodies of knowledge from the life histories of affected communities and share them with the internet activist community thereby fostering intellectual diversity and connecting campaigners, human and environmental rights advocates, engaged scholars and students and public officials. Gaventa and Cornwall [119] argue that this endeavor involves both horizontal and vertical networking. Contestation movements that "have gone to scale most effectively have done so horizontally ... they have included processes of peer-to-peer sharing"—but at the same time, to create new synergies with actors at different scales, "mediating organizations, processes and networks that vertically cut across hierarchies are critical [involving] processes of meaningful representation and voice from one scale to the other." [119] (p. 78. See also [145–150]).

Transnationally allied actors often engage with powerful domestic and international actors who share strategic interests, thereby extending the basis of shared knowledge and action. This extension and hybridization of grassroots knowledge frameworks offers important opportunities for strengthening the influence of local movements, but it also involves dangers. As Dupuits et al. [114] (p. 2) show, "the transnationalization of grassroots movements implies several transformations in the knowledge and languages they mobilize to defend their local common goods. They often have to appropriate and reframe expert knowledge and global norms to gain power and visibility in their mobilizations. This professionalization process is producing distinct effects back to the ground. On the one hand, it can mean more political and financial resources for grassroots movements and an increased recognition by States. On the other hand, it can also imply negative side effects in terms of exclusion of some actors at the margin of transnational processes, or resistance from actors who do not feel well represented". Grassroots transnationalization strategies, therefore, are complex and challenging, since knowledge 'adoption' and 'adaptation' may imply commensuration, institutionalization or normalization [151–158].

But this is not necessarily the case. Water grassroots collectives often react, modify and strategically use the ruling symbolic and political order. "Below [the] appearances of uniformity and formality, local collectives as trans-local networks strategize their ways to resist and construct their own, alternative orders, questioning the self-evidence of formal State, science or market-based frameworks for analyzing and regulating water flows and hydrosocial networks" [72] (p. 246). Here, economic-material and political-symbolic orders and struggles interweave in an effort to defend territories, water rights and livelihoods [10,76,110,121,143–152,159–161]. Following Foucault's remarks, the conceptual/intellectual and political challenge for multi-scalar, multi-actor grassroots networks seeking to defend the commons is to question the dominant epistemology of large dams in order to make space for "an autonomous, non-centralized kind of theoretical production, one whose validity is not dependent on the approval of the established regimes of thought" [94] (p. 81).

The contributions to this special issue present a diverse set of case studies and, at the same time, bring conceptual depth and intellectual rigor to explain these complex issues in their local, regional and global contexts. They highlight the importance of critically examining the (still) dominant regimes of representation that privilege mega-hydraulic projects and the attendant extractivist water policy and scientific regimes that deny the intrinsic complexity, social construction, and political ordering of local water management practices. In discussing the genealogy of dominant knowledge paradigms, the contributors to this special issue propose taking the societal coproduction of water, knowledge and governance as the point of departure, in order to build alternative water epistemologies and ontologies. Or, as Foucault [94] (p. 85) suggested, " . . . in contrast to the various projects which aim to inscribe knowledge in the hierarchical order of power associated with science, a genealogy should be seen as a kind of attempt to emancipate historical knowledge from that subjection, to render them capable of opposition and of struggle against the coercion of a theoretical, unitary, formal and scientific discourse". This implies including other types of knowledge about water control—for instance, that belonging to people who live in the environments that are threatened with being affected by large dams and mega-hydraulic development. This transdisciplinary co-creation of knowledge, which involves both confrontation and mutuality among the water-affected, water-users, activists, the policy and scientific communities—not only critically scrutinizes dominant water knowledge and the workings of neoliberal water culture but also enables the construction of alternative hydro-political knowledge regimes and more just hydro-territorial configurations.

Funding: This research received no external funding.

Acknowledgments: We would like to thank Nick Parrott for his careful and critical revision of this article.

Conflicts of Interest: The authors declare no conflict of interest.

References

1. Long, N. Actors, interfaces and development intervention: Meanings, purposes and powers. In *Development Intervention. Actor and Activity Perspectives*; Kontinen, T., Ed.; University of Helsinki: Helsinki, Finland, 2004; pp. 14–36.
2. Forsyth, T. Critical Political Ecology. In *The Politics of Environmental Sciences*; Routledge: London, UK; New York, NY, USA, 2003.
3. Horowitz, L.S. Power, profit, protest: Grassroots resistance to industry in the global north. *Cap. Nat. Soc.* **2012**, *23*, 20–34. [CrossRef]
4. Boelens, R. Water, Power and Identity. In *The Cultural Politics of Water in the Andes*; Routledge: London, UK, 2015.
5. Menga, F.; Swyngedouw, S. *Water, Technology and the Nation-State*; Routledge: London, UK, 2018.
6. Sneddon, C.; Fox, C. Struggles over dams as struggles for justice: The World Commission on Dams (WCD) and anti-dam campaigns in Thailand and Mozambique. *Soc. Nat. Resour.* **2008**, *21*, 625–640. [CrossRef]
7. Khagram, S. *Dams and Development. Transnational Struggles for Water and Power*; Cornell University Press: Ithaca, NY, USA; London, UK, 2004.
8. Bakker, K.; Hendriks, R. Contested Knowledges in Hydroelectric Project Assessment: The Case of Canada's Site C Project. *Water* **2019**, *11*, 406. [CrossRef]
9. Duarte-Abadía, B.; Boelens, R.; Pre, L.D. Mobilizing water actors and bodies of knowledge. The multi-scalar movement against the Río Grande Dam in Málaga, Spain. *Water* **2019**, *11*, 410. [CrossRef]
10. Dukpa, R.; Joshi, D.; Boelens, R. Contesting Hydropower dams in the Eastern Himalaya: The Cultural Politics of Identity, Territory and Self-Governance Institutions in Sikkim, India. *Water* **2019**, *11*, 412. [CrossRef]
11. Fox, C.; Sneddon, C. Political Borders, Epistemological Boundaries, and Contested Knowledges: Constructing Dams and Narratives in the Mekong River Basin. *Water* **2019**, *11*, 413. [CrossRef]
12. Hidalgo-Bastidas, J.P.; Boelens, R. Hydraulic order and the politics of the governed: The Baba Dam in coastal Ecuador. *Water* **2019**, *11*, 409. [CrossRef]
13. Hoogendam, P.; Boelens, R. Dams and Damages. Conflicting epistemological frameworks and interests concerning "compensation" for the Misicuni project's socio-environmental impacts in Cochabamba, Bolivia. *Water* **2019**, *11*, 408. [CrossRef]
14. Huber, A. Hydropower in the Himalayan Hazardscape: Strategic Ignorance and the Production of Unequal Risk. *Water* **2019**, *11*, 414. [CrossRef]
15. Lynch, B. What Hirschman's hiding hand hid in San Lorenzo and Chixoy. *Water* **2019**, *11*, 415. [CrossRef]
16. Teräväinen, T. Negotiating water and technology—Competing expectations and confronting knowledges in the case of the Coca Codo Sinclair in Ecuador. *Water* **2019**, *11*, 411. [CrossRef]
17. Warner, J.; de Vries, L.; Jomantas, S.; Jones, E.; Ansari, M.S. The Fantasy of the Grand Inga Hydroelectric Project on the River Congo. *Water* **2019**, *11*, 407. [CrossRef]
18. McCully, P. *Silenced Rivers: The Ecology and Politics of Large Dams*; Zed Books: London, UK; New York, NY, USA, 1996.
19. Steiner, A. Preface. *Water Altern.* **2010**, *3*, 1–2.
20. Duarte-Abadía, B.; Roa-Avendaño, T.; Boelens, R. Hydropower, encroachment and the re-patterning of hydrosocial territory: The case of Hidrosogamoso in Colombia. *Human Organization* **2015**, *74*, 243–254. [CrossRef]
21. Fainguelernt, M.B. The historical trajectory of the Belo Monte hydroelectric plant's environmental licensing process. *Ambiente Soc.* **2016**, *19*, 245–264. [CrossRef]
22. Johnston, B.R. Chixoy Dam Legacy Issues Study, Vol. 1–5. International Rivers Network. Available online: http://www.irn.org/programs/chixoy/index.php?id=ChixoyLegacy.2005/03.findings.html (accessed on 14 February 2019).
23. Martínez-Alier, J. *The Environmentalism of the Poor*; Edward Elgar: Cheltenham, UK, 2002.
24. Molle, F.; Floch, P. Megaprojects and Social and Environmental Changes: The Case of the Thai "Water Grid". *Ambio* **2008**, *37*, 199–204. [CrossRef]
25. Moore, D.; Dore, J.; Gyawali, D. The World Commission on Dams + 10: Revisiting the Large Dam Controversy. *Water Altern.* **2010**, *3*, 3–13.
26. World Commission on Dams (WCD). *Dams and Development: A New Framework for Decision-Making*; Earthscan: London, UK, 2000.

27. Ahlers, R.; Zwarteveen, M.; Bakker, K. Large Dam Development: From Trojan Horse to Pandora's Box. In *The Oxford Handbook of Megaproject Management*; Flyvbjerg, B., Ed.; Oxford University Press: Oxford, UK, 2017.

28. Baviskar, A. Waterscapes. In *The Cultural Politics of a Natural Resource*; Permanent Black: Delhi, India, 2007.

29. Kaika, M. Dams as Symbols of Modernization: The Urbanization of Nature Between Geographical Imagination and Materiality. *Ann. Assoc. Am. Geogr.* **2006**, *96*, 276–330. [CrossRef]

30. Shah, E.; Liebrand, J.W.; Vos, J.; Veldwisch, G.J.; Boelens, R. The UN World Water Development Report 2016, Water and Jobs: A Critical Review. *Dev. Chang.* **2018**, *49*, 678–691. [CrossRef]

31. Worster, D. *Rivers of Empire: Water, Aridity, and Growth of the American West*; Pantheon Books: New York, NY, USA, 1985.

32. Mehta, L. The Manufacture of Popular Perceptions of Scarcity: Dams and Water-Related Narratives in Gujarat, India. *World Dev.* **2001**, *29*, 2025–2041. [CrossRef]

33. Mehta, L. Whose scarcity? Whose property? The case of water in western India. *Land Use Policy* **2006**, *24*, 654–663. [CrossRef]

34. Swyngedouw, E. *Liquid Power: Contested Hydro-Modernities in Twentieth-Century Spain*; MIT Press: Cambridge, MA, USA, 2015.

35. Hommes, L.; Boelens, R. Urbanizing rural waters: Rural-urban water transfers and the reconfiguration of hydrosocial territories in Lima. *Political Geogr.* **2017**, *57*, 71–80. [CrossRef]

36. Lynch, B. Rivers of contention: Scarcity discourse and water competition in highland Peru. *Ga. J. Int'l. Comp. L.* **2013**, *42*, 69–92.

37. Aguilera-Klink, F.; Pérez-Moriana, E.; Sánchez-García, J. The social construction of scarcity. The case of water in Tenerife (Canary Islands). *Ecol. Econ.* **2000**, *34*, 233–245. [CrossRef]

38. Birkenholtz, T. Dispossessing irrigators: Water grabbing, supply-side growth and farmer resistance in India. *Geoforum* **2016**, *69*, 94–105. [CrossRef]

39. Boelens, R. *Rivers of Scarcity. Utopian Water Regimes and Flows against the Current*; Wageningen University: Wageningen, The Netherlands, 2017.

40. Mena-Vásconez, P.; Boelens, R.; Vos, J. Food or flowers? Contested transformations of community food security and water use priorities under new legal and market regimes in Ecuador's highlands. *J. Rural Stud.* **2016**, *44*, 227–238. [CrossRef]

41. Zwarteveen, M. *Regulating Water, Ordering Society: Practices and Politics of Water Governance*; Inaugural Lecture; University of Amsterdam: Amsterdam, The Netherlands, 2015.

42. Del Bene, D.; Scheidel, A.; Temper, L. More dams, more violence? A global analysis on resistances and repression around conflictive dams through co-produced knowledge. *Sustain. Sci.* **2018**. [CrossRef]

43. Hommes, L.; Boelens, R.; Maat, H. Contested hydrosocial territories and disputed water governance: Struggles and competing claims over the Ilisu Dam development in southeastern Turkey. *Geoforum* **2016**, *71*, 9–20. [CrossRef]

44. Dye, B. The return of 'high modernism'? Exploring the changing development paradigm through a Rwandan case study of dam construction. *J. East. Afr. Stud.* **2016**, *10*, 303–324. [CrossRef]

45. Hommes, L.; Boelens, R. From natural flow to 'working river': Hydropower development, modernity and socio-territorial transformations in Lima's Rímac watershed. *J. Hist. Geogr.* **2018**. [CrossRef]

46. Jasanoff, S. A New Climate for Society. *Theory Cult. Soc.* **2010**, *27*, 233–253. [CrossRef]

47. Johnston, B.R. Large-scale Dam Development and Counter Movements. Justice Struggles around Guatemala's Chixoy Dam. In *Water Justice*; Boelens, R., Perreault, T., Vos, J., Eds.; Cambridge University Press: Cambridge, UK, 2018; pp. 169–186.

48. Hidalgo-Bastidas, J.P.; Boelens, R.; Isch, E. Hydroterritorial Configuration and Confrontation: The Daule-Peripa Multipurpose Hydraulic Scheme in Coastal Ecuador. *Latin Am. Res. Rev.* **2018**, *53*, 517–534. [CrossRef]

49. Huber, A.; Joshi, D. Hydropower, Anti-Politics, and the Opening of New Political Spaces in the Eastern Himalayas. *World Dev.* **2015**, *76*, 13–25. [CrossRef]

50. Boelens, R.; Post Uiterweer, N.C. Hydraulic Heroes. The ironies of utopian hydraulism and its politics of autonomy in the Guadalhorce Valley, Spain. *J. Hist. Geogr.* **2013**, *41*, 44–58. [CrossRef]

51. Molle, F.; Mollinga, P.; Wester, P. Hydraulic bureaucracies and the hydraulic mission: Flows of water, flows of power. *Water Altern.* **2009**, *2*, 328–349.

52. Swyngedouw, E.; Boelens, R. " . . . And Not A Single Injustice Remains": Hydro-Territorial Colonization and Techno-Political Transformations in Spain. In *Water Justice*; Boelens, R., Perrault, T., Vos, J., Eds.; Cambridge University Press: Cambridge, UK, 2018; pp. 115–133.
53. Boelens, R. *Water Justice in Latin America. The Politics of Difference, Equality, and Indifference. Inaugural Lecture*; CEDLA & University of Amsterdam: Amsterdam, The Netherlands, 2015.
54. Blackbourn, D. *The Conquest of Nature: Water, Landscape and the Making of Modern Germany*; Norton: London, UK; New York, NY, USA, 2006.
55. Swyngedouw, E. Technonatural revolutions: The scalar politics of Franco's hydrosocial dream for Spain, 1939–1975. *Trans. Inst. Br. Geogr.* **2007**, *32*, 9–28. [CrossRef]
56. Nixon, R. Unimagined communities: Developmental refugees, megadams and monumental modernity. *New Formations* **2009**, *69*, 62–80. [CrossRef]
57. Nixon, R. *Slow Violence and the Environmentalism of the Poor*; Harvard University Press: Cambridge, MA, USA; London, UK, 2011.
58. Watts, M.J. 'Antinomies of Community: Some Thoughts on Geography, Resources and Empire'. *Trans. Inst. Br. Geogr.* **2004**, *29*, 195–216. [CrossRef]
59. Eisenstadt, S.N. Multiple modernities. *Daedalus* **2000**, *129*, 1–29.
60. Giddens, A. *Beyond Left and Right*; Polity: Cambridge, UK, 1995.
61. Latour, B. *We Have Never Been Modern*; Harvard University Press: Cambridge, MA, USA, 1993.
62. Rostow, W.W. *The Stages of Economic Growth: A Non-Communist Manifesto*; Cambridge University Press: Cambridge, UK, 1960.
63. Bauman, Z. *Modernity and the Holocaust*; Polity Press: Cambridge, UK, 1989.
64. Bauman, Z. Liquid Times. In *Living in an Age of Uncertainty*; Polity Press: Cambridge, UK, 2007.
65. Haraway, D. *Simians, Cyborgs and Women: The Reinvention of Nature*; Free Association Books: London, UK, 1991.
66. Polanyi, K. *The Great Transformation: The Political and Economic Origins of Our Time*; Farrar & Rinehart: New York, NY, USA, 1944.
67. Harvey, D. *Justice, Nature & the Geography of Difference*; Blackwell: Cambridge/Oxford, UK, 1996.
68. Gudynas, E. Buen Vivir: Today's tomorrow. *Development* **2011**, *54*, 441–447. [CrossRef]
69. Scott, J. *Seeing Like a State: How Certain Schemes to Improve the Human Condition Have Failed*; Yale University Press: New Haven, CT, USA, 1998.
70. Lynch, B.D. *The Chixoy Dam and the Achi Maya: Violence, Ignorance, and the Politics of Blame, Mario Einaudi Centres*; Cornell University: Ithaca, NY, USA, 2006.
71. Barnes, J.; Alatout, S. Water worlds. *Soc. Stud. Sci.* **2012**, *42*, 483–488. [CrossRef]
72. Boelens, R. Cultural Politics and the Hydrosocial Cycle: Water, Power and Identity in the Andean Highlands. *Geoforum* **2014**, *57*, 234–247. [CrossRef]
73. Boelens, R.; Hoogesteger, J.; Swyngedouw, E.; Vos, J.; Wester, P. Hydrosocial territories: A political ecology perspective. *Water Int.* **2016**, *41*, 1–14. [CrossRef]
74. Bijker, W.E. Dams and Dikes. Thick with politics. *Focus-Isis* **2007**, *98*, 109–123. [CrossRef]
75. Pfaffenberger, B. Fetishised objects and humanised nature: Towards an anthropology of technology. *Man* **1988**, *23*, 236–252. [CrossRef]
76. Sanchis-Ibor, C.; Boelens, R.; García-Mollá, M. Collective irrigation reloaded. Re-collection and re-moralization of water management after privatization in Spain. *Geoforum* **2017**, *87*, 38–47. [CrossRef]
77. Shah, E. Social Designs. Tank Irrigation Technology and Agrarian Transformation in Karnataka, South India. Ph.D. Thesis, Wageningen University, Wageningen, The Netherlands, 2003.
78. Verbeek, P. *Moralizing Technology: Understanding and Designing the Morality of Things*; University of Chicago Press: Chicago, IL, USA, 2011.
79. Zwarteveen, M.; Boelens, R. Defining, researching and struggling for water justice: Some conceptual building blocks for research and action. *Water Int.* **2014**, *39*, 143–158. [CrossRef]
80. Bakker, K. Privatizing Water. In *Governance Failure and the World's Urban Water Crisis*; Cornell University Press: Ithaca, NY, USA, 2010.
81. Mumford, L. *Technics and Civilization*; Harcourt and Brace: New York, NY, USA, 1934.
82. Mumford, L. *The City in History: Its Origins, Its Transformations, and Its Prospects*; Harcourt, Brace and World: New York, NY, USA, 1961.

83. Cronon, W. *Nature's Metropolis*; W.W. Norton & Company: New York, NY, USA; London, UK, 1991.
84. Angelo, H.; Wachsmuth, D. Urbanizing urban political ecology: A critique of methodological actiyism. *Int. J. Urban Reg. Res.* **2014**, *39*, 16–27. [CrossRef]
85. Menga, F. Building a nation through a dam: The case of Rogun in Tajikistan. *Natl. Papers* **2014**, *43*, 479–494. [CrossRef]
86. Swyngedouw, E. Power, nature, and the city. The conquest of water and the political ecology of urbanization in Guayaquil, Ecuador: 1880–1990. *Environ. Plan. A* **1997**, *29*, 311–332. [CrossRef]
87. Kearney, M. *Reconceptualizing the Peasantry. Anthropology in a Global Perspective*; Westview Press: Boulder, CO, USA, 1996.
88. Illich, I. *In the Mirror of the Past: Lectures and Addresses, 1978–1990*; Marion Boyars: London, UK, 1992.
89. Illich, I. *Shadow Work*; Boyars Publishers: London, UK, 1981.
90. Escobar, A. Culture sits in places: Reflections on globalism and subaltern strategies of localization. *Political Geogr.* **2001**, *20*, 139–174. [CrossRef]
91. Illich, I. *H_2O and the Waters of Forgetfulness*; Marion Boyars: London, UK, 1986.
92. Ingold, T. *The Perception of the Environment*; Routledge: London, UK, 2000.
93. Lefebvre, H. *The Production of Space*; Blackwell: Oxford, UK, 1991.
94. Foucault, M. *Power/Knowledge: Selected Interviews and Other Writings 1972–1978*; Gordon, C., Ed.; Pantheon Books: New York, NY, USA, 1980.
95. Foucault, M. *Society Must be Defended. Lectures at the College de France 1975–1976*, Picador: New York, NY, USA, 2003.
96. Anders, G. *Die Antiquiertheit des Menschen. Part 1 and 2*; Beck: Munich, Germany, 1980.
97. Warner, J. Contested Hydro-hegemony: Hydraulic Control and Security in Turkey. *Water Altern.* **2008**, *1*, 271–288.
98. Valladares, C.; Boelens, R. (Re)territorializaciones en Tiempos de 'Revolución Ciudadana': Petróleo, Minerales y Derechos de la Naturaleza en el Ecuador. *Revista de Estudios Atacameños* Forthcom. April 2019, in press. (In Spanish)
99. Anand, N. Ignoring power: Knowing leakage in Mumbai's water supply. In *Urban Navigations: Politics, Space & the City in South Asia*; Anjaria, J., McFarlane, C., Eds.; Routledge: Delhi, India, 2011; pp. 191–212.
100. Perrow, C. *Normal Accidents: Living with High-Risk Technologies*; Princeton University Press: Princeton, NJ, USA, 1984.
101. Huber, A.; Gorostizac, S.; Kotsilaa, P.; Beltránd, M.; Armieroe, M. Beyond "Socially Constructed" Disasters: Re-politicizing the Debate on Large Dams through a Political Ecology of Risk. *Cap. Nat. Soc.* **2017**, *28*, 48–68. [CrossRef]
102. Proctor, R.; Schiebinger, L. (Eds.) *The Agnotology: The Making and Unmaking of Ignorance*; Stanford University Press: Stanford, CA, USA, 2008.
103. Sismondo, S. *An Introduction to Science and Technology Studies*; Wiley-Blackwell: Sussex, UK, 2010.
104. Beck, U. *Risk Society: Towards a New Modernity*; Sage: London, UK, 1992.
105. Levitas, R. Discourses of Risk and Utopia. In *The Risk Society and Beyond: Critical Issues of Social Theory*; Adam, B., Beck, U., van Loon, J., Eds.; Sage: London, UK, 2000.
106. Scott, A. Risk Society or Angst Society? Two Views of Risk, Consciousness and Community. In *The Risk Society and Beyond: Critical Issues of Social Theory*; Adam, B., Beck, U., van Loon, J., Eds.; Sage: London, UK, 2000.
107. Espeland, W.; Stevens, M.L. Commensuration as a social process. *Annu. Rev. Sociol.* **1998**, *24*, 313–343. [CrossRef]
108. Duarte-Abadía, B.; Boelens, R. Disputes over territorial boundaries and diverging valuation languages: The Santurban hydrosocial highlands territory in Colombia. *Water Int.* **2016**, *41*, 15–36. [CrossRef]
109. Mollinga, P. Water and politics: Levels, rational choice and south Indian canal irrigation. *Futures* **2001**, *33*, 733–752. [CrossRef]
110. Valladares, C.; Boelens, R. Extractivism and the rights of nature: Governmentality, 'convenient communities', and epistemic pacts in Ecuador. *Environ. Politics* **2017**, *26*, 1015–1034. [CrossRef]
111. Baillie Smith, M.; Jenkins, K. Disconnections and Exclusions: Professionalization, Cosmopolitanism and (Global?) Civil Society. *Global Netw.* **2011**, *11*, 160–179. [CrossRef]

112. Roth, D.; Boelens, R.; Zwarteveen, M. Property, legal pluralism, and water rights: The critical analysis of water governance and the politics of recognizing "local" rights. *J. Legal Plur. Unoff. Law* **2015**, *47*, 456–475. [CrossRef]
113. Kauffman, C.M. *Grassroots Global Governance: Local Watershed Management Experiments and the Evolution of Sustainable Development*; Oxford University Press: New York, NY, USA, 2017.
114. Dupuits, E.; Baud, M.; Boelens, R.; de Castro, F.; Hogenboom, B. Transnational grassroots movements defending water commons in Latin America: Professionalisation, expert knowledge and resistance. Forthcom. September 2019, in press.
115. Dukpa, R.D.; Joshi, D.; Boelens, R. Hydropower development and the meaning of place. Multi-ethnic hydropower struggles in Sikkim, India. *Geoforum* **2018**, *89*, 60–72. [CrossRef]
116. Cruikshank, J. Melting glaciers and emerging histories in the Saint Elias Mountains. In *Indigenous Experience Today*; de la Cadena, M., Starn, O., Eds.; Berg: Oxford, UK, 2007; pp. 355–378.
117. Hale, C. Rethinking Indigenous Politics in the Era of the 'Indio Permitido'. *NACLA Rep. Am.* **2004**, *38*, 16–21. [CrossRef]
118. Laurie, N.; Andolina, R.; Radcliffe, S. Ethnodevelopment: Social Movements, Creating Experts and Professionalising Indigenous Knowledge in Ecuador. *Antipode* **2005**, *37*, 470–96. [CrossRef]
119. Gaventa, J.; Cornwall, A. Power and Knowledge. In *Handbook of Action Research*; Reason, P., Bradbury, H., Eds.; SAGE: London, UK, 2001; pp. 70–80.
120. Grande, S.M.A. Beyond the Ecologically Noble Savage. Deconstructing the White Man's Indian. *Environ. Ethics* **1999**, *21*, 307–320. [CrossRef]
121. Paredes Peñafiel, A.P.; Li, F. Nourishing Relations: Controversy over the Conga Mining Project in Northern Peru. *Ethnos* **2017**. [CrossRef]
122. Horowitz, L.S. Interpreting Industry's Impacts: Micropolitical Ecologies of Divergent Community Responses. *Dev. Chang.* **2011**, *42*, 1379–1391. [CrossRef]
123. Robbins, P. Beyond Ground Truth: GIS and the Environmental Knowledge of Herders, Professional Foresters, and Other Traditional Communities. *Hum. Ecol.* **2003**, *31*, 233–253. [CrossRef]
124. Kerschner, C.; Wächter, P.; Nierling, L.; Ehlers, M. Degrowth and Technology: Towards feasible, viable, appropriate and convivial imaginaries. *J. Clean. Prod.* **2018**, *197*, 1619–1636. [CrossRef]
125. Law, J.; Mol, A. *Complexities: Social Studies of Knowledge Practices*; Duke University Press: Durham, NC, USA, 2002.
126. Li, F. Relating Divergent Worlds: Mines, Aquifers and Sacred Mountains in Peru. *Anthropologica* **2013**, *55*, 399–411.
127. Baud, M.; Boelens, R.; Damonte, G. Introducción Nuevos capitalismos y transformaciones territoriales en la Región Andina. Special Issue, Revista de Estudios Atacameños "New capitalisms and territorial transformations in the Andean region". Forthcom. April 2019, in press.
128. Long, N.; Long, A. *Battlefields of Knowledge. The Interlocking of Theory and Practice in Social Research and Development*; Routledge: London, UK; New York, NY, USA, 1992.
129. Beck, U. Ecology and the Disintegration of Institutional Power. *Organ. Environ.* **1997**, *10*, 52–65. [CrossRef]
130. Collins, H.M.; Evans, R. The third wave of science studies: Studies of expertise and experience. *Soc. Stud. Sci.* **2002**, *32*, 235–296. [CrossRef]
131. Jasanoff, S. Breaking the Waves in Science Studies: Comment on H.M. Collins and Robert Evans, 'The Third Wave of Science Studies'. *Soc. Stud. Sci.* **2003**, *33*, 389. [CrossRef]
132. Lukes, S. *Power: A Radical View*; Macmillan Press: London, UK, 2005.
133. Lukes, S. Power and the Battle for Hearts and Minds. *Millennium* **2005**, *33*, 477–493. [CrossRef]
134. Gaventa, J. Finding the Spaces for Change: A Power Analysis. *Inst. Dev. Stud. Bull.* **2006**, *37*, 23–33. [CrossRef]
135. Gramsci, A. *Selections from the Prison Notebooks*; International Publishers: New York, NY, USA, 1971.
136. Foucault, M. *The Will to Knowledge: The History of Sexuality Volume 1*; Penguin Books: London, UK, 1998.
137. Foucault, M. *The Birth of Biopolitics*; Palgrave MacMillan: New York, NY, USA, 2008.
138. Foucault, M. Afterword: The subject in power. In *Michel Foucault: Beyond Structuralism and Hermeneutics*; Dreyfus, H.L., Rabinow, P., Eds.; University of Chicago Press: Chicago, IL, USA, 1982; pp. 208–225.
139. Feindt, P.H.; Oels, A. Does discourse matter? Discourse analysis in environmental policy making. *J. Environ. Policy Plan.* **2005**, *7*, 161–173. [CrossRef]

140. Atkins, E. Building a dam, constructing a nation: The 'drowning' of Capel Celyn. *J. Hist. Sociol.* **2018**, *2018*, 1–14. [CrossRef]
141. Keck, M.E.; Sikkink, K. *Activists beyond Borders: Advocacy Networks in International Politics*; Cornell University Press: London, UK, 1998.
142. Swyngedouw, E. Globalisation or 'glocalisation'? Networks, territories and rescaling. *Camb. Rev. Int. Affairs* **2004**, *17*, 25–48. [CrossRef]
143. Dupuits, E.; Bernal, A. Scaling-up water community organizations: The role of inter-communities networks in multi-level water governance. *Flux N°* **2015**, *99*, 19. [CrossRef]
144. Hoogesteger, J.; Verzijl, A. Grassroots Scalar Politics: Insights from Peasant Water Struggles in the Ecuadorian and Peruvian Andes. *Geoforum* **2015**, *62*, 13–23. [CrossRef]
145. Boelens, R.; Hoogesteger, J.; Baud, M. Water reform governmentality in Ecuador: Neoliberalism, centralization and the restraining of polycentric authority and community rule-making. *Geoforum* **2015**, *64*, 281–291. [CrossRef]
146. Boelens, R.; Perreault, T.; Vos, J. (Eds.) *Water Justice*; Cambridge University Press: Cambridge, UK, 2018.
147. Atzle, A. Transnational NGO Networks Campaign against the Ilisu Dam, Turkey. In *Evolution of Dam Policies*; Scheumann, W., Hensegerth, O., Eds.; Springer: Berlin/Heidelberg, Germany, 2014; pp. 201–228.
148. Bebbington, A.; Humphreys-Bebbington, D.; Bury, J. Federating and defending: Water, Territory and Extraction in the Andes. In *Out of the Mainstream*; Boelens, R., Getches, D., Guevara-Gil, A., Eds.; Water Rights, Politics and Identity, Earthscan: London, UK; Washington, DC, USA, 2010; pp. 307–327.
149. Hidalgo, J.P.; Boelens, R.; Vos, J. De-colonizing water. Dispossession, water insecurity, and Indigenous claims for resources, authority, and territory. *Water Hist.* **2017**, *9*, 67–85. [CrossRef]
150. Harris, L. M.; Alatout, S. Negotiating hydro-scales, forging states: Comparison of the upper Tigris/Euphrates and Jordan River basins. *Political Geogr.* **2010**, *29*, 148–156. [CrossRef]
151. Lynch, B.D. Vulnerabilities, Competition, and Rights in a Context of Climate Change toward Equitable Water Governance in Peru's Rio Santa Valley. *Glob. Environ. Chang.* **2012**, *22*, 364–373. [CrossRef]
152. Vos, J.; Boelens, R. Sustainability standards and the water question. *Dev. Chang.* **2014**, *45*, 205–230. [CrossRef]
153. Mehta, L.; Veldwisch, G.J.; Franco, J. Introduction, Special Isue: Water grabbing? *Water Altern.* **2012**, *5*, 193–207.
154. Perramond, E. Adjudicating hydrosocial territory in New Mexico. *Water Int.* **2016**, *41*, 173–188. [CrossRef]
155. Mosse, D. Is good policy unimplementable? Reflections on the ethnography of aidpolicy and practice. *Dev. Chang.* **2004**, *35*, 639–671. [CrossRef]
156. Jongerden, J. Dams and Politics in Turkey: Utilizing Water, Developing Conflict. *Middle East Policy* **2010**, *17*, 137–143. [CrossRef]
157. Li, T.M. Rendering Society Technical: Government Through Community and the Ethnographic Turn at the World Bank in Indonesia. In *Adventures in Aidland: The Anthropology of Professionals in International Development*; Mosse, D., Ed.; Berghahn: Oxford, UK, 2011; pp. 57–80.
158. Vos, J.; Boelens, R. Neoliberal water governmentalities, virtual water trade, and contestations. In *Water Justice*; Boelens, R., Perrault, T., Vos, J., Eds.; Cambridge University Press: Cambridge, UK, 2018.
159. Espeland, W. *The Struggle for Water. Politics, Rationality, and Identity in the American Southwest*; University of Chicago Press: Chicago, IL, USA, 1998.
160. Boelens, R.; Vos, J. Legal Pluralism, Hydraulic Property Creation and Sustainability: The materialized nature of water rights in user-managed systems. *COSUST* **2014**, *11*, 55–62. [CrossRef]
161. Schlosberg, D. Reconceiving environmental justice: Global movements and political theories. *Environ. Politics* **2004**, *13*, 517–540. [CrossRef]

Article

Contested Knowledges in Hydroelectric Project Assessment: The Case of Canada's Site C Project

Karen Bakker [1,*] **and Richard Hendriks** [2]

1 Department of Geography, University of British Columbia, Vancouver, BC V6T 1Z2, Canada
2 Department of Civil and Mineral Engineering, University of Toronto, Toronto, ON M5S 1A4, Canada;
   r.hendriks@mail.utoronto.ca
* Correspondence: karen.bakker@ubc.ca; Tel.: +1-604-822-2663

Received: 26 March 2018; Accepted: 3 January 2019; Published: 26 February 2019

Abstract: This paper analyzes contestation over aspects of the Site C Project on the Peace River in northeastern British Columbia, Canada. The $10.7 billion project, which is now under construction, has been vigorously debated for over 30 years. Initially proposed in the 1980s, project approval was not granted following review by the BC Utilities Commission, as the need for the project was not established. In 2010, the provincial government enacted legislation to exempt the project from future review by the BC Utilities Commission; an environmental assessment was initiated in 2012 and a constrained review by the Commission was undertaken in 2017, after construction had commenced. The paper explores key examples of contested knowledge regimes within the review process, focusing on debates over cumulative effects and greenhouse gas emissions. The analysis provides technical examples of the ways in which differing societal values are deployed and co-produced within regulatory processes.

Keywords: hydroelectric development; hydropower; dam; indigenous peoples; first nations; Canada; Site C; British Columbia; environmental impacts; socio-economic impacts

1. Introduction

Over recent decades, large-scale hydropower projects have sparked significant debates around the world [1]. Proponents and opponents differ on whether the benefits (e.g., flood control, expansion of irrigated land, improved drinking water supply, and hydropower production) outweigh the costs (environmental impacts, cost over-runs, and displacement (particularly of Indigenous peoples)). The World Commission on Dams emphasized these concerns over a decade ago, and many countries and international organizations (such as the World Bank) overhauled their hydropower policies as a result [2]. Some experts even called for a new post-dam era; in the United, States, a leader in building big dams in the 20th century, considerable attention has now been refocused on dam decommissioning and environmental restoration [3–5].

In Canada, recent large-scale hydropower projects have raised substantive concerns: significant adverse environmental effects, potential health effects due to methylmercury bioaccumulation and exposure (particularly through fish consumption), infringement on First Nations' Aboriginal and Treaty rights, questionable economics (due to construction cost overruns, declining cost of alternatives and low export market prices for surplus electricity), and constrained regulatory processes, which do not adequately consider many of the key issues [6–10]. The Site C Project, currently under construction on the Peace River in northern British Columbia, exemplifies many of these concerns, and similar issues have arisen in relation to two other major hydropower projects currently under construction in Canada: the Keeyask Generating Station in Manitoba and the Muskrat Falls Project in Labrador.

The issues arising from Site C, as from these other two projects, raise questions respecting the role of large hydropower in Canada's energy future, as well as the regulatory processes for approval of these projects, given what World Bank evaluator Besant-Jones has termed "pervasive appraisal optimism" [11,12] (p. 61)). Such optimism typically entails under-estimating risk by relying on overly positive assessments of future gains and benefits, while under-estimating and/or externalizing environmental and socio-cultural costs [2,12–17]. Mega projects, such as large dams, are also typically plagued by rent-seeking provoked by the enormous investments required, as well as principal-agent problems—in which the state makes decisions that may benefit certain actors, but negatively impact others with less political power [18,19]. In the regulatory decision-making process, multiple contestations over knowledge production arise between proponents and opponents. In this paper, we focus on contestations over knowledge regarding data sources, analytical methods, and underlying assumptions deployed in the technical assessment process.

Our analysis is based on a series of research papers that specifically addressed environmental, economic, and socio-legal issues that were not adequately studied during the regulatory review process due to governmental circumvention of comprehensive environmental assessment and economic evaluation of Site C [20–30]. All raw data was obtained from published sources, most of which were provided by the proponent, British Columbia Hydro and Power Authority (BC Hydro), the BC Utilities Commission (the utility board responsible for regulating BC Hydro), the federal government of Canada, or the provincial government of British Columbia. In our analysis, we found that Site C poses greater environmental risk, has equal or higher greenhouse gas emissions, and generates fewer jobs compared to the cost-comparable alternative portfolios of resources available for meeting needs (including wind generation, upgrades to existing hydroelectric facilities and additional conservation and demand management). Our research was submitted to the BC Utilities Commission during its review of the economic aspects of Site C, and our findings were cited more often than those of any other intervenor [31]. However, our focus in this paper is not on the technical details of the analysis, but rather on the knowledge regimes contested during the regulatory process.

The following section of the paper provides a brief overview of our conceptual framework underlying our presentation of the contested production of knowledge via the regulatory process. Section 3 discusses the historical and regulatory context of the Site C Project. Section 4 provides examples of the regulatory shortcomings and gaps in the assessment and review process. Section 5 concludes with reflections on further research into the technical dimensions of the contestation of knowledge in the evaluation of major hydroelectric projects.

2. Contested Knowledges: Conceptual Discussion

The focus of this paper is the contestation of knowledge within regulatory review processes of resource development. However, what do we mean by resource regulation? To begin with, we are referring to regulation in a practical sense: the rules and procedures whereby different orders of government (in this case, the federal government of Canada and the provincial government of British Columbia) conduct reviews of proposed resource projects, consult with stakeholders, and issue approvals or rejections. A study of resource regulation must include an analysis of the different arguments mobilized by specific stakeholders: the project proponent, government agencies, and affected constituents, including Indigenous communities and local landowners, some of whom may be opposed to a proposed project. In short, our definition of regulation refers to the set of rules for project evaluation, together with procedures for application and enforcement of these rules.

Our analysis is, however, more expansive than this procedural definition of regulation. Social scientists studying resource management often argue that formal rules are only a subset of the *institutions*—understood as 'norms, rules, and customs' [32]—governing resource decision-making. These institutions are relatively coherent albeit dynamic, and embedded within broader political economic processes. Resource regulation is, from this perspective, a socio-political practice of negotiating resource allocation. Within contested processes of regulatory review of hydroelectric projects, regulation is thus a strategic terrain, or what Jessop terms a 'site of struggle': An object and generator of strategies as well as the product of past political struggles [33]. As a 'key site in the strategic codification of power relations' [33] (p. 248), regulation is a deeply socio-political process in which competing interests are advanced and defended. Moreover, regulatory approval processes are also key sites of epistemological production. Regulatory decision-making is both analytical and representational: an act of interpretation as well as adjudication, involving categorization and assessment of analytical procedures, data, and outcomes. In other words, the production of knowledge in decision-making processes—such as those for large utility projects—combines an analysis of socio-environmental issues together with contestation over the underlying knowledge regimes at stake [33–35]. Resource regulation is thus inherently (but by no means solely) a socio-political practice, insofar as the institutional framework for conducting regulatory reviews includes rules that define knowledge and legitimize authority. This socio-political practice is enacted by individuals, or groups of individuals, who share common 'storylines'—sets of (often contested) ideas that unite people in particular ways of communicating and producing knowledge about a problem, issue, or event [36]. Regulation is thus inevitably inscribed within ideological allegiances, as well as political alliances. Defining regulation as the social negotiation of the exploitation of a dynamic resource landscape thus requires an analysis of both the discursive and the technical dimensions of regulatory decision-making. In other words, at the heart of the regulatory process is a contestation over knowledge which is simultaneously technical, sociopolitical, and discursive [37,38].

Within this conceptual framework, our specific goal in this paper is to provide examples of knowledge contestation concerning two critical environmental knowledge domains—cumulative effects and greenhouse gas emissions—that were at the core of debates over Site C; these issues are also highly relevant to international discourse over hydroelectric projects. In undertaking this analysis, our paper directly addresses some of the key questions posed in this special issue [39]. We provide examples of how these two knowledge domains were rigidly interpreted in favor of some stakeholders over others during the regulatory review process. Underlying this analysis is a split between distinct "epistemic communities" (BC Hydro on the one hand, and a coalition of environmental groups and Indigenous communities on the other). We explore how one dominant epistemic community successfully claimed legitimacy and authority, and how these dominant epistemologies were linked to norms, beliefs, and values about impacts and benefits of hydroelectric development. To frame this analysis, the next section provides a brief overview of the historical and regulatory context of the Site C Project.

3. Historical and Regulatory Context of British Columbia's Site C Project

Canada emerged from the Second World War as a hydroelectric superpower; only the United States generated more hydropower than Canada, and only Norway generated more per capita [40–42]. The majority of this hydropower development occurred in northern Canada, far from densely populated regions in the south, but often with significant impacts on local Indigenous communities; these impacts on Indigenous communities were politically controversial but relatively under-studied, in a pattern characterized as "Hydraulic Imperialism" [43]. The Canadian provinces of Quebec, Ontario, Manitoba and British Columbia emerged as hydropower leaders. Significant hydropower development took place from the 1930s onwards, creating disputes with Indigenous peoples across Canada due to negative effects from displacement, flooding of traditional hunting and harvesting territories, and infringement of Treaty rights [44–47]. In British Columbia (as in western Canada more generally), postwar debates over resource extraction and development intensified over the second half of the 20th century as Indigenous peoples gained recognition of their rights in the courts [48–51].

Site C is among the costliest infrastructure projects ever undertaken by the province of British Columbia: originally budgeted in 2011 at C$7.9 billion, the project cost is now estimated at C$10.7 billion, a real dollar increase of more than 25%, with a potential for further cost increases. The Project is currently under construction on the Peace River in northeastern British Columbia and is designed to provide 1145 MW of capacity and 5286 GWh/year of energy, starting in late 2024. Supporters of the project—notably BC Hydro, the provincial Crown corporation developing the project—have promoted it as "clean energy."

The genesis for the Site C Project lies in a policy first formulated more than a half-century ago, in the 1950s. The Two Rivers Policy was conceived by the then Premier of British Columbia, William Andrew Cecil (W.A.C.) Bennett, and called for large-scale hydroelectric development on both the Peace River and Columbia River watersheds. The result of the Two Rivers Policy was the development on the Peace River of the W.A.C. Bennett Dam and GM Shrum Generating Station in 1968 (2730 MW) and the Peace Canyon Dam and Generating Station in 1980 (694 MW) (see Figure 1). This enabled the creation of a large electricity supply that powered industrial growth, served growing demand in the rapidly developing southern part of the province of British Columbia, and provided revenues from export of electricity surpluses.

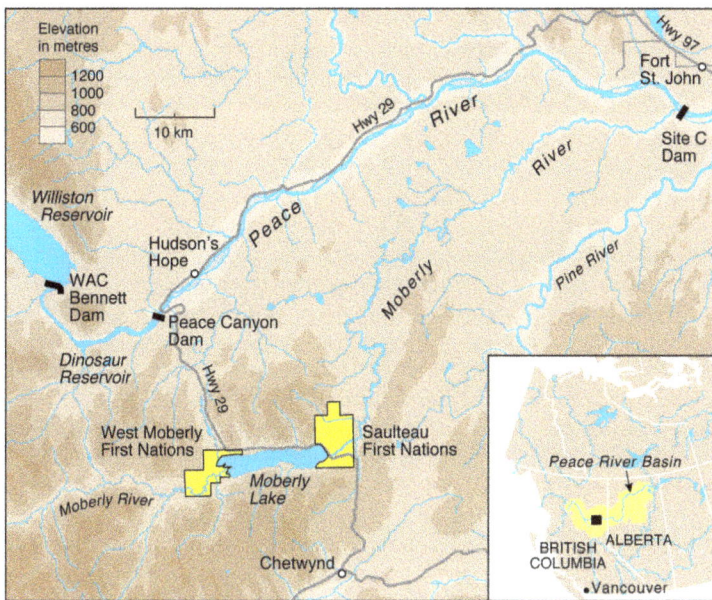

Figure 1. Hydroelectric developments on the Peace River [52].

By the early 1980s, BC Hydro was planning to move forward with its third hydroelectric facility on the Peace River, the Site C Project. The BC Utilities Commission was tasked with reviewing the project's justification, design, impacts and other relevant matters, and recommending whether and under what conditions to issue an Energy Project Certificate, which encompassed the issuance of a Certificate of Public Convenience and Necessity [53]. At that time, BC Hydro's original and final submissions to the Commission based the need for the Site C Project on the electricity requirements identified in its load forecasts [53]. BC Hydro's 1981 "probable" or "mid load" forecast was for BC Hydro system-wide energy demand to increase to 59,700 GWh/year by 1992–1993 [53]. Upon review, the Commission raised several "major issues" respecting the demand forecasts, as detailed in its report, including: forecast methodology, the role and forecast of key underlying variables, specific factors such as industrial sector growth, technological change, fuel shifting, conservation and self-generation, and prospects and potential in the export market.

Though the Commission's report was written over 30 years ago, the major issues raised at that time remain salient. The Commission concluded that: "Hydro's 'probable' load forecast should be considered as optimistic" [53] (p. 85) and recommended as follows:

> *The Commission recommends that Cabinet defer issuing an Energy Project Certificate for Site C until an acceptable load forecast demonstrates that construction of Site C must begin immediately in order to avoid supply deficiencies, and a comparison of alternative system plans demonstrates that Site C is the best project to meet the anticipated shortfalls [53] (p. 23).*

The conclusions reached by the Commission would prove to be illustrative of the value for the public interest of thorough, evidence-based consideration of proposed large-scale resource projects. The acceptance of the Commission's recommendations by the Government of the day would also prove to be prudent: as of 2017–2018, the system-wide energy demand forecasted in 1981 by BC Hydro for the year 1992–1993 has only just materialized, over 25 years later than initially forecast by BC Hydro [54].

The Site C Project was shelved for over two decades. However, in the mid-2000s, the Two Rivers Policy was revived when BC Hydro announced its intentions to reconsider developing Site C. This time, the provincial government was extremely supportive.

With the enactment of the *Clean Energy Act* in 2010, BC Hydro was no longer required to obtain a Certificate of Public Convenience and Necessity for the Site C Project, the legislation having exempted it from review and approval by the BC Utilities Commission. This exemption fast-tracked project development, despite the lack of growth in domestic electricity demand in British Columbia over the past decade [55], declines on the order of 60% in export market prices over that same period [56], and substantial reductions in the cost of alternative resources for meeting the electrical energy [57] and capacity [58] requirements of BC Hydro—all of which pointed to a need for policy reconsideration.

Site C was required to undergo a joint federal and provincial environmental assessment by an independent Joint Review Panel (relevant legislation includes the *Canadian Environmental Assessment Act* and the British Columbia *Environmental Assessment Act*) [59]. In February 2012, the federal and provincial Ministers of Environment finalized a Panel Agreement to conduct a cooperative environmental assessment, including the terms of reference for a Joint Review Panel (JRP), and guidelines for an environmental impact statement [60]. The assessment included an 8-month Joint Review Panel phase, in which the Panel was expected to: review nearly 30,000 pages of documentation; issue and review responses to information requests; determine the sufficiency of the proponent's environmental impact statement; consult with the public and Indigenous groups; hold public hearings; and prepare its final report and recommendations to government. The Panel's final report was issued in May 2014 and, following additional consultation with Indigenous groups concerning conditions for approval, environmental assessment approvals were granted in October 2014. These approvals were given despite recommendations from the Joint Review Panel that key matters related to the economic evaluation of Site C, including project costs and future electricity requirements, be referred to the BC Utilities Commission for further review prior to construction [61]. The provincial government rejected these recommendations and, as the shareholder for BC Hydro, issued a final investment decision in December 2014. By July 2015, construction commenced on the Site C Project on what was estimated at that time to be a 9-year construction period.

However, in May 2017, a new provincial government was elected and initiated the BC Utilities Commission Inquiry Respecting Site C. Pursuant to part 5 of the British Columbia *Utilities Commission Act*, the new government asked the Utilities Commission to inquire into Site C and advise on the implications of completing the project as scheduled, suspending the project, or terminating construction and remediating the site. The Inquiry focused on the Site C Project expenditures to that date, the likelihood of achieving the proposed budget, the implications to ratepayers of suspending or continuing the project, and the potential for an alternative portfolio of resources to meet BC Hydro's forecasted future requirements at a similar or lower cost [62]. The BC Utilities Commission Inquiry final report, issued in November 2017, provided a detailed analysis, determined that the project was unlikely to remain on schedule or on budget, and identified an alternative portfolio capable of meeting BC Hydro's needs at a similar cost to ratepayers [31]. Nonetheless, the provincial government decided to continue with construction of the project, with planned completion in 2025 (delaying the original schedule by one year).

4. Contesting Environmental Knowledge in Regulatory Decision-Making: Cumulative Effects and Greenhouse Gas Emissions

This section explores technical dimensions of the contestation of knowledge with respect to cumulative effects and greenhouse gas emissions within the regulatory decision-making process for the Site C Project. Prior to presenting this analysis, some context is necessary. The Panel Agreement between the federal and provincial governments concerning the environmental assessment of the Site C Project scoped the assessment to include environmental, economic, social, health and heritage effects, and included consideration of alternatives to the Project [60]. However, the Agreement imposed several time and resource constraints on the work of the three-person Panel, and left open to interpretation several pivotal aspects of the assessment, including in relation to cumulative effects and greenhouse gas emissions. With respect to time and resource constraints, while it is not unusual for environmental assessments to include large volumes of material, it *is* unusual (at least in the Canadian context) to limit a review of a large-scale hydroelectric project by an independent panel to be completed within 225 days, or less than eight months [60]. In contrast to the Site C process, the four-person Clean Environment Commission reviewing Manitoba's 695 MW Keeyask Hydroelectric Project was provided a total of 18 months, and the five-member Manitoba Public Utilities Board was provided an additional 14 months. The Lower Churchill Hydroelectric Generation Project in Labrador was reviewed by a five-person environmental assessment panel over 32 months, followed by a review over a nine-month period by the four-person Newfoundland and Labrador Utilities Board.

The implications of the Panel's time and resource constraints are evident in its findings related to Site C's cost—a central consideration in evaluating it relative to the available alternatives—a key mandate and responsibility of the Panel:

> *The Panel cannot conclude on the likely accuracy of Project cost estimates because it does not have the information, time, or resources. This affects all further calculations of unit costs, revenue requirements, and rates [61].*

The decision to impose significant constraints on the environmental assessment created a situation in which a comprehensive analysis of the relevant information could not be undertaken. Future environmental reviews of similar project proposals should be provided time and resources in order to ensure that the evidence can be properly gathered and analyzed in order to provide robust, evidence-based and defensible recommendations and conclusions to decision-makers.

4.1. Cumulative Effects

The first example of the contestation of knowledge that we present is that of "cumulative effects", defined by the federal government's Canadian Environmental Assessment (CEA) Agency as "changes to the environment that are caused by an action in combination with other past, present and future human actions" [63]. The question of cumulative effects became one of the focal points of the contestation of knowledge between BC Hydro and interveners in the regulatory process for Site C. The treatment of the effects of prior development within the region surrounding the project is of critical importance in this case, since development has been extensive [64], and includes the existing WAC Bennett Dam with the large Williston Reservoir, and Peace Canyon Dam with the Dinosaur Reservoir. These projects inundated more than half the length of the Peace River in British Columbia.

The Panel Agreement between the federal and provincial governments concerning the environmental assessment of the Site C Project scoped the assessment to include cumulative effects that are likely to result from the Project in combination with other projects or activities that have been or will be carried out [60]. According to the CEA Agency, a cumulative effects assessment includes a study area that is large enough to allow for the assessment of effects of the project or action being proposed, as well as identification of other projects or actions that have occurred, exist, or may yet occur and that may also affect the valued components of the ecosystem that are under study [63]. The cumulative effects assessment should address the incremental additive effects of the proposed action and the other actions, including against thresholds or policies. Following the consideration of proposed mitigation measures, the significance of residual effects is clearly stated and defended [63]. The CEA Agency guidance outlines the following options for scoping the temporal boundaries for the cumulative effects assessment, each of which progresses further back in time: (i) when impacts associated with the proposed action first occurred; (ii) existing conditions; (iii) the time at which a certain land use designation was made (e.g., lease of crown land for the action, establishment of a park); (iv) the point in time at which effects similar to those of concern first occurred; or (v) a past point in time representative of desired regional land use conditions or pre-disturbance conditions (i.e., the "historical baseline"), especially if the assessment includes determining to what degree later actions have affected the environment [63].

In preparing the draft guidelines for the preparation of the environmental impact statement (EIS) for the Site C Project, BC Hydro proposed a methodology for evaluating cumulative effects using a temporal baseline based on *existing* conditions, including the effects the two existing hydroelectric project and other prior development in its baseline case [65].

In contrast to the BC Hydro methodology, during the review of the draft guidelines, several First Nations suggested an alternative approach based on the last of the five options contained in the CEA Agency guidance.

> We also noted in our comments on the EIS Guidelines that in order to assess the cumulative environmental effects of the proposed Project and the cumulative implications for [Constitution Act] Section 35(1) rights, the initial case for consideration or the "baseline case" must include the historical circumstances, since these circumstances are essential to the understanding of the seriousness of the potential impacts on established Treaty rights, and which circumstances would include the WAC Bennett Dam, Peace Canyon Dam and the Peace Project Water Use Plan [66].

Parks Canada, a federal government department, raised concerns similar to those of the First Nations in its comments on the draft guidelines:

> BC Hydro's approach to cumulative effects assessment for the Site C project is based on accepting the present state of the Peace River and the Peace Athabasca Delta as the baseline condition upon which to add the incremental impacts from construction and operation of the Site C dam. This approach does not fully consider the cumulative impact from all BC Hydro's flow management operations against an unregulated, undammed river. The point here is that the WAC Bennett dam was proposed, and constructed, in a time when no environmental assessment legislation or process was in place. If the Bennett dam project was proposed today it is very unlikely that such dramatic regulation of the flow regime on the Peace River would be found to be justifiable in the circumstances. The project would then either be modified to limit the scope of the impact to the hydrology of the Peace River, or the project would be cancelled. Using the existing conditions as the baseline, conveniently incorporates the extensive impacts from the WAC Bennett Dam into the baseline, and avoids looking holistically at the collective impacts of all BC Hydro's flow management upon the flows and ecology of the Peace River downstream of the dams. The usual argument for not doing this in cumulative effects assessments is that it is unfair to saddle the current proponent or project with the responsibility to cumulatively assess the impacts of all relevant projects and activities on the receiving environment. This argument often succeeds and hence cumulative effects assessment typically becomes more an exercise in assessing incremental effects of the proposed project, than a comprehensive assessment of cumulative effects. The current circumstances before us are unique in that BC Hydro presently manages two dams on the Peace River and, in the event of the Site C project being approved, would manage all three dams on the Peace River. This provides a compelling argument for BC Hydro to assess the full impact of its operations on the Peace River. The WAC Bennett Dam damage is done and no one is going to ask for

that dam to be decommissioned. Given that is the case, the tolerance for accepting Site C incremental impacts to downstream environments should be correspondingly low [67].

BC Hydro responded to these concerns by noting that the reliability of the environmental assessment would be compromised by, in its view, the lack of available data concerning historical baseline conditions.

BC Hydro also believes that a pre-development case would be inherently unreliable. There are two methods by which a pre-development case could be developed. Firstly, if direct, reliable data about the pre-development state is available, that information could be used. BC Hydro is not aware of data from the pre-development era. Secondly, in the absence of data from the pre-development era, a model would have to be built based on various assumptions in order to emulate pre-development conditions. The longer the period of time between current conditions and the pre-development era, the greater the uncertainty would be [68].

BC Hydro's approach to the baseline case is identical to that used for the environmental assessment of Nalcor Energy's Lower Churchill Hydroelectric Generation Project. In that instance, the Joint Review Panel raised concerns about this approach that would have been known to the government agencies approving the guidelines for the Site C environmental impact statement, particularly the Canadian Environmental Assessment (CEA) Agency:

The Panel concluded that Nalcor's approach to cumulative effects assessment was less than comprehensive and that participants had raised valid concerns that contributed to a broader understanding of the potential cumulative effects of the Project. The Panel recognized the challenges involved, including limited information about past projects such as the Churchill Falls project, and the built-in disincentive for proponents to identify adverse cumulative effects when they are perceived as a potential threat to project approval [69].

In a critical review prepared in relation to Manitoba Hydro's Bipole III transmission line, Gunn and Noble argue the importance of this retrospective analysis, or the "historical circumstances" referred to above:

The development of a baseline for evaluation of cumulative effects is more than a description of current conditions, which alone can discount the effects of past changes as simply the 'new normal.' Baseline development requires a retrospective analysis of how VEC conditions have changed over time and whether that change is significant in terms of the sustainability of the VEC [70].

In finalizing the guidelines for the Site C Project, the CEA Agency and the British Columbia Environmental Assessment Office accepted BC Hydro's position concerning the temporal boundaries for the cumulative effects assessment, based on BC Hydro's position that there was insufficient data to prepare a pre-development baseline. As a compromise, the CEA Agency added the following requirement pertaining to previous developments:

The EIS will include a narrative discussion of existing hydro-electric generation projects on the Peace River (W.A.C. Bennett Dam and the Peace Canyon Dam). The narrative will include the description of any existing studies of changes to the environment resulting from those projects that are similar to potential changes resulting from the project, including any mitigation measures that were implemented, and any long term monitoring or follow up program that were conducted. The effectiveness of those mitigation measures and key results of monitoring or follow-up programs would be described. This narrative discussion should include historical data, where available and applicable, to assist interested parties to understand the potential effects of the Project and how they may be addressed [60].

BC Hydro tabled its environmental impact statement based on the approved guidelines. The Joint Review Panel, which was struck only after the impact statement was tabled, reviewed the documents and during the public hearings issued several requests for additional information related to cumulative effects, including to local, Indigenous and government participants in the environmental assessment process. During those hearings, the Panel queried BC Hydro concerning its approach to the temporal baseline in conducting the cumulative effects assessment:

> I'd like to know more about the arguments that you've used and managed; they must be magical. In addition, managed to convince the agency and environmental assessment office of the Province to go ahead and exclude the two dams. Because even if there is a narrative, it does not preclude the proponent to do a cumulative effect assessment, especially if in the narrative you acknowledge that the previous dams had effects.
>
> . . .
>
> Do I understand that the major argument was that you did not have the data? I mean, the Peace Canyon Dam had the Environmental Impact Assessment done. The Bennett Dam—when you build a dam, you have data. I mean, even if it is 1957, you would have data [71].

Upon receipt of the additional information from participants, and completion of the hearing, the Panel prepared its final report, which provided specific conclusions and recommendations regarding the environmental effects of the Site C Project. Table 1 presents a comparison of the number of significant environmental effects determined by the Joint Review Panel for the Site C Project compared to other projects assessed as having significant environmental effects under the *Canadian Environmental Assessment Act*. The table demonstrates that Site C was predicted to have more significant environmental effects than any project ever assessed under the *Act* [22,25,62]. This finding may seem surprising, and arises from the extensive biodiversity of the Peace River valley, which plays a key role in the ecology of the northeastern part of British Columbia, as well as being vital to the exercise of First Nations rights, including hunting, fishing and gathering.

Table 1. Significant environmental effects arising from projects assessed under the *Canadian Environmental Assessment Act* (1992–2017).

Projects Assessed under the Canadian Environmental Assessment Act	Number of Significant Environmental Effects
Site C Project	20
New Prosperity Gold and Copper Mine Project	5
Lower Churchill Hydroelectric Generation Project	5
Jackpine Mine Expansion Project	5
Pacific Northwest LNG	3
Encana Shallow Gas Infill Development Project	2
Cheviot Coal Project	2
Kemess North	2
Northern Gateway Project	1
White Pines Quarry	1
LNG Canada	1
Labrador-Island Transmission Link	1

Data source: CEA Agency, Environment and Climate Change Canada.

Of the numerous significant environmental effects identified by the Panel, many were cumulative effects, including on fish, vegetation and ecological communities, several wildlife species, heritage resources, and the current use of lands and resources for traditional purposes by Indigenous peoples. In addition to conclusions regarding the significance of the cumulative effects of the Site C Project, the Panel also drew more general conclusions respecting the cumulative effects assessment for the Site C Project.

While the Panel understands that, according to the CEA Agency Operational Policy Statement, past or existing physical activities may be helpful in predicting the effects of a designated project, it is not the sole intent of assessing past or existing projects. The Panel believes that providing a narrative with no analysis or conclusions on the cumulative effects of the existing hydroelectric facilities does not suit the needs of a cumulative effects assessment [61] (p. 259).

. . .

The Panel disagrees with BC Hydro's assertion that there was limited information available to conduct a cumulative effects assessment, particularly given the information from participants. The Panel received numerous testimonies from Aboriginal and non-Aboriginal participants about the effects of the Bennett and Peace Canyon Dams. This information was provided first-hand (by people who were alive at that time) or second-hand (by participants who learned of the effects from previous generations). The Panel understands that there is existing information in various formats such as air photos, environmental impact studies, research from various provincial and independent bodies, and historic maps of changing land tenure [61] (p. 259).

The Panel concludes that, whether the Project proceeds or not, there is a need for a government-led regional environmental assessment including a baseline study and the establishment of environmental thresholds for use in evaluating the effects of multiple, projects in a rapidly developing region [61] (p. 261).

. . .

Because of the importance of cumulative effects assessment, the Panel concludes that there is a need to improve and standardize cumulative effects assessment methods [61] (p. 262).

In summary, while the Panel was able to gather information to support its determinations respecting the significance of the cumulative effects of the Site C Project, this information was prepared and made available very late in the assessment process. As the Panel observed, additional information existed that was not made available to the Panel to support its assessment. In this instance, the contestation of knowledge could have been resolved by early provision of available knowledge, which would have supported a more comprehensive analysis. However, given that this analysis would have required scoping guidelines for the environmental impact assessment in a manner unfavorable for the proponent, the exclusion of this data by BC Hydro is perhaps unsurprising.

4.2. Greenhouse Gas Emissions

The second example of the contestation of knowledge pertains to greenhouse gas (GHG) emissions. The guidelines for the preparation of the environmental impact statement required: an estimate of the multi-year GHG emissions profile associated with the construction and ongoing operations of the Project; an estimate of the net change in GHG emissions from current conditions to post-inundation scenarios; and a comparison of the GHG profile of the Project with other electricity supply options [60].

The Site C Project was promoted, in part, for its purported reductions in GHG emissions relative to the available alternative portfolios of resources for meeting the energy and capacity requirements of BC Hydro [72]. While the generation of hydroelectricity results in no GHG emissions, the construction activities and reservoir creation for the Site C Project do result in meaningful emissions. In preparing its GHG emissions estimate, BC Hydro considered both "likely" (lower emission) and "conservative" (higher emission) scenarios.

Construction emissions from Site C result from fuel and electricity use associated with the dam and generating station, spillways, quarried and excavated materials, transmission lines, access roads, highway realignment and worker accommodation facilities. GHG emissions embedded in construction materials are also included in BC Hydro's analysis based on a life cycle assessment. The conservative scenario during construction assumes 15% greater fuel emissions and greater life-cycle emissions for construction materials than in the likely scenario [73].

Operations emissions from Site C result from the decomposition of flooded biomass within the reservoir, increasing emissions of both CO_2 and CH_4 over baseline conditions. The conservative scenario for operations assumes no storage of carbon (i.e., harvested timber stored as building materials for the construction industry) and no burial of biomass, while the likely scenario assumes that merchantable timber is converted entirely into stored carbon and that 30% of non-merchantable timber cleared would be buried (and therefore indefinitely stored) [73].

As presented in Table 2, the construction of the Site C Project and the inundation of biomass resulting from reservoir creation generate greenhouse gas emissions, estimated by BC Hydro at between 5.3 and 7.3 million metric tons of CO_2e over the first 100 years of project operations [74]. Both the likely and conservative estimates assume that the reservoir emissions occur mostly in the early years following inundation, and eventually decline to resemble emissions prior to reservoir creation.

Table 2. BC Hydro estimates of GHG emissions—Site C Project [74].

Activity	GHG Emissions Estimates	
	Conservative	Likely
	(tonnes CO_2e)	(tonnes CO_2e)
Construction (8 years)	1,483,708	997,225
Operations (100 years)	5,824,820	4,343,633
TOTAL (108 years)	7,308,528	5,340,858

During the environmental assessment of the Site C Project, BC Hydro developed an alternative portfolio of resources for meeting the energy and capacity needs of BC Hydro without the use of fossil fuels. This alternative "clean" portfolio consisted of available resources for meeting needs considering regulatory, planning and technical constraints, including the requirements of the *Clean Energy Act*. These resources included upgrades to existing hydroelectric facilities, municipal solid waste (MSW) generation, pumped storage hydroelectric, and wind.

Of potential concern in light of the *Clean Energy Act*, particularly the requirement to generate at least 93% of electricity from non-emitting sources, is that the MSW generation included in the alternative portfolio emits carbon dioxide at levels on par with diesel generation [74]. The analysis presented by BC Hydro during the environmental assessment did not seek to minimize the GHG emissions in this alternative portfolio by optimizing the selection and operation of the available resources. Our subsequent research illustrated that several options were available to optimize the greenhouse gas emissions of the alternative portfolios, including replacing the MSW generation with lower emitting resources [24].

During the Site C Joint Review Panel hearings, minimal attention was paid to the issue of GHG emissions. Over the course of the 25 days of hearings, the JRP dedicated one afternoon session to atmospheric and air quality issues, of which GHG emissions was one of five sub-topics [75]. No technical evidence concerning GHG emission estimates of Site C and the alternative portfolios was presented to the Panel during the hearings, other than by BC Hydro. The JRP undertook no independent analysis of the findings of BC Hydro, and solicited no additional evidence through undertakings by BC Hydro or other interveners. Yet, in its final report to the Ministers, the JRP reached the conclusion that the Site C Project "would produce a vastly smaller burden of greenhouse gases than any alternative save nuclear power, which B.C. has prohibited" [61].

In response to an information request from the JRP to estimate the GHG emissions that would be avoided by the Site C Project, inclusive of export of surplus electricity, BC Hydro provided the following table (Table 3), which it subsequently presented in its submission to the BC Utilities Commission Inquiry Respecting Site C. Table 3 presents the avoided GHG emissions of the Site C Project over 100 years compared to the alternative clean portfolio, as well as the additional avoided GHG emissions resulting from exporting the surplus energy created by Site C.

Table 3. BC Hydro's comparative GHG benefits of the Site C Project (2024–2124) [76].

Attribute	Units	Site C Energy Used in BC	Site C Surplus Energy Exported	Total
Generation (100 years)	(GWh)	476,300	33,700	510,000
Avoided GHGs—alternative "clean" portfolio	(Mt CO_2e)	19	15	34

BC Hydro then drew the following conclusions:

The portfolio including the Project has lower operational GHG emissions than both portfolios not including the Project. The Clean Generation portfolio selects a municipal solid waste resource option, which includes GHG emissions from fuel combustion [77].

In arriving at the conclusion that the Site C Project avoids 19 Mt relative to the alternative portfolio in relation to electricity used in BC, BC Hydro omits a key consideration. The MSW generation developed by BC Hydro and included in the alternative clean portfolio used in the Site C Project environmental impact statement was not ultimately selected in any of BC Hydro's portfolios developed for its submission to the BCUC Inquiry Respecting Site C. With the removal of MSW generation from the alternative clean portfolio, the GHG emission advantage of the Site C Project in terms of electricity sold in BC disappears entirely.

Secondly, as our research illustrated, the total energy surplus from Site C based on information presented by BC Hydro during the BCUC Inquiry Respecting Site C is approximately 20,400 GWh, much less than the 33,700 GWh reported in Table 3 [29]. With respect to the emissions avoided through the export of surplus electricity from the Site C Project, BC Hydro incorrectly compared electricity from Site C to the potential emissions from existing electricity resources rather than from other *new* resources that could also be developed to replace existing higher-emitting resources.

Finally, while Table 3 excludes construction phase emissions from Site C and the alternative portfolio, it also omits entirely the GHG emissions from the operations phase of the Site C Project, which were estimated at from 4.3 to 5.8 Mt, as shown in Table 2.

In summary, the alternative portfolio of clean resources produces fewer greenhouse gas emissions than the Site C Project, even after considering the emissions reductions resulting from the export of the energy surplus resulting from Site C. While the GHG emissions from both the Site C Project and the alternative portfolio are low compared to emissions from a natural gas facility of comparable capacity and energy production, our findings illustrate that there is no greenhouse gas emissions advantage to the Site C Project compared to alternative clean portfolios [29].

Indeed, in its terms of reference for the BCUC Inquiry respecting Site C, the provincial government included a requirement for the Commission to identify an alternative portfolio that could provide similar benefits to the Site C Project at similar or lower costs with "maintenance or reduction of 2016-2017 greenhouse gas emission levels" [31]. In preparing these terms of reference, the Provincial Government seemed unaware of the actual emissions profile of the Site C Project. In its final report, the Commission observed that while the alternative portfolio satisfied this requirement, the Site C Project did not [31].

In this instance, the underlying assumption that the Site C Project produces lower GHG emissions than the available and cost-comparable alternative portfolios resulted in a failure to properly contest the evidence during the environmental assessment. Time constraints on the regulatory process reduced the potential for the Joint Review Panel to undertake or solicit an independent review of BC Hydro's GHG emissions analysis. Based on our subsequent research, the findings of such an analysis would have contradicted the prevailing narrative surrounding the justification for proceeding with the Site C Project, including the incorrect postulate that Site C reduces GHG emissions more than the available alternatives.

5. Conclusions

Despite the deteriorating economic case for continuing with Site C, the provincial government rendered its decision in December 2017 to carry on with the project. In making this decision, the government acknowledged unresolved Treaty rights and economic risks but indicated that in the government's view, the project was too far advanced to halt [78]. The academic research conducted on Site C—including the project's implications for Treaty rights, significant adverse environmental effects, lack of greenhouse gas emissions reduction benefits, and lower employment benefits compared to the alternatives—was not referenced in the government's public announcement. In early 2018, an application for an injunction was launched by affected First Nations, which was later dismissed by the court [79–82].

Why was the Site C Project advanced to construction despite the numerous shortcomings? Additionally, why did the project continue despite the (albeit belated) BC Utilities Commission review challenging the "pervasive appraisal optimism" of the evidence offered by the proponent, BC Hydro? At first glance, the answer is simple: with the enactment of the *Clean Energy Act*, the government exempted the project from regulatory review by the Commission, and then—as publicly promised—pushed the project "past the point of no return" prior to the (belated and limited) review [83]. This was compounded by systemic shortcomings in the regulatory process for Site C, including a failure to consider the evolving framework of Indigenous rights (an important topic beyond the scope of this paper), and an overly constrained environmental assessment process.

The Site C Project is an example of what Boelens, Shah and Bruins term in the introductory article to this special issue [39] as "manufactured ignorance" via the exclusion of specific questions, analyses, data, and analytical methods from consideration. In the case of Site C, this "manufactured ignorance" was criticized by a wide range of stakeholders, including affected Indigenous communities, a previous CEO of BC Hydro, the former Chair of the Joint Review Panel (a former senior federal civil servant), and a federal government department [69,84–86]. Of course, this contestation of knowledge occurs in the context of uncertainties and unpredictability of planning mega-projects such as large dams, which inevitably understates the challenges and difficulties which arise in such projects. It is also important to emphasize the point made above that Indigenous treaty rights issues, while beyond the scope of this paper, are significant in the context of Canada's Truth and Reconciliation Commission, and its recent commitment to implementing the UN Declaration on the Rights of Indigenous peoples [87–90]. Nonetheless, our analysis reveals that the choices made by the government and BC Hydro regarding the evaluation of cumulative effects and GHG emissions favored a specific outcome: developing the Site C Project. The question thus arises: what forms of regulatory review could reduce the possibility for "manufacturing ignorance" in the future? How might legal, policy, and procedural changes create the space for more accurate, comprehensive, and inclusive evaluation? Our hope is that our analysis of the Site C case provides lessons for the future on these crucially important questions.

Author Contributions: K.B. and R.H. jointly conceived the initial study. R.H. was the lead author on the majority of background reports and submissions to the BCUC. K.B. provided funding and participated in the writing and editing of the background reports and BCUC submissions. K.B. was the lead author on this present publication; R.H. reviewed and edited this publication.

Funding: Academic funding was received by Bakker from the Social Sciences and Humanities Research Council of Canada and the University of British Columbia. Post-hoc funding for participating in the Site C Inquiry was received from the BC Utilities Commission (Participant Access Cost Award) for Hendriks (but not Bakker).

Conflicts of Interest: Hendriks provided environmental assessment advice to the Treaty 8 Tribal Association from 2010 to 2014. The funders had no role in the design of the study; in the collection, analyses, or interpretation of data; in the writing of the manuscript, and in the decision to publish the results.

Appendix A

All reports published by the Program on Water Governance (including those submitted to the BC Utilities Commission) are available online at www.watergovernance.ca/projects/sitec.

References

1. Zarfl, C.; Lumsdon, A.E.; Berlekamp, J.; Tydecks, L.; Tockner, K. A global boom in hydropower dam construction. *Aquat. Sci.* **2015**, *77*, 161–170. Available online: https://link-springer-com.ezproxy.library.ubc.ca/article/10.1007%2Fs00027-014-0377-0 (accessed on 10 January 2019). [CrossRef]
2. World Commission on Dams. *Dams and Development: A New Framework for Decision-Making*; Earthscan Publications Ltd.: London, UK, 2000; ISBN 978-185-383-798-2.
3. George, M.W.; Hotchkiss, R.H.; Huffaker, R. Reservoir sustainability and sediment management. *J. Water Resour. Plan. Manag.* **2016**, *143*, 04016077. Available online: https://ascelibrary-org.ezproxy.library.ubc.ca/doi/abs/10.1061/(ASCE)WR.1943-5452.0000720 (accessed on 10 January 2019). [CrossRef]
4. Hammersley, M.; Scott, C.; Gimblett, R. Evolving conceptions of the role of large dams in social-ecological resilience. *Ecol. Soc.* **2018**, *23*, 40–49. Available online: https://doaj.org/article/c1e9eb0446be41c683aee690258f273c (accessed on 10 January 2019). [CrossRef]
5. Ho, M.; Lall, U.; Allaire, M.; Devineni, N.; Kwon, H.H.; Pal, I.; Raff, D.; Wegner, D. The future role of dams in the United States of America. *Water Resour. Res.* **2017**, *53*, 982–998. Available online: https://agupubs-onlinelibrary-wiley-com.ezproxy.library.ubc.ca/doi/full/10.1002/2016WR019905 (accessed on 10 January 2019). [CrossRef]
6. Calder, R.S.D.; Schartup, A.T.; Li, M.; Valberg, A.P.; Balcom, P.H.; Sunderland, E.M. Future impacts of hydroelectric power development on methylmercury exposures of Canadian indigenous communities. *Environ. Sci. Technol.* **2016**, *50*, 13115–13122. Available online: https://pubs.acs.org/doi/abs/10.1021/acs.est.6b04447 (accessed on 10 January 2019). [CrossRef] [PubMed]
7. Doelle, M. The Lower Churchill Panel Review: Sustainability assessment under legislative constraints in Canada. In *Sustainability Assessment*; Gibson, R., Ed.; Routledge: London, UK, 2016; pp. 124–140.
8. Macdonald, D.; Lesch, M. Management of Distributive Conflicts Impeding Expansion of Interprovincial Hydroelectricity Transmission. *J. Can. Stud.* **2015**, *49*, 191–221. Available online: http://ezproxy.library.ubc.ca/login?url=https://search-proquest-com.ezproxy.library.ubc.ca/docview/1806500135?accountid=14656 (accessed on 10 January 2019). [CrossRef]
9. Mercer, N.; Sabau, G.; Klinke, A. Wind energy is not an issue for government: Barriers to wind energy development in Newfoundland and Labrador, Canada. *Energy Policy* **2017**, *108*, 673–683. [CrossRef]
10. Scott, D.N.; Smith, A.A. Sacrifice Zones in the Green Energy Economy: The New Climate Refugees. *Transnatl. Law Contemp. Probl.* **2016**, *26*, 371. Available online: http://link.galegroup.com.ezproxy.library.ubc.ca/apps/doc/A497798899/LT?u=ubcolumbia&sid=LT&xid=f939c3c5 (accessed on 10 January 2019).
11. Besant-Jones, J. A view of multilateral financing from a funding agency. In *Financing Hydro Power Projects '94, Proceedings of the International Water-Power and Dam Construction Conference, Frankfurt, Germany, 22–23 September 1994*; First European Communications: London, UK.
12. Usher, A.D. The Mechanism of 'Pervasive Appraisal Optimism'. In *Dams as Aid: A Political Anatomy of Nordic Development Thinking*; Usher, A.D., Ed.; Taylor & Francis: London, UK, 1997; ISBN 978-128-011-598-1.
13. Ansar, A.; Flyvbjerg, B.; Budzier, A.; Lunn, D. Should we build more large dams? The actual costs of hydropower megaproject development. *Energy Policy* **2014**, *69*, 43–56. Available online: https://www-sciencedirect-com.ezproxy.library.ubc.ca/science/article/pii/S0301421513010926?via%3Dihub (accessed on 10 January 2019). [CrossRef]
14. Flyvbjerg, B.; Bruzelius, N.; Rothengatter, W. *Megaprojects and Risk: An Anatomy of Ambition*; Cambridge University Press: Cambridge, UK, 2003; ISBN 052-180-420-5.
15. Khagram, S. *Dams and Development: Transnational Struggles for Water and Power*; Cornell University Press: Ithaca, NY, USA, 2004; ISBN 080-148-907-5.
16. Klingensmith, D. *One Valley and a Thousand: Dams, Nationalism and Development*; Oxford University Press: New Delhi, India, 2007; ISBN 019-568-783-3.

17. Zwarteween, M.; Ahlers, R.; Bakker, K. Large Dam Development: From Trojan Horse to Pandora's Box. In *The Oxford Handbook of Megaproject Management*; Flyvbjerg, B., Ed.; Oxford Handbooks: Oxford, UK, 2017; pp. 556–576. ISBN 978-019-873-224-2.

18. Flyvbjerg, B. What You Should Know About Megaprojects and Why: An Overview. *Proj. Manag. J.* **2014**, *45*, 6–19. Available online: https://doi-org.ezproxy.library.ubc.ca/10.1002/pmj.21409 (accessed on 10 January 2019). [CrossRef]

19. Grant, H.M. Resource rents from Aboriginal lands in Canada—Royal Commission on Aboriginal Peoples. 1994. Available online: http://publications.gc.ca/collections/collection_2016/bcp-pco/Z1-1991-1-41-83-eng.pdf (accessed on 10 January 2019).

20. Hendriks, R.; Raphals, P.; Bakker, K. Site C: Summary of Key Research Results. Available online: http://watergovernance.sites.olt.ubc.ca/files/2017/11/UBC-Report-Site-C-Key-Issues-Full-Report-1.pdf (accessed on 22 March 2018).

21. Bakker, K.; Christie, G.; Hendriks, R. Report #1: First Nations and Site C. Available online: http://watergovernance.sites.olt.ubc.ca/files/2017/11/Briefing-Note-1-First-Nations-and-Site-C.pdf (accessed on 10 January 2019).

22. Bakker, K.; Christie, G.; Hendriks, R. Report #2: Assessing Alternatives to Site C—Environmental Effects Comparison. Available online: http://watergovernance.sites.olt.ubc.ca/files/2017/11/Briefing-Note-2-Site-C-Environmental-Effects.pdf (accessed on 10 January 2019).

23. Bakker, K.; Christie, G.; Hendriks, R. Report #3: The Regulatory Process for the Site C Project. Available online: http://watergovernance.sites.olt.ubc.ca/files/2017/11/Briefing-Note-3-Regulatory-Process.pdf (accessed on 10 January 2019).

24. Hendriks, R. Report #4: Comparative Analysis of Greenhouse Gas Emissions of Site C versus Alternatives. Available online: http://watergovernance.sites.olt.ubc.ca/files/2017/11/Report-4-Site-C-Comparative-GHG-analysis.pdf (accessed on 10 January 2019).

25. Hendriks, R.; Raphals, P.; Bakker, K. Report #5: Reassessing the Need for Site C. Available online: http://watergovernance.sites.olt.ubc.ca/files/2017/11/2-Site-C-Economics-Report-FINAL.pdf (accessed on 10 January 2019).

26. Program on Water Governance, UBC. Report #6: Employment—Site C versus the Alternative Portfolios. Available online: http://watergovernance.sites.olt.ubc.ca/files/2017/11/UBC_Briefing_Note_Comparative_Employment_Assessment_of_Site_C_versus_Alternatives.pdf (accessed on 14 March 2018).

27. Hendriks, R.; Raphals, P.; Bakker, K. Submission F106-1/2. August 2017. Available online: http://www.bcuc.com/site-c-inquiry.html#Sub-1 (accessed on 10 January 2019).

28. Raphals, P.; Hendriks, R. Submission F106-5. October 2017. Available online: http://www.bcuc.com/site-c-inquiry.html#Sub-1 (accessed on 10 January 2019).

29. Hendriks, R.; Raphals, P. Submission F106-6. October 2017. Available online: http://www.bcuc.com/site-c-inquiry.html#Sub-1 (accessed on 10 January 2019).

30. Hendriks, R.; Raphals, P.; Bakker, K. Submission F106-7-11. October 2017. Available online: http://www.bcuc.com/site-c-inquiry.html#Sub-1 (accessed on 10 January 2019).

31. British Columbia Utilities Commission. British Columbia Utilities Commission Inquiry Respecting Site C—Project No. *1598922—Final Report to the Government of British Columbia*. BCUC: Vancouver, Canada. Available online: http://www.bcuc.com/Documents/wp-content/11/11-01-2017-BCUC-Site-C-Inquiry-Final-Report.pdf (accessed on 10 January 2019).

32. Ostrom, E. *Governing the Commons*; Cambridge University Press: Cambridge, UK, 2015.

33. Jessop, B. *State Theory: Putting the Capitalist State in Its Place*; Penn State Press: University Park, PA, USA, 1990.

34. Coutard, O. (Ed.) *The Governance of Large Technical Systems*; Routledge: London, UK, 2002.

35. Scott, J.C. *Seeing Like a State: How Certain Schemes to Improve the Human Condition Have Failed*; Yale University Press: New Haven, CT, USA, 1998.

36. Hajer, M.; Versteeg, W. A decade of discourse analysis of environmental politics: Achievements, challenges, perspectives. *J. Environ. Policy Plan.* **2005**, *7*, 175–184. Available online: https://www-tandfonline-com.ezproxy.library.ubc.ca/doi/abs/10.1080/15239080500339646 (accessed on 10 January 2019). [CrossRef]

37. Boelens, R. *Water, Power and Identity: The Cultural Politics of Water in the Andes*; Routledge: London, UK, 2015.

38. Menga, F.; Swyngedouw, E. (Eds.) *Water, Technology and the Nation-State*; Routledge: London, UK, 2018.

39. Boelens, R.; Shah, E.; Bruins, B. Contested Knowledges: Large Dams and Mega-Hydraulic Development. *Water* **2019**, *11*, 416. [CrossRef]

40. Evenden, M. *Allied Power: Mobilizing Hydro-Electricity during Canada's Second World War*; University of Toronto Press: Toronto, ON, Canada, 2015; ISBN 1442626259.

41. Webster, K.L.; Beall, F.D.; Creed, I.F.; Kreutzweiser, D.P. Impacts and prognosis of natural resource development on water and wetlands in Canada's boreal zone. *Environ. Rev.* **2015**, *23*, 78–131. [CrossRef]

42. Sandwell, R.W. (Ed.) *Powering up Canada: The History of Power, Fuel, and Energy from 1600*; McGill-Queen's Press-MQUP: Montreal, QC, Canada, 2016; Volume 6, ISBN 077-359-952-5.

43. Macfarlane, D.; Kitay, P. Hydraulic Imperialism: Hydroelectric Development and Treaty 9 in the Abitibi Region. *Am. Rev. Can. Stud.* **2016**, *46*, 380–397. Available online: https://www-tandfonline-com.ezproxy.library.ubc.ca/doi/abs/10.1080/02722011.2016.1228685 (accessed on 10 January 2019). [CrossRef]

44. Choquette, L.; Carlson, H.M. *Home is the Hunter: The James Bay Cree and Their Land*; UBC Press: Vancouver, BC, Canada, 2008; ISBN 9780774814942.

45. Jenson, J.; Papillon, M. Challenging the citizenship regime: The James Bay Cree and transnational action. *Polit. Soc.* **2000**, *28*, 245–264. Available online: https://journals-sagepub-com.ezproxy.library.ubc.ca/doi/abs/10.1177/0032329200028002005 (accessed on 10 January 2019). [CrossRef]

46. Manore, J. *Cross-Currents: Hydroelectricity and the Engineering of Northern Ontario*; Wilfrid Laurier University Press: Waterloo, ON, Canada, 1999; ISBN 088-920-317-2.

47. Schiehll, E.; Raufflet, E. Hydro-Québec and the Crees: The challenges of being accountable to First Nations–case and teaching notes. *Int. J. Teach. Case Stud.* **2013**, *4*, 243–258. [CrossRef]

48. Ariss, R.; Fraser, C.M.; Somani, D.N. Crown Policies on the Duty to Consult and Accommodate: Towards Reconciliation. *McGill J. Sustain. Dev. Law* **2017**, *13*, 1. Available online: https://mcgill.ca/mjsdl/files/mjsdl/2_volume_13_ariss.pdf (accessed on 10 January 2019).

49. Booth, A.; Skelton, N.W. "We are fighting for Ourselves"—First Nations' Evaluation of British Columbia and Canadian Environmental Assessment Processes. *J. Environ. Assess. Policy Manag.* **2011**, *13*, 367–404. Available online: https://www.jstor.org/stable/enviassepolimana.13.3.367?seq=1#page_scan_tab_contents (accessed on 10 January 2019). [CrossRef]

50. Peyton, J. *Unbuilt Environments: Tracing Postwar Development in Northwest British Columbia*; UBC Press: Vancouver, BC, Canada, 2017; ISBN 978-077-483-304-2.

51. Waldram, J. *As Long as the Rivers Run: Hydroelectric Development and Native Communities in Western Canada*; University of Manitoba Press: Winnipeg, MB, Canada, 1988; ISBN 978-088-755-631-9.

52. BC Hydro. *Site C Clean Energy Project Environmental Impact Statement*; Figure 4.2: General project location and regional topography; CEAR #63919-421; BC Hydro: Vancouver, BC, Canada, 2013.

53. British Columbia Utilities Commission. *Site C Report: Report & Recommendations to the Lieutenant Governor-in-Council*; British Columbia Utilities Commission: Vancouver, BC, Canada, 1983; pp. 1, 50, 85, 23.

54. BC Hydro. *Rate Design Application. Evidentiary Update on Load Resource Balance and Long Run Marginal Cost*; BC Hydro: Vancouver, BC, Canada, 2016; p. 12.

55. BC Hydro. *Electric Load Forecast Fiscal 2013 to Fiscal 2033*; BC Hydro: Vancouver, BC, Canada, 2012; p. 21.

56. BC Hydro. *BC Hydro Integrated Resource Plan*; Planning Environment, Table 5-5; BC Hydro: Vancouver, BC, Canada, 2013; pp. 5–37. Available online: https://www.bchydro.com/content/dam/BCHydro/customer-portal/documents/corporate/regulatory-planning-documents/integrated-resource-plans/current-plan/0005-nov-2013-irp-chap-5.pdf (accessed on 10 January 2019).

57. Lazard. *Lazard's Levelized Cost of Energy Analysis*; Version 11.0; Lazard: Hamilton, Bermuda, 2017; Available online: https://www.lazard.com/media/450337/lazard-levelized-cost-of-energy-version-110.pdf (accessed on 10 January 2019).

58. Lazard. *Lazard's Levelized Cost of Storage Analysis*; Version 3.0; Lazard: Hamilton, Bermuda, 2017; Available online: https://www.lazard.com/media/450338/lazard-levelized-cost-of-storage-version-30.pdf (accessed on 10 January 2019).

59. Canadian Environmental Assessment Act, 2012 (CEAA 2012). Available online: http://laws-lois.justice.gc.ca/eng/acts/C-15.2/20100712/P1TT3xt3.html (accessed on 10 January 2019).

60. Canadian Environmental Assessment Agency. Agreement to Conduct a Cooperative Environmental Assessment Including the Establishment of a Joint Review Panel, of the Site C Clean Energy Project. Available online: https://www.ceaa.gc.ca/050/documents/54272/54272E.pdf (accessed on 10 January 2019).

61. Site C Joint Review Panel. Report of the Joint Review Panel: Site C Clean Energy Project BC Hydro. Available online: https://www.ceaa-acee.gc.ca/050/documents/p63919/99173E.pdf (accessed on 10 January 2019).

62. Province of British Columbia. British Columbia Utilities Commission Inquiry Respecting Site C—Terms of Reference. *Order of the Lieutenant Governor in Council, Order in Council 244, Section 3.* Available online: http://www.bclaws.ca/civix/document/id/oic/oic_cur/0244_2017 (accessed on 10 January 2019).

63. Canadian Environmental Assessment Agency. Cumulative effects assessment practitioners guide. Prepared by The Cumulative Effects Assessment Working Group and AXYS Environmental Consulting Ltd. Available online: https://www.ceaa.gc.ca/43952694-0363-4B1E-B2B3-47365FAF1ED7/Cumulative_Effects_Assessment_Practitioners_Guide.pdf (accessed on 10 January 2019).

64. Lee, P.; Hanneman, M. Atlas of land cover, industrial land uses and industrial-caused land change in the Peace Region of British Columbia. Global Forest Watch Canada Report #4: International Year of Sustainable Energy for All. Available online: https://www.ceaa-acee.gc.ca/050/documents/p63919/96538E.pdf (accessed on 10 January 2019).

65. Canadian Environmental Assessment Agency. Site C Clean Energy Project—Draft Environmental Impact Statement Guidelines. Available online: https://www.ceaa-acee.gc.ca/050/evaluations/document/55123?culture=en-CA (accessed on 10 January 2019).

66. Treaty 8 Tribal Association. Site C Clean Energy Project Joint Review Panel Hearings—Summary Report Treaty 8 First Nations. 3 February 2014. Available online: https://www.ceaa-acee.gc.ca/050/documents/p63919/98286E (accessed on 8 January 2019).

67. Parks Canada. *Parks Canada Comments on the Draft Environmental Impact Statement (EIS) for the Site C Clean Energy Project*; CEAR #63919-922; Parks Canada: Gatineau, QC, Canada, 2013.

68. BC Hydro. Response to Public Comments Related to the EIS Guidelines (10 April 2012) for the Site C Clean Energy Project. Topic Summary Cumulative Effects Assessment. Available online: https://www.ceaa-acee.gc.ca/050/documents/p63919/57624E.pdf (accessed on 14 August 2018).

69. Nalcor Energy. Part A Project Planning and Description. In *Lower Churchill Hydroelectric Generation Project Environmental Impact Statement*; Nalcor Energy: St. John's, NL, Canada, 2010; Volume 1, pp. 9–28.

70. Gunn, J.; Bram, F.N. Conceptual and methodological challenges to integrating SEA and cumulative effects assessment. *Environ. Impact Assess. Rev.* **2011**, *31*, 154–160. [CrossRef]

71. Canadian Environmental Assessment Agency; British Columbia Environmental Assessment Office. In the Matter of the Joint Review Panel Established to Review the Site C Clean Energy Project Proposed by British Columbia Hydro and Power Authority. Volume 26, pp. 192–198. Available online: https://www.ceaa-acee.gc.ca/050/documents/p63919/98144E.pdf (accessed on 14 August 2018).

72. BC Government, BC Hydro. Province Announces Site C Clean Energy Project—News Release. Available online: https://www.sitecproject.com/news-and-information/province-announces-site-c-clean-energy-project (accessed on 31 July 2018).

73. BC Hydro. Greenhouse Gas Emissions. In *Site C Clean Energy Project Environmental Impact Statement*; BC Hydro: Vancouver, BC, Canada, 2013; Volume 2, Section 15.

74. BC Hydro. Greenhouse Gases Technical Report. In *Site C Clean Energy Project Environmental Impact Statement*; BC Hydro: Vancouver, BC, Canada, Appendix S; 2012.

75. Site C Clean Energy Project Joint Review Panel. Revised Public Hearing Schedule—Released 6 December 2013. Available online: http://www.ceaa-acee.gc.ca/050/documents/p63919/96899E.pdf (accessed on 31 July 2018).

76. BC Hydro. BC Hydro Submission to the British Columbia Utilities Commission Inquiry into the Site C Clean Energy Project. *Appendix G: Site C GHG Emission Reductions.* Available online: http://www.bcuc.com/Documents/wp-content/09/DOC_90101_F1-1-BCH_submission_SiteC_Public.pdf (accessed on 14 August 2018).

77. BC Hydro. *Site C Clean Energy Project Environmental Impact Statement*; BC Hydro: Vancouver, BC, Canada, 2013; Volume 1, Section 5. 5-70.

78. Office of the Premier. Government will complete Site C construction, will not burden taxpayers or BC Hydro customers with previous government's debts—News Release. Available online: https://news.gov.bc.ca/releases/2017PREM0135-002039 (accessed on 24 April 2018).

79. British Columbia Supreme Court. West Moberly First Nations v British Columbia, BCSC 270 (CanLII). Available online: http://canlii.ca/t/hql7n (accessed on 25 April 2018).

80. British Columbia Supreme Court. Yahey v. *British Columbia, BCSC 1302 (CanLII)*. Available online: http://canlii.ca/t/gkd05 (accessed on 25 April 2018).
81. Kurjata, A. West Moberly and Prophet River First Nations file court claim to stop Site C. Available online: http://www.cbc.ca/news/canada/british-columbia/site-c-dam-court-case-1.4489679 (accessed on 24 April 2018).
82. Sage Legal. West Moberly First Nations application for Interim Injunction. Available online: https://www.sagelegal.ca/injunction-application/ (accessed on 24 April 2018).
83. Palmer, V. Getting Site C to point of no return a damning progress report, so far. Available online: http://vancouversun.com/opinion/columnists/vaughn-palmer-getting-site-c-to-point-of-no-return-a-damning-progress-report-so-far (accessed on 24 April 2018).
84. Eliesen, M. Submission F13-1, F13-2. August & October 2017. Available online: http://www.bcuc.com/site-c-inquiry.html (accessed on 14 March 2018).
85. Swain, H. Submission F36-1, F36-2. August & October 2017. Available online: http://www.bcuc.com/site-c-inquiry.html (accessed on 14 March 2018).
86. Prophet River and West Moberly First Nations. Submissions F28-1, F28-2, F28-3. August & October 2017. Available online: http://www.bcuc.com/site-c-inquiry.html (accessed on 14 March 2018).
87. United Nations. *United Nations Declaration on the Rights of Indigenous Peoples*; United Nations: New York, NY, USA, 2008.
88. Askew, H.; Snelgrove, C.; Wrightson, K.; Couturier, D.; Koebel, A.; Nowlan, L.; Bakker, K. Between Law and Action: Assessing the State of Knowledge on Indigenous Law, UNDRIP and Free, Prior and Informed Consent with Reference to Fresh Water Resources. West Coast Environmental Law, University of British Columbia: Vancouver, BC, Canada.
89. Slattery, B. Aboriginal Rights and the Honour of the Crown. *Supreme Court Law Rev.* **2005**, *29*, 433–445.
90. Department of Justice Canada. Principles respecting the Government of Canada's relationship with Indigenous peoples. Available online: http://www.justice.gc.ca/eng/csj-sjc/principles-principes.html (accessed on 15 March 2018).

water

MDPI

Article

What Hirschman's Hiding Hand Hid in San Lorenzo and Chixoy

Barbara Deutsch Lynch

Sam Nunn School of International Affairs, Georgia Institute of Technology (retired), Atlanta, GA 30332, USA; bdl5@cornell.edu

Received: 29 August 2018; Accepted: 3 January 2019; Published: 26 February 2019

check for updates

Abstract: Implementation of big water projects requires that their funders, contractors, and government officials will move projects forward ignorant of their potential social and environmental costs. Economist Albert O. Hirschman raised the issue of ignorance in a widely-read analysis of the factors driving the project process in Latin America, Africa, Asia, and southern Europe. This ignorance, which Hirschman referred to as 'the hiding hand,' led to creativity in the case of the San Lorenzo irrigation system in northern Peru, but had lethal consequences in the case of Guatemala's Chixoy dam project. While Hirschman saw what he called 'the hiding hand' as accidental, examination of documents related to large hydraulic infrastructure projects in Peru and Guatemala suggests that in the late twentieth century it was systematically produced by resistance on the part of international financial institutions to addressing the broader political context for project development, or to adequately addressing potential social and environmental impacts early in the project process.

Keywords: hiding hand; A.O. Hirschman; irrigation; hydraulic projects; San Lorenzo irrigation project; Chixoy irrigation project; Peru; Guatemala

1. Introduction

Large hydraulic works and river basin development programs, artifacts, and expressions of high modernization as a social and political project require an institutional array of funders, contractors, and government officials who will move projects forward with scant knowledge about their potential social and environmental costs (see [1]). Heterodox development economist Albert O. Hirschman raised the issue of ignorance in a widely-read analysis of the factors driving the project process in Latin America, Africa, Asia, and southern Europe. This ignorance, which Hirschman saw as benevolent on the whole when he studied World Bank funded projects in the 1960s, would prove nastier than he anticipated. In the case of big water projects, it would be lethal. In addition, while Hirschman saw what he called 'the hiding hand' as accidental, examination of documents related to large hydraulic infrastructure projects in Peru and Guatemala suggests that it may be systematically produced.

Hirschman's experience with the Marshall Plan drew him into the field of development economics in its infancy. Perhaps his major contributions to the study and practice of development were to critique the approach of World Bank economists to project planning and evaluation, and, more importantly, to insist on the importance of talking to people on the ground, whether local project staff or intended beneficiaries. He played a pioneering role in the field at a time when big water projects assumed a privileged position on the development agenda, due in part to the apparent success of the Tennessee Valley Authority in 'jumpstarting' development in the southern United States. Politicians, planners, and analysts who favored the hydraulic megaproject as an instrument for international development in the global south included President Harry S. Truman, former TVA administrator David Lilienthal, World Bank president Eugene Blank [2], Nehru, and Nasser, as well as Hirschman. Lilienthal spent the 1950s sharing his river basin development recipe with national governments in the Indus Valley,

Iran, Colombia, and the Dominican Republic [3,4]. It was in this context that Hirschman, who admired Lilienthal's work, moved to Colombia to work for the World Bank.

Support for big hydro in the postwar period was not just about identifying a new cold war weapon or opportunities to deploy 'mobile fixed capital' in new settings. It also had a moral dimension: global inequality was a concern, and hydropower development—even in authoritarian states—was seen as a motor for economic growth, and an instrument for poverty reduction and democratization. By the 1970s, hydropower development was also justified as a 'clean' energy alternative. Lilienthal captured this moral dimension when he said,

> If a great dam or new system or roads inspires people in a country with a feeling that this is theirs, and that it provides an opportunity, a leverage by which they and their young people can look to the future with hopefulness in specific ways, then that great dam as an inspiration will produce more than electricity and irrigation, the road network more than transport. It will produce a change in spirit, a release of energies and self-confidence which are the indispensable factors in the future of that country [5] (p. 13)

Hirschman shared Lilienthal's optimism about the promise of infrastructure projects. His long-term involvement with Latin American development began with his move to Colombia in 1952. Reflecting on his early work for the World Bank consultant during La Violencia [6] (pp. 80–81), he noted,

> At one time I was actively involved in the attempt to develop a regional authority on the model of the Tennessee Valley Authority. The idea to create a multifunctional entity was then quite widespread. This entity would provide irrigation, electric power, and even land reform. This kind of work gave me the desire to begin to know in depth the reality of this country, and it put me into contact with many people. Now it hardly ever happened that I would take a plane without meeting this or that minister or corporate executive whom I knew personally. I felt positive about all this because I had the feeling that the country was moving forward. However, I don't want to deny the tremendous problems the country was going through—we must not forget that a civil war was still going on—but in any event we had the perception that the country was progressing.

Hirschman did not view the project as a weapon of counterinsurgency or military control, nor did he display willful ignorance of Colombia's political realities. Still, his faith in human progress and the democratic potential of public works withstood whatever doubts he may have had about the efficacy of infrastructure projects in a climate of civil unrest and state repression.

Jeremy Adelman does a masterful job of contextualizing Hirschman's thinking. He writes that Hirschman's Colombian experience informed his decision to study the World Bank projects in 1964 in order to learn more about how massive development projects worked [7] (p. 385). His research, supported by the Brookings Institution and the Carnegie Foundation, took him to eleven countries where he interviewed World Bank staff and borrowers. Still optimistic about the democratizing potential of irrigation and river basin projects which, as he saw it, compelled water users to build new institutions for water allocation and for system maintenance, he became increasingly dubious about their representing triumphs of rational planning. His general optimism, coupled with his doubts about the rational nature of the project process, informed a 1967 *Public Interest* article [8] and a 1967 book, *Development Projects Observed* [9].

To challenge prevailing assumptions about the rational nature of project planning, Hirschman referred to the 'hiding hand,' a feature of the project process that fosters creativity by concealing the difficulties inherent in a project [9] (p. 13). In a 1995 reflective work, *The Propensity to Self-Subversion*, he described his principle of the hiding hand as speculation intended to "endow and surround the development story with a sense of wonder and mystery" [10] (p. 129). Hirschman argued that if planners knew all the obstacles to a project's successful implementation, they would not undertake it in the first place, but when facing difficulties in the field they would respond with creative solutions.

He stressed that "in developed countries less hiding of the uncertainties and likely difficulties is required than in underdeveloped countries where confidence in one's creativity is lacking, and where new tasks harboring many unknowns must be presented as though they were all 'cut and dried' in order to be undertaken" [8] (p. 14). For Hirschman, creativity did not simply mean the invention of new solutions to technical problems. he explains that he saw *Development Projects Observed* as a critique of the overly technocratic approaches to development and to project evaluation promoted by Robert McNamara when he was president of the World Bank. Looking back at the earlier book, Hirschman notes that, in order to challenge the 'scientific' rationality prevalent in World Bank circles, he expanded the definition of obstacles to development to include social as well as technical factors, and, citing the San Lorenzo case, argued that the hiding hand fostered creative social policies as well as technical solutions [8] (p. 128).

Hirschman goes on to argue that river basin development projects are excellent examples of projects that appear deceptively easy, because they are cast as TVA clones. In other words, the hiding hand makes proposed projects look like simple replications of past successes. It also exaggerates benefits, making it possible to swallow high costs and to achieve positive intermediate outcomes.

One of the projects that Hirschman analyzed for his book was the San Lorenzo irrigation project in Peru's Piura region. Although he found the project plagued by delays and opposition, he celebrated its contribution to land redistribution and agricultural experimentation. Drawing upon his understanding of this project and several others, Hirschman concluded that the hiding hand was largely benevolent. His findings in *Development Projects Observed* were generally favorable to the World Bank, although he did argue for better project monitoring and evaluation. However, the Bank's response to his recommendations was chilly. Staff objected both to his call for systematic project evaluation and his recommendation that the distributional consequences of projects be assessed [7] (pp. 398–403). They also faulted the book's lack of concrete recommendations.

Hirschman's upbeat view of the hiding hand has continued to provoke criticism over the half century since the publication of *Development Projects Observed*. Referring to his support for a Biafra transport project when the region was on the cusp of civil war, Adelman, his sympathetic biographer, concludes,

> Either Hirschman's optimism blinded him to the simmering tensions, or the evidence of this tension was still muted. The latter is not plausible; Hirschman's notes are filled with the grouchy testimonies of his witnesses. It is more likely that his wish for surprising, positive effects overwhelmed what he saw and heard. Either way, he failed to predict that this was one project that would have disastrous consequences and contribute to the devastating civil war in Nigeria not long after Hirschman toured the region [7] (p. 393)

Flyvbjerg and Sunstein [11] conclude that if Hirschman had expanded his sample, he would have found that, more often than not, the hiding hand results not in creativity, but in poor project design and cost overruns that could have been prevented. They argue that in such cases the hiding hand is malevolent rather than benign. Anheier [12] offers a typology of different types of possible hiding hands. Underestimation of task complexity in a context of ignorance, he argues, produces the hiding hand defined by Hirschman, one that makes for creative approaches to problem solving. In contrast, underestimation of task complexity in a state of awareness, argues Anheier, produces an 'information asymmetry' that leads to profiteering. This is Flyvbjerg and Sunstein's malevolent hiding hand. Where task complexity is overestimated in a context of ignorance, Anheier sees a protecting hand that he equates with the precautionary principle. Lastly, overestimation of task complexity in a context of knowledge may lead to an excess of caution or risk avoidance that can stifle development, as Hirschman predicted.

Hirschman's hiding hand metaphor interested me when in 2003, as a volunteer working with the International Rivers Network, I examined documents in the NGO's extensive files related to the Guatemalan Chixoy project, seeking to learn whether or not the World Bank and the InterAmerican

Development Bank approved loans for a big dam project during Guatemala's 30 year long civil war, knowing that the Guatemalan government agency implementing the project was implicated in human rights abuses. Although state-sanctioned violence against Maya communities in the project area was well documented, I could not find in the project documentation produced by the banks or international contractors evidence that they knew that activity directly related to building the dam would contribute to what was later determined to be genocidal behavior. This may have been due simply to a lack of transparency, or it may have been the work of a hiding hand—far more destructive than Flyvbjerg and Sunstein's [11] malevolent hiding hand—that deliberately produced ignorance of both the violent acts committed by the Guatemalan state and the devastating social and cultural impacts of displacement caused by the project, and led to outcomes that Hirschman would certainly have deemed objectionable. If the lending agencies had a clear understanding of the political environment in which the project process was unfolding, it can be argued that they were knowingly complicit in a campaign of repression. If, on the other hand, they were ignorant of what was going on, it becomes imperative to ask how and why ignorance was produced.

Inquiring about the role of international development actors in concealing the horrific side effects of project development led to some basic questions about Hirschman's optimism: Was it delusional at best, or did the probability of positive outcomes associated with big dam development simply diminish over time? Or was the expectation that the TVA model could be adapted to vastly different natural, political, and cultural environments without provoking serious harm simply untenable? This led me in turn to an epistemological question about whether the hiding hand is an accidental phenomenon, as Hirschman implied, or whether ignorance is systematically produced. To address these questions, I refer to World Bank and InterAmerican Bank project documentation, to historical analyses of the San Lorenzo Project discussed in Development Projects Observed, and to the extensive body of analytical and testimonial literature on the Chixoy dam project.

2. Hirschman's Hiding Hand and the San Lorenzo Project

The San Lorenzo or Quiroz-Piura irrigation project was one of several designed to deliver water to Piura's export cotton producers and rice growers, and to open new areas on Peru's arid north coast to agriculture. Hirschman saw it as a positive example of how proceeding with project development despite an absence of information could lead to creativity and positive, if unanticipated, outcomes. This raises two questions: Did project implementation in the face of ignorance of potential obstacles to its success lead to positive outcomes, and did the hiding hand operating in the San Lorenzo project represent a simple absence of information, or was it a case of ignorance produced in a specific cultural and political context? Recent irrigation development in the area dates to the late nineteenth century, and in 1902 a privately funded irrigation canal was built on the right bank of the Río Chira, expanding the area under cotton cultivation by more than 4000 ha [13] (pp. 56–57). Water conflict in the early twentieth century was severe, pitting large cotton producers in different parts of in the Piura valley against one another. These powerful growers managed to contain their problems by buying lands in the upper watershed—allowing them to cultivate when water was abundant—and by setting up municipal water rationing systems [14]. However, demographic growth and the opening of new lands increased water demand, and, given the role of cotton in the generating foreign exchange, improving water delivery was deemed to be in the national interest in the mid-1940s.

Shortly after coming to power in 1948, Peru's president Manuel Odría authorized construction of the Quiroz project, which, when completed in 1953, irrigated 31,000 ha of cotton lands. At the time, it was the largest water management project in Peru; it diverted water from the Río Quiroz into the new San Lorenzo dam, and ultimately into the Rio Piura. The diversion was intended to serve some 50,000 ha of newly colonized lands in San Lorenzo. In 1955, the Peruvian government obtained a $18 million loan from the World Bank to finance construction of a dam on the Quiroz and 85 km of main canals and diversion structures. In principle, this would have permitted irrigation of 50,000 ha of uncultivated land, and provision of supplementary irrigation to another 31,000 ha in the Rio Piura valley [15]. However, this estimate may be overoptimistic, given that the new system probably incorporated previously irrigated areas, but I have no data to confirm this. According to a 1955 World Bank report [16], 20,000 ha of the irrigated area would be planted in cotton, with the remainder in food crops and pasture. Newly irrigated lands would be sold to farmers in plots ranging from 15 to 100 ha.

In 1965, the government of president Fernando Belaunde Terry, seeking to sustain the power of agrarian elites while stemming the tide of growing peasant unrest, applied to the World Bank for additional project funding. With World Bank funding, ORDEN, a newly created planning agency, conducted a hydrological study of the Piura department in 1967. Feasibility studies followed, and in 1968 the newly installed military government of President Juan Velasco Alvarado requested an additional $109 million in loans from the World Bank and other sources, but, despite positive assessments of the project's technical mission, the World Bank pulled the plug in 1970. Cleaves and Scurrah's [14] interviews indicate that the bank's withdrawal was due to its discomfort with Velasco's decision to nationalize Peru's oil industry.

Hirschman visited San Lorenzo in 1964, the year in which the Belaunde government passed a weak agrarian reform law [17]. Hirschman found his visit to the project exhilarating "because of the strong reform wind that was blowing about the place" [9] (p. 159). In introducing the San Lorenzo experience, Hirschman argued that development projects are accompanied by a series of unanticipated threats not just to their performance, but to their very survival, and by a counterbalancing set of "remedial actions that can be taken should a threat become real" [9] (p. 11). He noted that the San Lorenzo project experienced significant delays due to political changes and a fundamental reassessment of the kind of irrigated agriculture that the program ought to support. This rethinking may well have been occasioned by a drop in world cotton prices that affected Peru in the 1960s [13]. Hirschman found that farmers in newly irrigated areas saw the adoption of modern cotton production as risky, so they turned to familiar crops with a known market, thus reducing anticipated benefits. He also cited the problem of excess demand for newly diverted waters, but he noted that, faced with these obstacles, San Lorenzo was transformed into a pilot project for subdividing land into small family farms and for offering credit and technical assistance to once landless cultivators, thereby establishing a new pattern for Peruvian agriculture and providing a training ground for a new breed of agricultural administrators able to apply lessons learned in Piura in other places [9] (p. 12).

For Hirschman, San Lorenzo had become a victim of its own success when, following completion of the project's first phase, all of the water diverted from the Río Quiroz went to irrigate cotton plantations in the lower Piura valley. Because the powerful cotton producers in the lower Piura valley enjoyed an ample water supply in the years before development of new irrigated areas, they objected to programmed reductions in their water supply and sought to block planned expansion. The 1955 World Bank Report noted that droughts in 1950, 1951, and 1954 and frequent water shortages "prevented the effective use of land suited for the cultivation of Pima cotton, a major export product" [16] (p. 3). Expansion of the irrigated area by 50,000 ha would further reduce the amount of water for cotton. According to Hirschman, cotton growers also feared the loss of labor with land colonization and promotion of small holder agriculture.

In their effort to speed expansion of the San Lorenzo irrigated area, project administrators portrayed the presence of goat herders and landless cultivators living and working in the area to be colonized as a threat to project completion, and they referred to this 'threat' when lobbying for the project in Lima [9] (p. 109), [14]. These references to the threat of squatting by the landless, according to Hirschman, allowed administrators to respond creatively in the face of opposition from the cotton elite by changing the sequence of project development and providing water to a highly contested downstream area, thus clearly defining the outer edge of the project before it could shrink.

Hirschman also argued that opposition from the cotton *hacendados* would vanish if the water supply were increased, either through tubewell introduction or river diversion from the Chira as well as the Quiroz. And he lauded the experimentation with alternate forms of water acquisition taking place in the project area:

> The river that is being tapped is frequently found not to have enough water for all the agricultural, industrial, and urban uses which had been planned or which are staking claims, but the resulting shortage can then often be remedied by drawing on other sources which had not been within the horizon of the planners: ground water can be lifted by tubewells, the river flow can be better regulated through upstream dams, or the water of more distant rivers can be diverted. At present, such plans are underfoot for the San Lorenzo irrigation scheme in Peru [9] (p. 10)

It could be argued that in the San Lorenzo case the hiding hand produced malevolent as well as benevolent results. It is quite probable that the 'creative solutions' that Hirschman identified as responses to unforeseen obstacles came at the expense of similarly hidden, but longer-term environmental problems that would inevitably result from aquifer mining and interbasin water transfers, problems that have in this century reached a critical phase in Peru.

Hirschman did worry, however, that the establishment of a new authority responsible to the executive and independent of the bureaucracy's line agencies would create administrative headaches. He also worried that San Lorenzo would lose its soul when waters were initially diverted to the lands of wealthy cotton producers rather than to the family farms and landless cultivators who were the intended project beneficiaries. These unanticipated threats, argued Hirschman, led to unanticipated but positive responses. He saw the hiding hand operating through a process that he called 'coat-tail riding': governments and international lending institutions privilege problems like inflation and balance of payments and neglect social problems like land reform, and problems that involve public works get more attention than those that do not. For this reason, Hirschman concluded that "the San Lorenzo irrigation project would hardly have been started in the early 1950s in Peru if it had been anticipated that the project would one day become a training and testing ground for agrarian reform" [9] (p. 172). Here, Hirschman finds that social change was smuggled in along with the project, and that "happily for the smuggling-in act, the changes that are most subversive of the existing order are often hardest to detect for the simple reason that the more fundamental the change, the more ramified and hence innocuous-looking will be its beginnings" [9] (p. 173).

In the book's conclusion, however, Hirschman qualifies his argument about benefits of the hiding hand in San Lorenzo by stressing the importance of the larger political context in determining whether creative solutions are possible. He suggests that the decision by project staff to prioritize the project did require some foresight as to its probable outcomes for cultivators in the project area, and for Peruvian social and economic development. This caveat became increasingly important for him as he came to understand how the struggle over Biafra made it difficult, if not impossible, to carry out development programs [7] (Ch. 13, 15).

That said, both World Bank project documents and Hirschman's conclusions suggest that, in addition to fostering creativity, the hiding hand obscured the likelihood of problematic societal and environmental outcomes. Neither Hirschman nor the World Bank paid attention to the problems that landless cultivators and pastoralists in the projected irrigated area would face as the project moved forward. In contrast, Cleaves and Scurrah [14] find that as agricultural modernization took place and cotton, rice, and corn production expanded, Piura saw two migrant flows: One consisted of immigrants from Europe and from elsewhere in Peru, who became small and medium scale farmers. A second stream of immigrants from highland Peru worked as farm labor, rented land, or established *minifundia*. Simultaneously, with the development of irrigation infrastructure for market agriculture, "local peasants began to lose their land and, more important, their water rights" [14] (p. 137). Land concentration in the Piura and Chira valleys was extreme, although in the Piura Valley there were an unusually high number of owner-operated modern farms ranging in size from 100 to 500 ha. Cleaves and Scurrah conclude that "displacement of the traditional peasantry and concentration of landownership were not checked until the 1969 Agrarian Reform Law" (p. 138). Their findings are particularly interesting in light of the fact that the 1955 World Bank report [16] called for spending $500,000 US on land expropriation for the project, but left to the imagination the question of who would be expropriated. Cleaves and Scurrah report that in the 1960s an entire community was displaced by construction of the Poechos Dam, only to be resettled in an area with poor, sandy soils. The credit provisions of the project may have resulted in additional displacements in cases where small farmers were unable to repay their loans. Lastly, as was so often the case in the 1950s and 1960s, the environmental changes that inevitably accompany development of irrigation infrastructure and the introduction of crop varieties with higher water requirements received no mention either from Hirschman or World Bank project documents.

While the hiding hand in San Lorenzo was not entirely benign, it was not wholly malevolent. The second phase of system construction, which Hirschman studied, took place as land reform was finding its way onto the development agenda, along with agricultural modernization. This accounted in large measure for Hirschman's optimism. While Peru had experienced coups and peasant rebellions during the construction period, it would not be engaged in a full-scale civil war until the 1980s. Second, it is not possible to assess even the medium-term social impacts of the project, because in 1969 the Velasco government enacted a sweeping agrarian reform and expropriated the large cotton *haciendas*. The reform radically transformed agriculture and land tenure in the project area. Further profound environmental and social change accompanied 1983 floods. It may be, however, that the small and medium fruit and vegetable producers who rose up in 2001 to block proposed mine expansion in Tambogrande were inheritors of social as well as the land-tenure changes produced by the hiding hand [18,19]. It is not possible to determine with any confidence the secondary impacts of the hiding hand in Piura.

That said, ignorance was not purely accidental. It was, at least to some extent, produced by the failure of project personnel to ask about the project's impacts on *campesinos* in the project area, and the ways in which their local economies depended upon water that was slated to be diverted to other parts of the project area. Potentially displacing effects of the project were not addressed in World Bank project documents. This failure could be attributed to the institutional culture of the World Bank and Peruvian implementing agencies. Ore and Rap [20] trace this failure to Peru's 1917 water law which placed irrigation engineers in a dominant position. Over the years, engineering perspectives shaped thinking about projects, obscuring agronomic, social, and environmental concerns. However, Hirschman also helped to produce ignorance by limiting his interviews to World Bank and project personnel. His optimism might have been tempered had he interviewed the small cultivators, pastoralists, or landless workers whose livelihoods were likely to be affected by dam and irrigation system construction.

To conclude, it is clear that Hirschman saw the San Lorenzo experience as a positive example of creativity enabled by the hiding hand. He observed the project's progress at a time when enthusiasm for land reform was growing in Peru and in the international development community, so the political context was more supportive of the experiments undertaken in Piura than it would have been in the 1950s or in the 1970s, when opposition to the Velasco land reform began to consolidate. That said, the cultural biases and political norms that occluded the interests of smallholders and landless rural workers, particularly those who had migrated from the highlands, produced an ignorance that concealed problems related to their displacement and resettlement. The hiding hand in this instance had both positive and negative consequences.

3. The Chixoy Project: A Nasty Hiding Hand in Guatemala

The Chixoy project—officially known as the Pueblo Viejo-Quixal Hydroelectric Project—was a component of a coordinated effort by international donors to jump start development in Guatemala by using grants and loans to build a TVA clone. The dam was intended to supply some 60 percent of Guatemala's electricity and provide energy for copper, nickel, and possibly oil extraction. The project area straddled Baja and Alta Verapaz, provinces with a large number of Maya communities that suffered since the nineteenth century from policies favoring land concentration, and from a state ideology that had associated them with backwardness [21]. By the 1970s, conflict between the military and civil patrols on one side and Maya communities on the other had become severe [22,23]: Guatemalan project personnel were complicit in the massacre of over 400 Achi Maya men, women, and children living in the dam catchment area, and the destruction of communities whose existence was predicated on an intimate relationship to place. Johnston [24,25] provides excellent summaries of how the project moved forward in a violent context and of the reparations campaign that revealed a history of genocidal violence. The history of the Chixoy project reveals a hiding hand that seems less a happy accident and the product of unforeseen events than the outcome of the deliberate production of ignorance by state and international development actors.

Details of the Chixoy project process and its impacts are well documented [24–26]. Suffice it to say here that the political context in which the two projects evolved differed fundamentally—in part as a function of timing. While the World Bank loan approval process for the second phase of San Lorenzo began during the Eisenhower administration, the Peruvian project as Hirschman knew it was implemented at a time when state and international actors were beginning to see land reform, land colonization by poor and landless cultivators, and the creation of a class of small and medium scale producers as drivers of economic growth as well as alternatives to revolution, but before the Velasco government nationalized the Talara oil fields and carried out its massive agrarian reform, which confiscated foreign-owned plantations.

The initial river basin development project that would include the Chixoy dam construction was first proposed in 1951 by Guatemalan president Jacobo Arbenz, who was also responsible for a major land reform enacted in 1952, one that included expropriation of United Fruit Company plantations. The 1951 proposal Arbenz submitted to the World Bank was basically a TVA clone that envisioned the movement of small farmers out of wheat agriculture and into manufacturing [24]. Nonetheless, the World Bank, responding to U.S. opposition to the Arbenz land reform, rejected the proposal. In 1954 Arbenz, who had been labeled as a communist sympathizer, was toppled by a U.S. Central Intelligence Agency-orchestrated coup, initiating a long period of counterrevolution and military control over the countryside, and more generally over the Guatemalan economy. In 1961, during the increasingly violent administration of the rightwing president Miguel Ydígoras Fuentes, World Bank interest in Guatemalan river basin development resurfaced. With World Bank, InterAmerican Development Bank (IADB), and support from bilateral assistance agencies, the project moved through its planning phase.

Thus, while Phase II of the San Lorenzo irrigation project evolved in a climate of openness to inclusive structural reform, Chixoy planning and execution took place in a context of increasingly violent repression framed in terms of the Cold War. The project's gestation coincided with the most violent period (1975–1985) of Guatemala's 30 year civil war, a period when individual assassination plots gave way to massacres, the principal victims of which were the Maya [27,28]. Construction on the Chixoy dam began in 1974, and the dam began to fill in 1983. In 1976, the IADB offered INDE a $105 million loan for the project; the World Bank lent the government of Guatemala an additional $72 million in 1978. The loans were replenished in 1978, 1981, and 1985 [26].

Project supervision was militarized over the course of the planning process. The Instituto Nacional de Electrificación (INDE), a parastatal agency created in 1963, acted as a pass-through for international loans and grants, and as an intermediary between engineering firms and the Guatemalan government. In 1982, during the bloody dictatorship of General Efraín Ríos Montt, the agency was placed under military control, and in the mid-1990s it was privatized. It should be noted that the principal victims of state violence during the project planning phase were Maya communities. Project finance came largely in the form of loans from the IADB and the World Bank. Responsibility for project design and implementation lay with private contractors. Fifteen firms from nine countries participated in the project. The governments of West Germany, Italy, and Canada made large grants for the project, and firms from these countries were well represented. Most of the work was done by these firms or contracted out to private consultants by the firms or directly by INDE.

The hiding hand helped the project to move forward not only in the face of technical obstacles, but in an increasingly problematic political environment. While the larger electrification effort was probably part of a broader effort to assert state control in the highlands, continued international interest in implementation may have been due at the outset to the shrinking number of sites available for hydropower development at a global level, which encouraged the construction sector and its financial backers to focus their efforts more intensely on the few sites remaining that appeared attractive from an engineering standpoint. In addition, the 1973 oil crisis generated new pressures for hydropower development as an alternative energy source. These pressures may have discouraged the IFIs from carrying out environmental or social impact assessments, although, as Partridge [29] remarks, this was still seen as a marginal activity within the banks.

In 1972, the West German government made a grant to the Guatemalan government to draft a hydropower development plan [30], and a consortium headed by the engineering firm Lahmeyer International (LAMI) conducted a hydrological study of the Chixoy watershed, financed in part by a 1967 World Bank Loan. The study did not address potential social impacts, nor did community consultation take place. However, according to a LAMI document, helicopter inspection of the Chixoy valley above the dam site showed that a large amount of the area on the valley walls had been cultivated [30,31].

INDE's 1974 development plan called for feasibility studies for four possible dam sites, one of which was Chixoy. That year, a tropical ecologist and a medical doctor conducted a two-week environmental reconnaissance for LAMI. Their report contained a two-page social impact assessment human ecology and public health. It concluded,

> The region is remote from population centers and comparatively few people will have to be relocated; only 210 dwellings will be affected. Public health in the area is reasonable at present and should improve with completion of the project. No major adverse effects on the plants and animals are predicted [32] (p. 7)

A section on archaeology and history listed ancient Maya sites threatened by dam construction, but did not treat recent Maya history in the area or Maya relations with the Guatemalan state. Colajacomo [33] (p. 2) reported that consultation with people living in the affected area did not occur until 1976, after construction began, and that talks took place in an environment of intimidation. There is no evidence that either funders or the implementing agencies sought input from social scientists who had worked with the Maya and could speak local Mayan languages. The next year, INDE approved the Chixoy project. The German and Guatemalan governments signed an agreement to draft a master plan for the project with the aid of foreign consultants. When, in 1975, INDE applied for a project loan from the Inter-American Development Bank, LAMI prepared bidding documents for construction and equipment, evaluated tenders, carried out financial and engineering studies, and developed the project design and specifications, but contracts were apparently put out to bid before the engineering study was complete [30] (p. 8). According to a 1991 INDE report, "this was done in order to create pressure for rapid completion by using time during the construction phase for studies, research, trials, and direct observation of hydrological conditions [34] (p. 76). A 1991 World Bank report concluded that "the project preparation process appears to have been hurried in 1975 and 1976 due to a growing sense of urgency reflecting the fear of further 'oil' shocks and of growing needs for future generating capacity" [35] (p. 48). The Guatemalan Institute of Anthropology and History conducted an archaeological survey and a salvage program [29], but there is no record of a social impact assessment or a resettlement plan. In January 1976, the IDB and the government of Guatemala signed a loan contract for $105 million, or about a third of the project's estimated costs.

Construction began in 1978, and INDE told communities in the watershed that they would have to move. A month later, a severe earthquake struck Guatemala; international development agencies poured aid into the country, much of which came in the form of additional funding for projects seen as contributing to economic revitalization. Chixoy was one such project, and the World Bank added $72 million to the IADB loan, but as a loan condition it required a resettlement plan. INDE submitted a resettlement plan in 1979, estimating that 450 residents would be displaced by the reservoir. In 1985, the World Bank approved a second loan of $44.6 million to cover cost overruns. By 1988, Chixoy loans represented 40 percent of Guatemala's debt [36] (p. 271). INDE failed to meet its resettlement obligations, a failure amply discussed in the reparations campaign literature. Equally worrisome were decisions by local firms to buy security from the military, INDE's close ties to the military, and the willingness of the IFIs to continue project funding as systematic state-sponsored violence against Guatemala's rural indigenous population was escalating in the project area.

The hiding hand, as Hirschman predicted, hid cost overruns, delays, engineering errors, and routine bureaucratic incompetence [37]. It also hid the almost inevitable consequences of displacement for the Maya families living in the project area. In 1983, when the dam was already scheduled to fill, the IADB for the first time engaged a social scientist familiar with Maya culture and languages to assess the project process [38]. However, the full extent of the project's social impacts was revealed only when investigation into atrocities associated with the project was made possible by the 1994 Oslo Accord. Massive and carefully documented studies by the Commission on Historical Clarification, established by the treaty and by the Archdiocese of Guatemala, concluded that Guatemalan government forces (including the army and civil patrols) were responsible for 93 percent of human rights violations and acts of violence [22]. In a period roughly coinciding with the planning and implementation phases of the Chixoy project, violence escalated in Guatemala's Maya Highlands, a region that included Alta Verapaz and Baja Verapaz. Estimates of death vary widely, but human rights groups estimate about 50,000 were killed and up to a million were displaced. Over 83 percent of the victims of violence between 1962 and 1996 were Maya [27].

Those directly harmed by the Chixoy project were largely Achi Maya, and communities that opposed their displacement suffered disproportionately. The hardest hit was Río Negro. From 1980 to 1982, INDE and the Guatemalan government responded to Río Negro's reluctance to move with revocation of their title to their lands, theft of documents proving their title, and the massacre of some 440 Río Negro community members—men, women, and children [33,37,39]. Those who survived were reduced to abject poverty. Lands received in compensation for those flooded were generally unsuited for agricultural production, and hunger became endemic. Housing was insufficient and poorly built, commitments to provide public facilities were generally honored in the breach, as were agreements to provide free electricity [24,37]. Moreover, Rio Negro survivors who refused to move to INDE's resettlement site were subjected to intimidation by paramilitaries and military police [24,37].

Drawing on evidence from the 1999 report of the Commission for Historical Clarification, environmental and human rights groups attributed the killing of Río Negro community members to INDE. Their position is supported by the oral histories of Río Negro survivors [25,37] and by a ten-volume report on the violence, based heavily on forensic anthropology and oral history, prepared by the archdiocese [40]. Evidence from exhumations corroborated local testimonies about the Río Negro massacres. Drawing on these findings, a coalition of local and international NGOs and Achi Maya, including Río Negro survivors who were still very much in danger, began the reparations campaign outlined by Johnston: In 2010 a reparations plan was finalized, and in 2014 the Guatemalan government apologized to dam-affected communities and began to make reparations payments.

However, even as Chixoy human rights violations gained international attention and evidence of atrocities in the project area mounted, the hiding hand continued to hide. In response to a 1996 Witness for Peace Report that documented dam impacts, and a strong letter from an NGO coalition seeking reparations, the World Bank sent an inspection team to investigate causes of violence and implementation of resettlement plans. In a letter to the NGOs summarizing the results of the investigation, Bank President James Wolfensohn [41] summarized its findings as follows:

> Although team members had read about the events in Guatemala, and in some cases had worked there, they were deeply affected by their experience and the account of the events which they heard. The widespread destruction of indigenous organizations in Guatemala, the murders and repression were vividly recounted and have made a lasting impact. What

happened is not questioned. In 1982, women and children from Rio Negro were brutally murdered by civil patrols from a neighboring village. Why they were murdered is less certain. Some people attributed the deaths to counterinsurgency efforts, others to the fact that the people of Rio Negro were politically organized, and some to the fact that they were opposed to resettlement. Others saw a confluence between these forces. It is evident, however, that the civil disorders which wracked Guatemala in the late 1970's and 1980's were not focused on or confined to the population displaced by the Chixoy Hydroelectric Project. Most resettled communities were not subject to violence and many communities in the vicinity, with no connection to Chixoy, experienced murder and repression. In 1982, the year of the massacre, neither the Bank, nor other observers, knew the extent of the violence and terror that were occurring in Rabinal, nor did Bank staff associate the violence, of which it had only general and limited knowledge, with resettlement activities. The Bank at the time attributed these actions to the ongoing insurgency/counterinsurgency struggle. To this day there are still varying and conflicting interpretations of the causes of the violence which occurred. (pp. 1–2)

Wolfensohn acknowledged the massacre, but, aided by a hiding hand that was produced by an unwillingness on the part of lenders and international project staff to question INDE's role in the project area, he attributed it to generalized violence unrelated to dam construction. He admitted to ignorance of the extent of the violence in areas where the Achi Maya were resettled, but did not account for the absence of staff on the ground to monitor INDE's behavior. In any case, it would prove difficult to hold INDE accountable for its actions, because the agency was privatized in the late 1990s. With privatization, INDE's Resettlement Agency closed its doors, and INDE distanced itself from its past performance.

In addition, nothing in the project documentation demonstrates that IFI staff and international contractors knew that INDE was complicit in violence against residents of the dam area, despite mounting evidence to the contrary [39] (pp. 37–44). INDE's complicity may have been informally noted and dismissed as irrelevant, but it may well be that the epistemological and social preferences of IFI and contractor staff, as well as their institutional practices, prevented the construction of a knowledge base that would have led international project actors to question INDE's behavior and, more broadly, the wisdom of implementing a hydraulic project in a conflict zone.

When, in 1983, anthropologist William Partridge presented the results of a study for the Inter-American Development Bank comparing the Arenal Hydroelectric project in Costa Rica with Chixoy to World Bank staff responsible for Guatemala, he discovered that

The World Bank team responsible had not once visited the resettlement operation, and the unfolding disaster was never mentioned in the team's supervision reports to management over the previous four years. The bank project team was incensed; they angrily defended themselves and attacked me [29] (p. 153)

At one point in the meeting the vice president for the Latin American and Caribbean region strongly advocated that staff fix the problem, but, as Partridge noted, fixing the resettlement problem faced resistance because it would mean holding up a new project loan.

4. How Ignorance Is Produced

Hirschman argued that, while it could on occasion conceal social problems too profound to be addressed through creativity on the part of project staff, he saw the hiding hand as largely benevolent. I would argue that his optimism was in part due to the fact that he saw the hiding hand largely as an absence of knowledge—that which cannot be foreseen. In contrast, ignorance is often socially produced, either quite deliberately or as an inevitable outcome of political and social preferences, values, and assumptions. This was true in the San Lorenzo case, to the extent that staff ignored the potential impacts of water diversion on peasant cultivators and herders. The ignorance that allowed the Chixoy project to advance in a far more toxic political environment can be attributed in part to the interests of the IFIs, INDE, and project contractors. Firms hired to do pre-feasibility and feasibility studies, and who seek contracts to implement the same projects, were disinclined to draw attention to factors that might slow or stop funding. In the 1980s, the World Bank project cycle did not call for social and environmental impact assessment until the appraisal phase. These tasks were considered peripheral to the Bank's mission until the 1990s, when massive protest against the Indian Narmada Dam and Brazilian Polonoreste projects, coupled with the "Fifty Years is Enough" NGO campaign to abolish the bank, drew attention to these issues. As Wade [42] concluded, based on his thorough analysis of World Bank performance in the environmental arena, staff during the 1970s paid only minimal attention to environmental issues, and were hesitant to raise these with borrowers. In the 1980s, the relationship between the World Bank's Office of Environmental Affairs, which held responsibility for addressing displacement and impacts on indigenous peoples as well as environmental issues, and the rest of the bank was highly contentious. Other lenders, subcontractors, and bilateral assistance agencies shared the World Bank's interest in marginalizing social and environmental concerns.

Ignorance was also produced when lenders expressed reluctance to intervene in the domestic affairs of member nations, or to render judgment on particular governments or their leaders, although this was not the case when the World Bank withheld support for the Guatemalan River Basin project in 1951, and for Peru's Chira-Piura project in 1970. Equally important, however, were optimistic assumptions about modernization and the value of hydraulic megaprojects in 'jumpstarting' development.

Several studies have addressed the types of knowledge gaps that probably enabled the World Bank and the IDB to continue supporting the Chixoy project. Scott, for example, argues that standardization of the subjects of development associated with high modernist development entails a refusal to consider what he calls "situated and contextual attributes" [43] (p. 346). Mitchell [44] described the politics of development in Egypt as framed in an economic rationality that emphasized calculability and rules, but obscured the extent to which force and violence were deployed to advance a development agenda as conceived by international financial institutions and bilateral assistance agencies. In his analysis of World Bank practice in Laos, Goldman [45] identifies elements in the project process that constrain and direct knowledge production. These include the 'terms of reference' which specify the kinds of information to be collected, the time allowed for field work, and reporting deadlines. Other elements in project culture that foster the production of ignorance include pressures to move money out of the door, reliance on short-term consultants, and the close ties between IFI staff and international contractors.

Three elements in the Chixoy project process enabled the hiding hand to hide: timing, reliance on contractors and IFI staff as producers of knowledge, and a framing of the public good in terms of high modernist development practice. The long gestation of the project, coupled with a sense of urgency stemming from the oil crisis and earthquake, meant that the forces favoring rapid implementation were powerful incentives for funders. Both the sense that the path to the project was inevitable and the perceived need to 'just do something' created an environment where the careful research needed to reveal potential social and environmental impacts was unlikely to happen. In both Peru and Guatemala, the technical biases of the institutions conducting the project identification and feasibility studies reflected a project culture in which engineering and macroeconomic concerns were routinely accorded priority over social and environmental issues. For example, Oré and Rap [20] show how irrigation engineers trained at the National Agrarian University at La Molina, lacking training in either agronomy or the social sciences, came to Peruvian irrigation bureaucracy. As noted above, social impact analysis for the Chixoy project was relegated to a biologist and a physician, and no consideration was given to social and environmental impacts until significant project investments had already been made. Language skills were not seen as particularly relevant to the data gathering process, which in any case favored quantitative over qualitative data.

In both San Lorenzo and Chixoy, the kind of economic calculation that Hirschman criticized in *Development Projects Observed* obscured the impacts of construction on people whose economies were not fully monetized or integrated into the national economy, systematically undervaluing the contribution of crops and livestock produced for subsistence or local markets. Nor did these studies take into account the importance of resource complementarity in rural livelihood strategies. For example, a study recommending what should be covered in an impact assessment for INDE concluded that negative impacts "can essentially be summarized as the loss of agricultural production in flood zones. This loss will be of little importance owing to the small extent of the cultivated area and the scant value of the products of the zone" [34] (p. 9). The technical bias resulted in the assignment of multiple mandates to the wrong project actors. For example, the 1973 environmental impact reconnaissance for LAMI, which purported to identify social impacts, was performed by consultants who did not speak Maya languages [32]. Furthermore, the IFIs put INDE in charge of resettlement even though, by its own admission, INDE lacked the capacity to perform this function [34].

Another element in the institutional culture of development actors that helped to produce ignorance was the equation of modernization with the public good. The ethos that underlay support for TVA-like projects in the global south was grounded in an idea of progress viewed as the progression from pastoralism and subsistence agriculture to urbanization and industrial development. A corollary is the idea that projects in the public interest can be planned and executed by technical experts acting independently of political context. This idea of progress, which underlay Hirschman's assessment of his Colombian experience, informed the development of Peru's irrigation bureaucracy, as well as the World Bank's decision to finance San Lorenzo and to undertake its massive hydroelectric effort in Guatemala. However, the two projects took place in vastly different political contexts. In Piura, the hand that hid the project's political context played a somewhat positive role: it allowed project staff to invent creative ways in which to promote small and medium scale agriculture in the face of opposition from an entrenched *hacendado* elite. In Guatemala, it appears to have blinded both IFI staff and contractors to the severity of the human rights abuses that were already occurring when project planning took place. In making its case for reparations, a 2004 mission report prepared by the Centre on Housing Rights and Evictions [39] suggests that the IADB and the World Bank probably did know about INDE's aggression against Maya communities, but that if the banks were indeed ignorant, this represented gross negligence given the evidence at their disposal.)

As Hirschman predicted, the hiding hand hid technical problems that resulted in huge cost overruns for the Chixoy project. Incorrect assessments of the Middle Chixoy Basin's geology and seismicity were partly to blame, but poor management also contributed to cost overruns. However, writing in the 1960s, Hirschman did not recognize that the hiding hand could also obscure the existence of peoples whose culture depended upon a strong relationship to place. With the exception of Partridge's 1983 study for the IADB [38], project documents for both San Lorenzo and Chixoy portrayed residents of the project area as scattered, backward, and resistant to change. San Lorenzo project documents identified goat herders and landless cultivators mainly as potential threats to colonization efforts, rather than as people facing displacement. Of the raft of IADB and World Bank documents, only the Partridge report considered what the loss of place would mean for Maya family livelihoods and community survival.

In large part, official Chixoy project documents ignored, belittled, and misconstrued the concerns of those affected by the dam. For example, a 1981 ex-post evaluation of the dam's social and economic impacts [34] made no mention of ethnicity, the cultural value of landscapes, or even Maya archaeological sites. By 1991, the IDB was somewhat more aware of dam affected people, yet a 1991 loan proposal for management and conservation of renewable natural resources in the Upper Chixoy Valley reflected a somewhat more benign, yet still essentialist, view of area residents. It stated,

> In the world view of the native peoples, traditional lifestyles and agricultural practices are expected to remain changeless for evermore, which explains why native campesinos fitting the traditional mold have proven resistant to change and novelty and prefer to stick to subsistence agriculture [35] (Annex II-2, p. 1)

In contrast, a proposal for indemnification made by the Community of Pacux to the World Bank in the mid-1990s makes this poignant assessment of the effects of deracination on the elderly:

A majority of elderly lost their family even when young since the massacres of Pacoxom, Xococ, El Naranjo and also Rio Negro. Now they are already growing old and cannot go to gather firewood. Some have been abandoned by their families, today they have no money for health care. Youth don't respect them, the committee has not been able to do anything to protect them. Authorities don't take an interest. There are no programs to support them either on the part of the municipality or the department (my translation) [46] (p. 5)

The hiding hand operating during the Chixoy Project process enabled the erasure of the connection of people to place, first conceptually and then literally. Components of place that were obliterated in the transformation of the project area included connection to ancestors, sacred elements in the landscape, the knowledge that resides in landscape and its features, relations and networks of economic interaction, and knowledge about safety and danger. Río Negro inhabitants were not only reluctant to lose these connections, but their testimonies also indicate their fear that they would be moving to a site of danger under surveillance of Guatemalan military and civil patrols. Additionally, as the above quote indicates, displacement contributed to cultural loss. Even as they reflected critically on the performance of INDE with regard to resettlement, World Bank staff barely acknowledged the place-based nature of Maya Achi concerns when it cited the need to allocate resettlement lands "along kinship lines" [47] (p. 50). For the most part, however, Bank criticism addressed INDE's failures to comply with its own standards, but allowed the hiding hand to conceal the extreme risks faced by those displaced by the Chixoy dam.

Lastly, Hirschman's thoughts on Colombia notwithstanding, war zones are poor environments for infrastructure projects, however well intentioned. Transfer of INDE to the Guatemalan military should have raised serious concerns for the lending agencies, given the escalation of human rights abuses in the countryside. Thorough investigation of conditions in the project area was called for, but by the early 1980s, as the World Bank planned its second loan, it was no longer sending teams into the countryside even for brief periods, citing generalized violence as the reason. This vastly increased the likelihood that INDE's activities would go undetected by the bank, although the Witness for Peace report argues that it would be reasonable to assume that bank staff were aware of the violence against Río Negro as early at 1982 [37]. The first mention of the civil war that I found in the official project documentation came from INDE's 1991 ex-post evaluation [34] (p. 81) carried out by an economist, a civil engineer, a geologist, and a public accountant. It stated that

Especially in El Quiche and part of Alta Verapaz, where the presence of security forces (army, Guardia de Hacienda, and paramilitary groups) as well as subversive cells provokes an instability in the communities who see the need to keep moving (perigrinar) and at times leave the region seeking refuge on the Mexican border (my translation).

The Chixoy project is not simply an isolated a cautionary tale about the unanticipated side effects of well-intended international intervention. Exclusion of the voices of those affected by the project was a manifestation of fundamental flaws in the way knowledge was defined within the IFIs. It is by no means clear that prior consultation would have prevented displacement, but willful ignorance of local concerns and values enabled violation of human rights. The hiding hand made it difficult for the lending agencies to define the point where human rights should have taken precedence over hydropower development; this would have required accurate information about the state and its presence in the project area.

Even if the IFIs and contractors had known about the genocide in the Maya highlands and about INDE's complicity, it does not necessarily follow that they would have withdrawn support for the project. Fox and Brown [48] call our attention to World Bank emphasis on 'the counterfactual'—if we didn't participate in the process, it would be worse. There is some truth to this in the case of Chixoy. As noted above, when in 1975, the Guatemalan government applied to the IADB for funds, it did so without offering a plan for resettlement. When the World Bank negotiated its loans in the period following the earthquake, INDE was forced to make a plan for resettlement, although this plan was inadequate and largely honored in the breach. (On the other hand, even in the absence of other information, Río Negro's resistance to resettlement should have raised concern. Of the four reasons to resist resettlement offered by Oliver-Smith [49], three are highly germane. First is the relationship of the target population to its environment, a relationship having to do with factors that include soil fertility, resource availability, territoriality, inter-group relations, cosmology, world view, and individual and cultural identity. A second is relations between the target population relationship to the resettlement agent. Oliver Smith finds that resistance to resettlement occurs where there are ethnic differences between those who control the state and those subject to eviction, and where the state's resettlement history is bad. A third factor, he argues, is the quality of the resettlement plan. All three were at issue in the Chixoy case.)

In the San Lorenzo case, the hiding hand both fostered creativity and obscured some potentially displacing effects. In contrast, the hand that propelled the Chixoy project forward did not just obscure engineering obstacles and allow for unreasonably optimistic cost-benefit calculations; more importantly, it concealed a political reality that should have prevented the project from moving forward. The hiding hand appeared to guide not only the pre-feasibility and feasibility studies, but the entire course of the project process well into its evaluation phase. We still see evidence of its operation in 1996, after the restoration of democracy, although by this time local and international organizations had begun a lengthy, risky effort to uncover what the hiding hand hid and to seek reparations for the damages caused as a result.

5. Conclusions: The Hiding Hand and Its Potential for Mischief

With the benefit of hindsight, we see that the hiding hand's potential for mischief has been far greater than Hirschman realized when he wrote *Development Projects Observed*, although even when he wrote the book he harbored a certain skepticism about the potential of the development project writ large. The optimism that underlay Hirschman's understanding of the hiding hand metaphor may have been warranted when he assessed the progress of Peru's San Lorenzo project in the mid-1960s, an unusually optimistic moment in that nation's history. Interest in land reform on the part of state and international institutions led to what, for Hirschman, was a highly positive unintended outcome: the project became a site for experimentation with agrarian reform and food crop production. On the other hand, the hiding hand hid what were probably serious social problems related to displacement and resettlement both from Hirschman and from the project's promotors. However, the magnitude of these problems is hard to assess, as the agrarian reform program enacted in 1969 brought with it massive changes in Piura's agrarian sector.

Hirschman had moved on to other pursuits in the 1970s, so we don't know how he would have judged the impact of the hiding hand in the Chixoy case. That said, optimism about its role fostering creativity would have been delusional even at the project's outset: militarization of the Guatemalan countryside was well underway in the 1960s when project planning took place, and rural Maya communities had already become victims of policies designed to undo the Arbenz agrarian reform. Had serious social and environmental impact assessment been undertaken early in the project process, IFI staff would have known that the project would be likely to have seriously negative outcomes for Maya communities in the area. By continuing to support the project without monitoring social impacts on site and questioning the role of the implementing agency, the IFIs were in effect condoning some of the most severe human rights abuses of the Guatemalan civil war.

Regardless of political context, the likelihood that the positive outcomes of big hydroelectric projects would outweigh their negative societal and environmental impacts had diminished sharply by the 1970s and 1980s. Enthusiasm for the TVA model meant that it was exported to natural, political, and cultural environments where it was entirely inappropriate. In the global context, the number of sites where dams could be built without displacing substantial numbers of people or wreaking serious environmental damage had by this time become vanishingly small. Combined with population growth in dam catchment areas, the number of people negatively affected by resource development projects would increase.

This meant that, regardless of whether it fostered creativity, the hiding hand encouraged the funding of ill-conceived water projects with highly displacing effects and seriously detrimental environmental impacts. In the Chixoy case, ignorance of a political context that should have prompted the IFIs and contractors to withdraw support for the project was produced by the failure to include more than perfunctory social and environmental assessment in the planning process, and disinterest in monitoring the political activities of implementing agencies. These failures ensured that the export, a U.S.-based model, would allow a malevolent hiding hand to operate.

In contrast, Hirschman's later work shows increased attention to the role of the state and the broader political environment in governing the shape of development projects. By the early 1980s, Hirschman, ever the heterodox economist, had shifted his research agenda in ways that would yield far more information about the impacts of development on Latin American peoples. Rather than learning about the development process from World Bank and government officials, in research that would inform his 1971 book *A Bias for Hope*, he chose poor urban dwellers, fishers, teachers, and cultivators as teachers. He still saw a creative hiding hand at work in the projects he examined. However, as a result of his interviews, he concluded that overreliance on quantitative methods and economic analyses to guide projects would produce ignorance of critical factors in project success or failure [7].

The learning curve for the international institutions, as we see from the Chixoy case, was far slower. Ignorance was produced both by the timing and attention paid to social and environmental assessment. Well into this century, World Bank outlines of the project process relegated social and environmental impact assessment to a fourth appraisal phase of project development, well after the commitment of substantial amounts of money to feasibility studies and cost-benefit analyses. Regional development banks like the IDB were even more reluctant to address these concerns. In the Chixoy case, ignorance was also produced by the unwillingness of the World Bank as an institution to address human rights concerns in the borrowing country, or to put people on the ground where they could monitor the actions of the INDE, the implementing agency. It is unfortunate that Hirshman's awakening to the importance of non-economic elements in determining the impacts of development projects was not matched by a similar interest in the lives of project-affected people on the part of the IFIs and the contractors working on Chixoy and other big hydraulic projects.

Funding: This research received no external funding.

Acknowledgments: The author wishes to thank colleagues at PUCP for the enthusiasm with which they have shared their knowledge of Peruvian development and water management.

Conflicts of Interest: The author declares no conflict of interest.

References

1. Boelens, R.; Shah, E.; Bruins, B. Contested Knowledges: Large Dams and Mega-Hydraulic Development. *Water* **2019**, *11*, 416. [CrossRef]
2. Ekpladh, D. "Mr. TVA": Grass-roots development, David Lilienthal, and the rise and fall of the Tennessee valley authority as a symbol for U.S. overseas development, 1933–1973. *Dipl. Hist.* **2002**, *26*, 335–374. [CrossRef]
3. Neuse, S.; David, E. *Lilienthal: The Journey of an American Liberal*; University of Tennessee Press: Knoxville, TN, USA, 1996; Chapter 12.
4. Khagram, S. *Dams and Development: Transnational Struggles for Water and Power*; Cornell University Press: Ithaca, NY, USA, 2004.
5. Lilienthal, D. The road to change. *Int. Dev. Rev.* **1964**, *6*, 9–14.
6. Hirschman, A.O. *Crossing Boundaries: Selected Writings*; Zone Books: New York, NY, USA, 1998.
7. Adelman, J. *Worldly Philosopher: The Odyssey of Albert O. Hirschman*; Princeton University Press: Princeton, NJ, USA, 2013.
8. Hirschman, A.O. The Principle of the Hiding Hand. *Public Interest* **1967**, *6*, 10–23.
9. Hirschman, A.O. *Development Projects Observed*; Brookings Institution: Washington, DC, USA, 1967.

10. Hirschman, A.O. *A Propensity to Self-Subversion*; Harvard University Press: Cambridge, MA, USA, 1995.

11. Flyvbjerg, B.; Sunstein, C. The Principle of the Malevolent Hiding Hand or, the Planning Fallacy Writ Large. *Soc. Res.* **2016**, *83*, 979–1004. [CrossRef]

12. Anheier, H. Infrastructure and the principle of the hiding hand. In *The Governance of Infrastructure*; Wegrich, K., Genia Kostka, G., Hammerschmid, G., Eds.; Oxford University Press: Oxford, UK, 2017; Chapter 4; pp. 979–1004.

13. Thorp, R.; Bertram, G. *Peru 1890–1977: Growth and Policy in an Open Economy*; Columbia University Press: New York, NY, USA, 1978.

14. Cleaves, P.S.; Scurrah, M. *Agriculture, Bureaucracy and Military Government in Peru*; Cornell University Press: Ithaca, NY, USA, 1980.

15. Projects Department, International Bank for Reconstruction and Development and International Development Association. San Lorenzo Irrigation and Land Settlement Project (Stage III), Peru. World Bank. Unpublished work, 26 March 1965.

16. Department of Technical Operations, International Bank for Reconstruction and Development. Report on Quiroz-Piura Irrigation Project (Second Stage, Peru). Unpublished work, 25 March 1955.

17. Mayer, E. *Ugly Stories of the Peruvian Agrarian Reform*; Duke University Press: Durham, NC, USA, 2009.

18. Slack, K. *Mining Conflicts in Peru: Condition Critical*; Oxfam America: Boston, MA, USA, 2009. Available online: https://www.oxfamamerica.org/static/media/files/mining-conflicts-in-peru-condition-critical.pdf (accessed on 8 January 2019).

19. Bebbington, A.; Humphreys-Bebbington, D.; Bury, J. Federating and defending: Water, territory and extraction in the Andes. In *Out of the Mainstream: Water Rights, Politics and Identity*; Boelens, R., Getches, D., Guevara Gil, A., Eds.; Earthscan: New York, NY, USA, 2010; Chapter 16; pp. 307–327.

20. Oré, M.T.; Rap, E. Políticas neoliberales de agua en el Perú. Antecedentes y entretelones de la Ley de Recursos Hídricos. *Debates Sociol.* **2009**, *34*, 32–66. (In Spanish)

21. Grandin, G. *The Blood of Guatemala: A History of Race and Nation*; Duke University Press: Durham, NC, USA, 2000.

22. Comisión para el Esclaramiento Histórico. *Conclusiones y Recomendaciones: Guatemala Memoria del Silencio*; F&G Editores: Guatemala City, Guatemala, 2004. (In Spanish)

23. Sanford, V. *Buried Secrets: Truth and Human Rights in Guatemala*; Palgrave MacMillan: New York, NY, USA, 2003.

24. Johnston, B.R. Chixoy dam legacies. *Water Altern.* **2010**, *3*, 341–361.

25. Johnston, B.R. Large-scale dam development and counter movements: Water justice struggles around guatemala's Chixoy dam. In *Water Justice*; Boelens, R., Perreault, T., Vos, J., Eds.; Cambridge University Press: Cambridge, UK, 2018; Chapter 9.

26. Lynch, B.D. *The Chixoy Dam and the Achi Maya: Violence, Ignorance, and the Politics of Blame*; Working Paper Series 10-06; Mario Einaudi Center for International Studies; Cornell University: Ithaca, NY, USA, 2006. Available online: https://einaudi.cornell.edu/sites/default/files/Lynch_WP10-2006.pdf (accessed on 8 January 2019).

27. Sanford, V. *Violencia y Genocidio en Guatemala*, 2nd ed.; F&G, Editores Colonia Centro America: Guatemala, Guatemala, 2004.

28. Thorp, R.; Corinne Caumartin, C.; Gray-Molina, G. Inequality, ethnicity, political mobilization and political violence in Latin America: The cases of Bolivia, Guatemala, and Peru. *Bull. Lat. Am. Res.* **2006**, *25*, 453–480. [CrossRef]

29. Partridge, W. Multilateral governmental organizations. In *Handbook of Practicing Anthropology*; Nolan, R.W., Ed.; John Wiley & Sons: Hoboken, NJ, USA, 2013; Chapter 14.

30. Johnston, B.R. *Chixoy Dam Legacy Issues Document Review: Chronology of Relevant Events and Actions. Volume Two: Chronology of Relevant Events and Actions*; Center for Political Ecology: Santa Cruz, CA, USA, 2005.

31. LAVALIN (Lammarre Valois International Limitée). *Estudio de Desarrollo de la Cuenca del Río Chixoy, Segunda Etapa. Metodología para la Evaluación "Ex-Post" del Impacto Económico y Social de la Central Hidroeléctrica Pueblo Viejo-Quixal*, Instituto Nacional de Electrificación (INDE). Unpublished work, June 1981. (In Spanish)

32. Goodland, R.; Pollard, R. *Chixoy Development Project: Environmental Impact Reconnaissance*; Cary Arboretum, Environmental Protection Program; New York Botanical Garden. Unpublished report, 1974.

33. Colajacomo, J. *The Chixoy Dam: The Maya Achi' Genocide: The Story of Forced Resettlement*; Contributing Paper submitted to the World Commission on Dams; Prepared for Thematic Review 1.2: Dams, Indigenous People and Vulnerable Ethnic Minorities; World Commission on Dams: Cape Town, South Africa, 1999.
34. INDE (Instituto Nacional de Electrificación). *Informe de Evaluación Ex-Post Proyecto Hidroelectrico*, Pueblo-Viejo Quixal. Unpublished report, August 1991. (In Spanish)
35. InterAmerican Development Bank. *Management and Conservation of Renewable Natural Resources in the Upper Chixoy Valley*, Loan Proposal (GU-0064); InterAmerican Development Bank. Unpublished document, 1991.
36. McCully, P. *Silenced Rivers (Enlarged and Updated Edition)*; Zed: London, UK; New York, NY, USA, 2001.
37. Witness for Peace. A People Dammed—The Impact of the World Bank Chixoy Hydroelectric Project in Guatemala. 1996. Available online: www.witnessforpeace.org/ (accessed on 26 September 2003).
38. Partridge, W. Comparative Analysis of BID Experience with Resettlement, Based on Evaluations of the Arenal and Chixoy Projects; Consultancy Report to the Banco Interamericano de Desarrollo [IBD]. Unpublished report, December 1983.
39. COHRE (Centre on Housing Rights and Evictions). *Continuing the Struggle for Justice and Accountability in Guatemala: Making Reparations a Reality in the Chixoy Dam Case*; Mission Report; Centre on Housing Rights and Evictions: Geneva, Switzerland, 2004.
40. Oficina de Derechos Humanos del Arzobispado de Guatemala. Guatemala, Nunca Mas. Informe del Proyecto Interdiocesano de Recuperación de la Memoria Histórica. 1998. Available online: http://www.odhag.org.gt/pdf/Guatemala%20Nunca%20Mas%20(resumen).pdf (accessed on 8 January 2019).
41. Wolfensohn, J. Letter to Paul Scire, Executive Director, Witness for Peace and Owen Lammers, Executive Director, International Rivers Network. Berkeley, CA. International Rivers Archives. Unpublished work, 27 September 1996.
42. Wade, R. Greening the Bank: The struggle over the environment, 1970–1995. In *The World Bank: Its First Half Century, Volume 1: Brookings Institution*; Kapur, D., Lewis, J.P., Webb, R., Eds.; The World Bank: Washington DC, USA, 1997; Chapter 13.
43. Scott, J. *Seeing Like a State: How Certain Schemes to Improve the Human Condition Have Failed*; Yale University Press: New Haven, CT, USA, 1998.
44. Mitchell, T. *Rule of Experts: Egypt, Techno-Politics, Modernity*; University of California Press: Berkeley, CA, USA, 2002.
45. Goldman, M. *Imperial Nature: The World Bank and Struggles for Social Justice in the Age of Globalization*; Yale University Press: New Haven, CT, USA, 2005.
46. Comunidad de Pacux. Propuesta de la Comunidad de Pacux al Banco Mundial. Unpublished work, International Rivers Archives. n.d. (In Spanish)
47. World Bank. Project Completion Report on Guatemala Chixoy Hydroelectric Power Project (Loan 1605-GU) Report No. 10258; Latin America and Caribbean Regional Office. Unpublished document, 31 December 1991.
48. Fox, J.; Brown, L.D. (Eds.) *The Struggle for Accountability: The World Bank, NGOs, and Grassroots Movements*; MIT Press: Cambridge, MA, USA, 1998.
49. Oliver-Smith, A. Involuntary resettlement, resistance and political empowerment. *J. Refug. Stud.* **1991**, *4*, 132–149. [CrossRef]

water

MDPI

Article

Hydropower in the Himalayan Hazardscape: Strategic Ignorance and the Production of Unequal Risk

Amelie Huber

Institut de Ciència i Tecnologia Ambientals, Universitat Autònoma de Barcelona, 08193 Cerdanyola del Vallès, Barcelona, Spain; amelie.huber@gmx.de

Received: 13 August 2018; Accepted: 3 January 2019; Published: 26 February 2019

check for updates

Abstract: Rapidly expanding hydropower development in areas prone to geological and hydro-climatic hazards poses multiple environmental and technological risks. Yet, so far these have received scant attention in hydropower planning processes, and even in the campaigns of most citizen initiatives contesting these dams. Based on qualitative empirical research in Northeast India, this paper explores the reasons why dam safety and hazard potential are often marginal topics in hydropower governance and its contestation. Using a political ecology framework analyzing the production of unequal risks, I argue that a blind-eye to environmental risks facilitates the appropriation of economic benefits by powerful interest groups, while increasing the hazardousness of hydropower infrastructure, accelerating processes of social marginalization. More specifically, this paper brings into analytical focus the role of strategic ignorance and manufactured uncertainty in the production of risk, and explores the challenges and opportunities such knowledge politics create for public resistance against hazardous technologies. I posit that influencing the production of knowledge about risk can create a fertile terrain for contesting hazardous hydropower projects, and for promoting alternative popular conceptions of risk. These findings contribute to an emerging body of research about the implications of hydropower expansionism in the Himalayan hazardscape.

Keywords: large dams; dam safety; hazard risk; environmental governance; uncertainty; knowledge politics; marginalization; political ecology; Himalayas; India

1. Introduction

"[Dam safety is a matter of] calculation. When we design we take all these things into consideration (. . .) we assume that what we have designed will not fail. Generally, you might never have heard that a dam has failed. Maybe in Europe, but in India dam breakages are very few" (Vice-President, private hydropower company, Sikkim, 19 April 2015).

On 22 July 2018, an under-construction hydropower dam in southern Laos collapsed, killing at least 40 people, while hundreds went missing and nearly 10,000 were displaced in Laos and downstream Cambodia. Investigations in the aftermath of the disaster found faulty construction and operation and the authorities' failure to heed early warning signs to be responsible for the catastrophe [1]. Barely one month later, the South Indian state of Kerala experienced unprecedentedly severe flooding. For the first time in history, 35 out of 54 dams in the state had to be opened for safety reasons. Nearly 500 people were killed. Although various experts held the dams responsible for aggravating the floods, a report by the country's Central Water Commission quickly asserted the opposite, absolving the dams and their operators from any blame [2].

These disasters shine light on some often-neglected facts. Large dams are risk-laden artifacts. Exposed to earthquakes, floods, extreme rainfall, avalanches and landslides, and able to cause an equal number of environmental hazards, their functioning and (in)stability is ultimately a product of human excellence and error. As these catastrophes and numerous others in the past months, years and decades remind us, dams do sometimes fail or otherwise produce large-scale hazards. The history of the modern large dam includes a long list of dam-related disasters with a substantial human death toll [3]. This fact is often forgotten or negated, as in the above quote, partly owing to "hydro hubris" [4]—unwavering faith in the godlike power and brilliance of modern hydraulic engineering and in large dams as infallible human creations. But there are also economic and political reasons behind attempts to ignore or negate the fallibility of dams, and to erase such "accidents" from collective memory [5,6]. This paper analyzes the structural mechanisms and power relations behind policies and discourses, which sideline the obvious risk management challenges posed by hydropower dams.

Following the fall and resurgence of dams as "green" energy solutions and objects of financialization [3,7–9], the public debate on large dams has entered a new phase. Dominant pro-dam discourse celebrates hydropower as an uncomplicated, sustainable, and renewable source of energy indispensable to development objectives, such as green growth, climate change mitigation, and poverty alleviation [7,10,11]. While today the dam lobby more readily discusses mitigation strategies for contentious "externalities" like social displacement or ecological impacts, the delicate question of environmental and technological risks emanating from hydropower infrastructure rarely figures in public narratives. And yet, as the global hydropower frontier is expanding into many highly hazard-prone river basins [12,13], risk management emerges as a major challenge of environmental governance. With private and public corporations vying to tap the world's remaining unexploited rivers [8,14], located often in ecologically sensitive forest and mountain areas where climate change increasingly destabilizes precarious local environmental equilibria [15–19], the question of how new hydro infrastructure interacts with environmental hazards, and how decision-makers act to mitigate and adapt to associated risks has become acutely relevant.

This study looks at risk governance in one of these new hydropower hotspots, the Eastern Himalayan region of Northeast India. Over the last two decades the rapid proliferation of new hydropower infrastructure has exacerbated ecological precarity in this seismically and geologically active mountain range [20–23]. In addition, climate change and other anthropogenic pressures are expected to accelerate the frequency and intensity of landslides, flash floods, and seasonal droughts in the coming years [15,16]. Nevertheless, climate adaptation, risk management, and disaster preparedness have received only scant attention in hydropower planning processes across the Himalayas [20,24–28]. Even for most affected communities and civil society organizations contesting hydropower development, safety risk has not been a preferred mobilizing concept – with few notable exceptions, such as the South Asia Network on Dams, Rivers and People (SANDRP), or the movements against the Tehri Dam in Uttarakhand and the Lower Subansiri Hydroelectric Project (HEP) in Arunachal Pradesh/Assam (see Section 5.2). In this article I explore the reasons why hazard potential and dam safety are often marginal topics in discussions about Himalayan hydropower governance and its contestation, and how this has helped produce and exacerbate uneven risks and vulnerabilities.

The bulk of empirical research was carried out between 2011 and 2015 in the small mountainous state of Sikkim, a late entrant to the Indian Union and a forerunner in the Himalayan race for hydropower exploitation. I conducted semi-structured and informal interviews with experts and hydropower professionals from various state departments, a German development agency, three nongovernment organizations working in the fields of environment and development, one state-owned and two private hydropower companies, as well as with local activists and rural households who deal with the ground realities. I also carried out on-site observations in seven hydropower project areas and consulted available policy documents and environmental reports, including the Sikkim State Action Plan on Climate Change [29], the Sikkim State Disaster Management Plan [30], and ten Environmental Impact Assessment (EIA) reports for hydropower projects available in the public domain [31].

Combining insights from political ecology and critical hazards geography, and from literature on the "strategic unknowns" [32], I argue that ignoring environmental and technological risks in the planning and implementation of hydropower projects is a central mechanism in the production of unequal risk. Turning a blind-eye to risk enables the shifting of risks and hidden costs, thereby facilitating the appropriation of economic benefits from hazardous hydropower infrastructure by political and corporate powers, and accelerating processes of social marginalization among already vulnerable social groups. Political ecologists have pointed out similar generative patterns in the production of unequal risk, as well as their underlying institutional and discursive drivers [33–37]. What has not been explored in as much depth is why and how processes of marginalization and facilitation are met with acquiescence and/or resistance by the affected public.

I try to explore the latter question by paying attention to how knowledge politics mediate the production and contestation of risk [32,38,39]. I show how certain experts and hydropower professionals instrumentalize and manufacture scientific uncertainty and controversy to depoliticize and conceal the subject of risk in dam conflicts [40]. By contrasting the Sikkim experience with a second case study from the Eastern Himalayas—the protracted conflict over the 2000 MW Lower Subansiri Hydroelectric Project on the Assam-Arunachal Pradesh border—I discuss how knowledge politics can serve to curb resistance, while at the same time providing a fertile terrain for contestation. Thus, the conflict over the Lower Subansiri project—one of India's largest hydropower projects under construction to date—turned into a highly politicized public controversy precisely because civil society groups were able to draw on both vernacular knowledge and scientific expertise to challenge techno-scientific hubris and knowledge politics with powerful counter-claims.

The paper is structured as follows. Section 2 theoretically situates the paper. Section 3 discusses the inherent and aggravated risks associated with the Himalayan dam building spree. Section 4 looks at the hydropower governance process, analyzing institutional mechanisms and policy lacunae, which facilitate "risky" hydropower projects and foster relational processes of facilitation and marginalization. Section 5 first illustrates how these processes are reinforced through a politics of ignorance and neglect, which is legitimized through the mobilization and manufacture of uncertainty. It then discusses the challenges and opportunities for publicly resisting such discursive strategies, making a case for lay-expert knowledge co-production. Section 6 concludes with theoretical and policy implications.

2. Theoretical Framework

Political ecologists and critical hazard geographers have long argued against hazard-centric and techno-managerial approaches to the study and management of environmental risks, which locate the blame for calamity in nature [41–43], pointing instead to the social, structural, political, and institutional dynamics, which produce risk, disaster, and differential vulnerabilities [34,35,44]. Often those further down the social ladder have to pay for natural disasters—a result of power relations, structural inequalities and exploitative processes, which allow those who create or decide for risk, intentionally or through ignorance, to shift it onto others who lack the power to influence these decisions [45]. A common frame in hazards geography thus defines risk as the combination of the probability of a biophysical or technological hazard event (e.g., an earthquake or a dam failure), hazard exposure and social vulnerability ("the ability to anticipate, respond to, and recover" from the inflicted damage [33,44,46] (p. 589).

Recent studies have moreover emphasized the importance of analyzing the generative processes, systemic drivers and institutional decisions, which produce unequal risk and differential vulnerabilities, particularly those enhanced by neoliberalism and capitalism [47–49]. Baldwin and Stanley [46], for example, conceptualize environmental risks and hazards not merely as by-products of capitalism, but as integral to the circulation, viabilities, and crises of capital. Huber et al. [6] suggest that in order to give greater visibility to the role of capital in driving "risky" development decisions, dam failures and other so-called "socially constructed disasters" should better be characterized as "capital-driven destructions".

Similarly, research on environmental justice and environmental racism has shown how racially uneven geographies of risk and vulnerability are the product of colonial ecological violence during the early days of modern capitalism and its postcolonial iterations today [25,50]. In the context of the Eastern Himalayas, Gergan [25] argues that hazardous hydropower infrastructure is built on historical terrains marked by the relationship of dependency, exploitation and negation between the Indian state and its northeastern frontier, and by generations of regional marginalization, uneven development, and racialized, exclusionary state-building practices.

A useful analytical frame to understand how capital interests and elite social groups benefit from the creation of unequal risk is Collins' [33] relational concept of marginalization and facilitation. It highlights multi-dimensional, mutually constitutive and materially inseparable social constructions of nature, which turn environmental risks simultaneously into amenities for some, and externally imposed threats for others. Specifically, facilitation denotes "how powerful groups are provided privileged access to institutional resources" to exploit the environmental rewards associated with hazardous places, with deleterious socioenvironmental outcomes (p. 589). This definition emphasizes the integral role of state and market institutions in unevenly allocating protective resources, such as insurance, land-use and disaster relief subsidies [33,36,37]. In Collins' [33] example, wealthy US residents appropriated the environmental rewards of living in flood-prone neighborhoods (scenic views and being amidst nature) by securing privileged access to flood recovery resources and institutional support. Risks were thus externalized and shifted onto poor migrant workers (marginalization) who occupied flood-prone neighborhoods for want of options, while lacking access to a similar safety net.

But processes of marginalization and facilitation and the uneven geographies of risk and vulnerability they create are also a product of analytical lenses and discursive formations—often based on expert knowledge systems and technocratic managerial discourses—through which they are viewed, represented, and contested [33,34]. Mustafa [35], for example, in his work on the technocratic production of an urban flood "hazardscape" in Pakistan argued that the authorities' material interventions in the watershed were heavily influenced by a narrow technocratic view of hazard problems and solutions, which was largely incongruent with the lived hazardscape reality of the flood victims. Such narrow technical framings of contested environmental problems may also be used strategically to depoliticize and cover up value conflicts, to justify decisions already made [51], or to facilitate the shifting of the harmful effects of accumulation through hazardous technologies [52].

Similarly, one of the main objectives for scholars of ignorance studies has been to show how ignorance, knowledge gaps, or "undone science" are used strategically to preclude, obfuscate, deflect and insulate against unsettling information, magnifying what remains unintelligible [32,38,39,53,54]. As McGoey [32] argues, "the cultivation of strategic unknowns remains (…) perhaps the greatest resource for those in a position of power" (p. 1), a "productive asset helping individuals and institutions to command resources, deny liability in the aftermath of crises, and to assert expertise in the face of unpredictable outcomes" [28,55] (p. 553). Industries, for example, increasingly take advantage of uncertain evidence—or question the validity of existing evidence—to shroud claims of causal linkages and to protect financial interests, as was the case with the tobacco industry, climate skeptics or the 2008 financial crash [40,55,56].

In the Himalayan hydropower sector, too, the strategic mobilization of the "unknowns" about environmental and technological risks is a pervasive practice employed both by state and corporate actors [32]. Lord [28,57] and Butler and Rest [24] explore how the "speculative logics" of private financial interests and the state's ambition to meet demands of domestic electricity and revenue propel state and corporate actors within Nepal's hydropower community into "environmental denial" [24] (p. 15), "perpetuat(ing) a "strategic ignorance" [32] of palpable environmental and infrastructural risks" [57]. In a "conjuring trick (and) spectacle" aimed at gathering investments and maintaining the promise of a "hydropower nation", hydropower proponents gloss over inherent environmental and technological uncertainties (e.g., that of seismic risk), while championing "an understanding of risk as objectively calculable" [9,24] (pp. 21–22).

This paper delves deeper into the material and discursive process enrolled to invisibilize the hydro-climatic, geological, and technological risks of large-scale hydropower development in Sikkim.

While many of these risks are known in principle, the incalculability of their occurrence, timing and scale allows experts and political decision-makers to overlook, ignore and deliberately conceal them, much like the knowledge politics surrounding climate change allow some to argue climate change is a hoax.

A further concern of this paper is the role of risk conflict and public resistance in processes creating unequal risk, and in attempts to challenge manufactured uncertainty. The "hazardscape" concept invoked by Mustafa [35] and Collins [33] frames geographies of uneven risk as products of contestation between competing social groups. However, the discussion on what determines public resistance or compliance with the policies and discourses responsible for risk creation has remained relatively thin. Leaning on Gramsci [58], Collins [33] (p. 600) argues that some hegemonic discourses are invoked "even by people who appear to be poorly served by them," because marginalization—being predicated on unequal power relations—is often legitimized ideologically.

Alternatively, people may be aware of and challenge the technocratic gaze and the power dynamics that put them at risk; yet, they fail to transform existing configurations of power and injustice due to the differentially powerful epistemic authority of popular epistemologies vis-à-vis policy and science-based knowledge claims. In a "globalizing world, characterized by the hegemony of technocratic and social modernity", it is this "power of modern institutions to limit debate and discussion" that prevents a more democratic approach to risk management [35] (pp. 567, 582).

The literature on risk conflicts paints a more hopeful picture for the contestation of dominant risk discourses from below. Contrary to popular claims about science being indispensable for understanding many of today's "invisible", diffuse, and difficult to perceive risks [59–61], scholars argue that increasing awareness among the lay public and reduced appreciation for science as a privileged, authoritative source of knowledge create new opportunities to influence definitions of risks. As Cooper and Bulmer [62] (p. 264) argue, hegemonic expert discourses about risk can be hijacked and altered by counter-hegemonic forces, promoting "contradictory popular conceptions of risk". While the greater contestability of knowledge also implies that politically powerful interest groups can reject valid scientific evidence as "fake news" [40,56], it has also given rise to new risk conflicts [59,60,63–65], allowing grounded, material, and embodied experiences of environmental precarity to converse with and challenge scientific representations of risk [25].

This paper seeks to contribute to these debates about acquiescence and resistance to the production of unequal risk, by making a case for knowledge controversy as a major site of political struggle and contestation. It contrasts two cases of risk conflict over hazardous hydropower infrastructure: in the first, scientific uncertainty was effectively mobilized and reinforced by government experts, decision-makers and power developers to maintain the status quo of undefined liabilities; in the second, the ambiguity and malleability of science was exposed and exploited to stage alternative risk claims and to re-politicize the risk question.

3. Hydropower Risks in an Intensifying Hazardscape

The Himalayas are naturally hazard-prone. As one of the world's most geologically and seismically active mountain ranges transected by a multitude of steep, fast-flowing, silt-laden rivers, earthquakes, landslides, and floods are recurrent phenomena. Yet, anthropogenic activities including urbanization, deforestation and infrastructure development have led to an intensification of hazard potential in recent decades [66,67]. Further, severe climate change effects on the weather-climate and hydrological regimes of the Himalayan region have been observed and predicted for the coming decades [15,68–70], increasing in particular the risk of landslides and large-scale hydrogeological hazards, such as "landslide dam outburst floods", "glacial lake outburst floods" (GLOFs) and other erosive flash floods [16,27,71].

Recent disasters hint at how vulnerable Himalayan hydropower projects are. In 2013, an extreme flash flood in the Indian state Uttarakhand caused extensive damage to the state's hydropower infrastructure as an excess of water, boulders, debris, and silt choked the floodgates of hydropower stations, leading to overtopping [72]. A similar problem was caused by a landslide dam outburst flood in Nepal's Sunkoshi river basin in 2014 [73]. During the devastating 2015 Nepal earthquakes over 30 hydro projects were damaged, mostly by earthquake-triggered landslides, causing the loss of 34% of Nepal's installed hydropower capacity and USD 200 million estimated losses for its hydropower industry [28,74]. Schwanghardt et al. [74] estimate that ~25% of existing and planned Himalayan hydropower projects "have high probabilities of moderate to severe damage during future earthquakes".

But hydropower infrastructure is not only *at risk*. It also contributes to the intensification of hazard potential, often as a consequence of political and economic decisions about its siting, construction and operation. Schwanghardt et al. [75] (p. 1) observe a systematic push of hydropower activities into the headwaters of Himalayan river basins, closer to glacial lakes and on potential GLOF tracks, estimating that a third of the sampled sites "could experience GLOF discharges well above local design floods." Likewise, investigations identified hydropower infrastructure as one of the main contributors to the Uttarakhand flood damage [72]. Damage was greater near existing and under-construction hydro-projects—a result of how these projects manage destructive water and sediment flows. Construction debris was inadequately disposed of and washed into reservoirs, obstructing dam/barrage gates and leading the river to overflow and laterally outflank the dams. Excessive siltation had reduced the carrying capacity of rivers and increased their erosive capacity.

Dam-induced flash floods are a recurrent problem, caused by sudden releases of water from hydropower stations. Designed primarily for power generation, most run-of-the-river projects today lack adequate flood cushions, as Das [27] explains. Especially at times when reservoirs are full (e.g., at the end of the monsoon season), flood absorption is not guaranteed—a fact often brushed over by hydropower proponents. When excessive inflows from floods or heavy rainfall exceed storage capacity, it is standard practice to release water to ensure dam safety. But such patterns of water release can be highly disruptive, accentuating flood impacts downstream (ibid.), as has been most cruelly demonstrated by the massive flood disaster in Kerala this year [2].

Finally, what is easily overlooked, especially when thinking in terms of large-scale dam disasters, is the "slow violence" of everyday ecological precarity accompanying the construction of hydropower infrastructure in fragile geological settings [25,76]. Phenomena reported from hydropower-affected areas across the Himalayas, such as the sudden appearance of cracks in houses, the activation of landslide zones, or water resources running dry may represent more tangible and cumulatively impactful hazards to the lives and livelihoods of rural Himalayan communities [26,57,77–79]. Excavation works for hydropower infrastructure tend to destabilize fragile mountain slopes, with impacts often felt for months or years post-construction, exacerbated by natural hazard activity. For instance, following a 6.9 magnitude earthquake in Sikkim in 2011, a particularly large concentration of earthquake-induced landslides and damaged buildings was found in vicinity of the 1200 MW Teesta III HEP under construction at the time [22,80]. As a prominent Sikkimese activist commented: "It is any one's guess that the severely disturbed area just needed another jolt to cause devastation as that happened on the 18th of Sept 2011." [81].

Despite these obvious environmental risks, prevalent hydropower governance approaches in Himalayan states have brushed risk and safety considerations under the carpet, often due to economic considerations [24,26,28,82]. Sikkim, given its reputation as one of India's most environmentally conscious states would appear more likely to approach hydropower governance in a more holistic manner [11]. However, as I discuss in the following sections, analysis of governance approaches with respect to hydropower risks exposes serious shortcomings behind Sikkim's progressive "green" politics façade. The government's "quick-business" approach to hydropower development and efforts at providing favorable investment conditions have not only invited "risky" hydropower projects but have also accentuated vulnerabilities in marginal rural areas.

4. Hydropower Governance and the Production of Unequal Risk

4.1. Facilitating Risk through Privatization Policies

Large-scale privatization gave the impetus for the Himalayan hydropower boom [83]. With the bad fame large dams had acquired globally by 2000—following persistent civil society activism and the World Commission on Dams' staggering report—international finance for large dams had become scarce [3,84,85]. Privatization was widely embraced to revive an ailing hydropower industry [85]. In India, liberalization was gradually initiated with the Mega Power Policy 1995 and a number of other policies followed suit. The launch of the 50,000 MW Hydro Initiative in 2003 promulgated a discourse of sustainable hydropower as imperative for satisfying India's escalating energy demand and identified the country's exploitable hydropower reserves, predominantly located in difficult to access borderland mountain regions [83]. To attract private investments, the policy framework pledged financial support; fostered deregulation of the renewable energy market, minimizing state intervention; relaxed clearance procedures; reduced the minimum threshold capacity for so-called 'mega' power projects to increase the number of projects eligible for the attendant benefits"; and gave state governments the power to allot projects [86,87]. In Sikkim, investments were further incentivized by extending a 10-year tax and import duty exemption and by facilitating land acquisition and accelerated clearances through the government [86].

The Eastern Himalayas were labelled "India's future powerhouse" [27]. Here, privatization kick-started a race by the states of Sikkim and Arunachal Pradesh to contract out public rivers to private power producers [27,86]. These being predominantly rural, non-industrial states, hydropower was presented as a major source of revenue, greater financial and political autonomy, urgently needed infrastructure, employment, and regional development [11]. By 2007, Sikkim had signed 24 Memoranda of Understanding (MoUs) with selected private and public-sector undertakings for developing 5000 MW within five years [86]. Private undertakings were allocated on "Build–Own–Operate–Transfer" basis for 35 years (after which projects are handed back to the state in good operating condition), with the state entitled to 12–15% of the generated power/revenue and 26% equity for projects above 100 MW developed in joint venture [88]. Arunachal Pradesh signed 130 MoUs by 2010 for 40,140.15 MW installed capacity and 26% equity committed to each [27].

With neither a formal hydropower policy nor written rules for public-private financial transactions [87,89], this so-called "MoU virus" thrived on speculative investments and political brokering [27] (p. 3). Agreements with private sector companies allowed for "greater flexibility in negotiating financially lucrative deals" [83] (p. 15), and involved large upfront premiums and individual commissions [27]. Critics lamented that projects were allocated arbitrarily and at "throwaway charges", lacking transparency and competitiveness [90] (p. 738).

A consequence of this open-arms approach has been what Hill [90] (ibid.) calls "frontier capitalism": the entrance of private investors with minimal accountability and experience in hydro, including from the courier and logistics, real estate, steel fabrication, and tourism sectors [91]. As a result, many projects under construction were eventually stalled or abandoned due to financial insolvency (in Sikkim only five private-held projects have been commissioned as of 2018) [20]. The hydropower sell-out has also invited public suspicion about a massive scam involving "private deals, covert decision-making and corruption", as well as the sale of the procured memoranda, clearance papers, and licenses for profit [92] (p. 56). As Rahman [91] (p.19) notes about the similar situation in Arunachal Pradesh:

> "There have been allegations that many dubious private companies are raising huge capital in the stock market, increasing their market profile and bagging infrastructure projects in other under-developed countries on the back of such hydroelectric project allocations (. . .) The local perception is that many of these small and medium dam projects are 'paper dams' or 'MoU dams' and will not see any construction on the ground, as there have been no signs of urgency in ground assessment and feasibility studies."

By incentivizing and facilitating this fast-tracked and intransparent hydro-business model, privatization policies have prepared the ground for inequitable and "risky" dam projects. For example, India's Hydropower Policy 2008 contains a number of generous concessions, which insulate private companies from the majority of risks inherent to the sector (e.g., hydrological risk), shifting these onto the public, while maximizing the margin for profits and ultimately raising the costs of power [27]. Thus, the imposed tariff regime gives power producers no economic incentive to optimize project designs according to comprehensive hydrological data. Producers are paid for full "design energy" generation, even when water is scarce and power generation is low, while buyers pay more for less. The result, Vagholikar [93] argues, are "over-designed" dams based on unrealistic data.

Moreover, joint ventures "reduce the distance between project regulators and implementers" [89] (p. 118), affecting enforcement and compliance with environmental clearance and environmental management requirements [27]. Vagholikar and Das (ibid. p. 3) note that projects allocated through preliminary payments tend to be seen by both parties as a fait accompli. Even the Ministry of Environment and Forests (MoEF) is known to proactively grant clearances to 95% of the appraised projects, including for EIA reports of extremely poor quality, or which have been "sanitized by developers (...) to weed out problematic portions" (p. 5), ignoring concerns raised by civil society. In Sikkim, lack of regulatory oversight and willful ignorance have permitted environmental decision-making frameworks to be easily sidestepped, resulting in "regulatory collapse" [22] (p. 20). By 2011, at least 17 HEPs had received environmental clearance despite warnings, improper assessments, and without meeting negotiated conditions or addressing regulatory violations (ibid.).

The following sub-section looks more closely at the ground effects of existing policy frameworks and state-level governance decisions, illustrating how these have obscured liabilities, thereby producing increased ecological precarity and social vulnerability in hydropower areas.

4.2. Producing and Shifting Risks and Costs

The project giving the best insights on long-term hydropower impacts is the 510 MW Teesta V HEP in Central Sikkim, one of the first large dam projects in Sikkim, commissioned in 2008 by the public-sector National Hydroelectric Power Corporation (NHPC). During household interviews in the area directly affected by the project's infrastructure, environmental degradation was reported by various communities in the project area—spanning a 20 km-long stretch of the Teesta river—but was found to be most severe above the reservoir and dam site. The villages Jang, Aapdara, and Phidang have been struggling with perpetual sinking of the mountain slope on which they are located, likely a consequence of the cyclical release of impounded water [94]. Visible damage included cracks in residential buildings and agricultural land, as well as enhanced landslide activity.

This has created a situation of unanticipated displacement during and sometimes long after project construction. In Aapdara and Jang several residential buildings have collapsed, and families been relocated from the area. The safety risk has become so pronounced that plans are underway to move the entire village. In Phidang the reservoir backwater has come dangerously close to the settlement, and lower lying areas of the village are at risk of toe erosion and flash floods, including sudden water releases by planned and existing dams upstream. Here, too, respondents were considering abandoning their properties. In Dipudara, located above the vertical tunnel carrying water to the powerhouse, water was repeatedly found flowing from 'cracks' in the mountain within one year of commissioning, raising fears that the tunnel system may be faulty and risk collapsing. The villagers have since been demanding compensation and resettlement.

Negligent construction practices can partly be blamed for increased ecological precarity. For example, despite knowledge about the fragile geology NHPC didn't implement effective precautionary measures, such as reservoir rim treatment [94]. In Dipudara, slope stabilization and protective concreting works to mitigate tunnel leaks were reportedly ineffective, rendering the land increasingly unsuitable for agriculture and construction. Excessive amounts of explosives used to accelerate tunnel construction was another problem cited (Interview, Mines, Minerals & Geology Department, Government of Sikkim, 26 April 2011).

Such corporate negligence, however, has also been institutionally facilitated, notably by failing to hold developers accountable. First, India's rehabilitation and resettlement policy considers only land users whose lands are acquired prior to project construction as project-affected and entitled to compensation and rehabilitation. These provisions may work for dam projects with 'traditional' storage designs. Modern large-scale run-of-the-river hydropower projects—the new standard dam design in the Himalayas—have smaller submergence zones but require an extensive underground tunnel system carrying river water to a powerhouse located several kilometers downstream [27]. These tunnels affect places far away from reservoir and power house, often in ways that are not immediately obvious (see Section 5.1). In Sikkim, entire hillslopes are pierced by tunnels or nibbled at by water-level fluctuations in the reservoir, but the often-irreversible damage and displacement caused, sometimes years after project completion, is systematically ignored in EIAs because policy does not mandate land acquisition in such areas. (For an alternative rehabilitation and resettlement model see Lord's discussion of the Upper Tamakoshi HEP in Nepal [28].

Second, for impacts not accounted for in the EIA it is difficult to claim compensation and "project-affected person" status later. Without clearly formulated guidelines defining liabilities, and authorities unwilling to hold power developers accountable, the latter can easily shirk responsibility. The burden of proof is shifted onto the victims, but laypeople have been unable to prove that the observed impacts are caused by the project. While victims receive financial support through a 'natural calamity fund' from the district administration, access to these resources often hinges on social status and connections to influential decision-makers. As one interviewee tellingly suggests: "Compensation is not dealt out to all households, only to the rich. After all, cracks cannot be seen in simple huts" (24 February 2011).

The authorities' failure to efficiently handle rehabilitation claims further propounds the vulnerability of those affected. In Aapdara and Jang inhabitants still lived in visibly damaged buildings since their compensation payments had not yet been settled. One displaced landowner received Rs 10,000 (€160) on the spot for immediate relief but was told that a full damage assessment, to be conducted by a different department was required to compensate his entire loss. After ten months and 19 visits to different offices he still had no news. Such bureaucratic limbo hits economically weak households unable to self-fund safety measures particularly hard. Lengthy administrative procedures, trips to offices and possible legal fees cut further into tight household budgets, and struggling for rehabilitation also requires a certain level of education and political clout.

This is not to say developers are totally uncooperative. In some cases, NHPC agreed to compensate the damages incurred—either through voluntarily ex gratia payments to the victims, or by repairing or constructing public infrastructure. Such voluntary rehabilitation assistance appeases public relations and avoids negative publicity, but it also undermines a more drastic reconfiguration of the rules of the game. As one Sikkimese journalist relates:

> "NHPC is generous in dealing out compensation money and undertaking protection works but refuses to admit on paper that the project works are responsible for the damage. If they would, the calculations for future projects would have to be expanded. However, in this way the same underrating of impacts is likely repeated with the next project or EIA. NHPC is also very quick in doing the repairs because in this way they control the works and the contractors too" (24 February 2011).

A more impactful way to address the vulnerabilities created by hydropower development would be policy amendments, which establish on paper the range of (hydro-)geological disturbances hydropower infrastructure can cause; and which define corporate liabilities for social and environmental rehabilitation at any project stage, preventively or through aftercare. But institutional consequences are not in the economic interest of the developers, nor of the local elites, who gain kick-backs and contracts through the hydropower business. Thus, avoiding that certain environmental risks and impacts become officially recognized and their assessment and mitigation institutionalized is key to cost-cutting and a continuation of business-as-usual. The following section explores this process of "strategic denial" by government experts, politicians and power developers within Sikkim's hydropower community further [24,57]. It illustrates how scientific uncertainty about hydropower risks and impacts is fostered through individual and institutional complacency, and by manipulating knowledge production.

5. Risk, Knowledge, and Resistance

5.1. The Politics of Knowledge around Hydropower Risks

Dam sanctioning guidelines in India do contain provisions to assess risks and ensure dam safety. Detailed project reports (DPRs) drafted during the dam planning stage must contain seismic studies and dam design and safety parameters in accordance with the Bureau of Indian Standard's civil and hydraulic engineering codes [87]. Further, states must prepare dam operation manuals for individual dam projects, which contain an emergency action plan [2]. However, there are other risks—many—which are not or insufficiently accounted for through the national policy framework [87]. For example, a major risk management challenge, given the fast proliferation of hydropower projects in a context of heightened ecological precarity and climatic uncertainty, is reconciling hydropower development with climate action and disaster management. Yet, since not mandated, none of the publicly available EIA reports and Environmental Management Plans for hydropower projects in Sikkim contain climate change considerations. Even the newly named national Ministry of Environment, Forests, and Climate Change does not have any written positions on climate change adaptation for the hydropower sector [87].

Likewise, Sikkim's in many ways progressive environmental policy framework largely ignores the critical hydro–climate–disaster intersection, failing to keep in mind inherent ecological precarity [20,25]. Sikkim's State Action Plan on Climate Change elaborates at length on water resource vulnerability and water security but steers clear of hydropower and associated environmental risks [29]. Investigations with the State Disaster Management Authority (SDMA) found that so far neither are there risk assessments for dam-induced disasters, nor disaster management provisions to deal with hydropower-related risks. Even the department head admitted that dam safety standards were exclusively monitored by the project developers themselves, while the state was ill-prepared to deal with hydropower-related hazards:

> "We are not monitoring [the power companies] (. . .) As a disaster management department, we should be going to the dam, but now they themselves have scientists and technical persons (. . .) We should also have a connection, but now we are busy with some other things. Later we'll be doing that also" (7 May 2015).

Interviews with experts and hydro professionals reflected a striking disregard and ambiguity about dam safety, even among senior government officials. Several respondents tried to downplay or deny the hazard potential of hydropower, or to retreat from responsibility. An interviewee in the Power Department suggested I better go ask these questions to the Forest Department, since they are the ones dealing with environmental impacts and climate change (11 May 2015). There seemed to be a general lack of mandate by the authorities to look at the effects of climate change on hydropower infrastructure. An interviewee at the German development agency supporting the drafting of a state climate change policy said he was aware of climate change impacts on hydropower, and hydropower impacts on the environment, but explained that his agency only works "on whatever the government will request" (2 December 2013).

To justify this lack of engagement with hydropower risks several interviewees invoked scientific uncertainty and knowledge controversy. One NHPC official explained that in the absence of scientific climate data, climate change was a factor too uncertain for consideration in his project's EIA report:

"This is something that has to be established first (scientifically) (. . .) Somebody else has to do it" (8 November 2013). Some respondents disputed that climate change poses a problem at all. The head of the disaster management department, for example, opined that for him as a geologist, climate change was "sort of a political issue", rather than a "real" scientific phenomenon to be taken seriously (7 May 2015).

Uncertainty regarding the intensity, timing and spatial effects of climate change is undoubtedly more pronounced in the Himalayas, not least due to the lack of basic, reliable, long-term climate data for the region [27]. However, responses by project developers indicated that neglecting rather than mitigating potential environmental risks and their social ramifications was at least partly an economic strategy for safeguarding the financial viability of hydropower investments. Since the companies usually try to recover their investments as quickly as possible, and since from an economic perspective structural safety loses its importance as time advances (especially for licenses lasting merely 35 years), investing in costly precautionary measures not prescribed in project plans cuts into narrowly calculated returns. Responses also reflected that for the developers, environmental risks and their mitigation were calculable problems. Asked how his company deals with natural hazards, the director of a private power company commented: "You have to put up with it. If it harms the project, we have to bear it. We take insurance. (. . .) That is a must. That is a condition put by the lenders. Because after all they get the money back" (15 May 2015).

The strategic use of knowledge gaps also became evident from the rather selective framing of uncertainty. For example, despite the inherent and inevitable uncertainties associated with complex technological systems like dams [38], dam safety concerns were either brushed aside, or exalted with "hope" or "faith" in state-of-the-art technology, and the blunt denial of historical dam failures. The following statements were made by engineers and scientists in the Power and Disaster Management departments, the Central Water Commission and by a private hydropower company director:

"We expect earthquakes (. . .) and damage is inevitable. Some damage happens, it is designed that way. But to totally fail and all, that's not going to happen. *Hopefully*" (11 May 2015).

"There are so many components, which should be earthquake resilient features. If that is done, then there is no issue that it shouldn't hold the intensity of an earthquake with a magnitude of around 8. I *think* they are made to hold this 8" (7 May 2015).

"The geology is very fragile here and the conditions of the rock are not very good. But everything has engineering solutions and that's what we do provide. From the geological and engineering point of view, there are no problems. For every geological problem there is an engineering solution available." (8 May 2015).

"Dam break analysis has been done by the Institute of Roorkee based on which the dam has been designed. (. . .) In India, it has never happened. No dam has broken till now. Once the dam break analysis is done, it is ensured that the reinforcement and construction methodology used are fool-proof" (5 May 2015)

But uncertainty was not only discursively framed. Its active reinforcement, both by dismissing non-scientific sources of information and by obstructing the generation of scientific evidence is illustrated by the controversy about the hydrogeological impacts of hydropower tunnels. Local people in various project-affected villages in Sikkim have noted declining agricultural productivity, along with the sinking and degradation (cracking) of land, the depletion of soil moisture, and the drying and disappearance of springs—predominant source of domestic and agricultural water. Especially farmers who observed such changes daily were unanimous that these problems were caused by hydropower tunnels, which they argued "swallow" and divert soil moisture and spring water, evidenced by rivulets emerging from the tunnel access points. Likewise, they associated land degradation with tunnel blasting, (which "made the land shake just like during an earthquake,") and recalled starting to witness these changes roughly when project construction began.

While the locals believed to have an answer, science did not. A spring expert explained the difficulties in scientifically pinpointing the causes of hydrogeological change, with current scientific understanding of Himalayan hydrogeological systems being too coarse and qualitative, due to their complexity and location-specificity, and available research methods being too resource-intensive (2 May 2015). He nevertheless suggested that a multi-site study analyzing the state of the environment before and after project construction could explore scientifically valid linkages between tunneling and spring decline. However, when I enquired about such studies with the Department of Mines and Geology, the state agency responsible for (hydro)geological assessments, the geologists there were disheartened. While their team had carried out damage assessments in three project-affected areas, confirming many of the locals' assumptions, their findings couldn't be considered "legitimate" evidence, since the assessments were exclusively carried out *after* the changes had occurred. A systematic baseline inventory of landslide zones and spring discharge *prior* to tunneling had never been commissioned:

> "We tried to have baseline data earlier, but we could not do it. I tried to convince the government, this is the main requirement. Because a lot of comparisons will come up at a later stage. (...) I insisted to carry out such studies, either they should be financed by the power projects, or they [themselves] should carry out those studies."

> *"So, you couldn't carry out these studies even though you wanted to?"*

> "These power developers are such influential people, they can influence the government also."

> *"But isn't there any incentive for the government to carry out these kinds of studies?"*

> "They have not initiated. [Why] I don't know exactly. Because the government is interested in harnessing all the power" (26 April 2011)

Even four years later, despite further ex-post reports confirming extensive environmental degradation in project-affected areas, the government had still not made any efforts to establish the extent and causes of these damages. Only one project developer had apparently agreed to collect baseline data before starting tunnel excavation (6 May 2015).

In the absence of sufficient scientific-experimental data and for system dynamics difficult to explain with scientific methods, vernacular forms of knowledge, based on historic memory, long-term observations and an intimate understanding of the local environment have become important sources of information to better understand patterns of climate and environmental change in the Himalayas [95,96]. Even Sikkim's acclaimed rural development program "Dhara Vikas", which revives springs to ensure water security in drought-prone villages uses local perceptions as a main data input [97].

In most expert interviews about hydropower impacts, on the other hand, vernacular knowledge was categorically dismissed as insufficient or invalid, rumors and "misconceptions" resulting from ignorance, illiteracy, or political motivations, and pitted against science as the only valid source of information. A common argument was that Sikkim's geology is "naturally" fragile and multiple other factors may influence hydrogeological processes. An official in the Rural Management and Development Department suggested that locals blame the tunnels because they produce sudden changes and are highly visible, but how sure can the locals be? "Actually, we need some scientific evidence" (8 May 2015). An engineer with the Central Water Commission got rather worked up by my question and the account of the villagers' perceptions:

> "After making the tunnels, it has not been observed anywhere that the resources have dried up. (...) How can it be? It's basically a foolish question. Those who are asking those questions, they don't have a basic knowledge of engineering. (...) There is no valid reason,

no engineering reason that water will be depleted. (. . .) People are either making false statements or it is just against hydropower. Unnecessary rumors are being created (. . .) Because in our village areas (. . .) the literacy rate [is] less than 10 per cent, people have not studied even 4th and 5th standard. (. . .) If you tell them that the sun rises in the west, they will be convinced by that" (8 May 2015).

5.2. Contesting Dominant Risk Discourses

Quantitative, techno-scientific representations of risk often have a depoliticizing effect on environmental governance, compromising public political deliberation and democratic decision-making. As Anderson [98] (p. 41) argues, by employing an "end-eschewing mode of argument" and presenting fundamentally political and ideological choices as technically complex, scientific issues, such discourses disqualify the scientifically illiterate from debates about risk, making risk "the exclusive province of the expert". Expert representations of environmental and technological risk may also serve hidden agendas of profit and power:

"When a certain (warped) technical rationale is presented as the standard of rationality it is really an unacknowledged form of political domination, one that facilitates the selection of a particular kind of society as it obscures this end from public political discourse. Naturally, we find this mode of argumentation advanced by those who stand to lose the most from free democratic discussion—those who represent corporate and state power" (p. 43)

Equivalent arguments by political ecologists speak about the capture of environmental discourse and risk controversies in a "post-political consensus", concealing that what really is at stake in environmental politics is a democratic-ideological struggle between different socio-environmental futures [99,100].

Similarly, in India, the high degree of epistemic authority awarded to science vests experts with the exclusive power to determine which risks are real or unreal, allowing no room for lay knowledge claims to contribute even to problems associated with great scientific uncertainty. This erasure of risk by invisibilizing and excluding certain risks from the terms of the debate presents a major obstacle to public resistance against hazardous hydropower projects. The entire burden of proof is shifted onto affected households—often the most vulnerable social groups who have little to no means to consult expert opinions, mount campaigns or enter litigation and exert political pressure.

Yet, while uncertainty and knowledge politics lend themselves to manipulation, they also represent a fertile terrain for contestation. Research suggests that in the age of manufactured, diffuse or invisible risks [59–61], the reduced appreciation of science as an exclusive, authoritative source of knowledge gives alternative knowledge claims more influence over definitions and debates about risks—one reason for the growing number of risk conflicts [59,60,63–65]. As the struggle against the Lower Subansiri mega-hydro project on the Assam-Arunachal Pradesh border illustrates, influencing the production of knowledge about risk can be a most effective strategy for challenging expert representations of risk and the top-down imposition of hazardous technologies.

NHPC's 2000 MW Lower Subansiri HEP is among the largest of Arunachal's 160+ hydropower projects to be constructed on various tributaries of the Brahmaputra river. While most of these projects cause only limited displacement in sparsely populated Arunachal, the strongest resistance has come from civil society and political groups in downstream Assam, a densely populated state occupying the Brahmaputra floodplains. Conflict over Lower Subansiri started in the early 2000s. River-dependent communities in Assam were concerned about the dam impacting river flows and the floodplain ecology—livelihood base for a vast, rural population. They were also familiar with the dams' unintended, incalculable and uninsurable risks. The collective memory of the 1950 Assam earthquake, when the failure of a natural landslide dam on the Subansiri river caused catastrophic floods is still strong [101]. Since 2004, floods have been aggravated repeatedly by sudden excessive releases of water from hydropower projects in Arunachal Pradesh, Bhutan and Nagaland [27,102]. However, since Indian environmental policy includes no mechanism to account for downstream impacts of large dams, downstream-affected people were systematically excluded from negotiations over rehabilitation and benefit-sharing.

By 2006, an alliance of affected people and local student unions succeeded in pressuring the Government of Assam to commission a downstream impact study, conducted by an interdisciplinary eight-member expert committee from reputed Assamese universities. Their report, completed in 2010, gave scientific endorsement to the activists' concerns: operating the dam would cause dramatic daily flow fluctuations to be felt for hundreds of kilometers downstream. Another alarming discovery were flaws in the seismic-geological aspects of the project. Despite poor geological conditions and a highly earthquake-prone location, NHPC had made several modifications to the dam's original design parameters, which would reduce the costs of the project but increase its risks —including reducing the width and depth of the foundation, shifting the powerhouse underground, choosing concrete over rock-fill technology, and minimizing the project's flood cushion. The report triggered massive public support and the formation of a mass movement comprising affected communities, student and farmer organizations, and the state's major opposition party. In 2011, a road blockade temporarily shut down construction works. The project has remained suspended ever since.

The expert committee suggested to simply change the dam's design parameters. But NHPC, possibly concerned about additional costs and its reputation, mobilized its own group of experts from reputed national scientific institutions (some of whom had helped design and clear the project in the first place) to delegitimize the concerns raised against the dam and the validity of the claims by the Assamese experts [101]. This led to a highly politicized stand-off between the corporation and the movement and a more ambiguous group of national and regional politicians, bureaucrats and scientists between the two ends of the spectrum.

The government, responsible to take a final call on the fate of the project has since been under immense political pressure from both sides. The stakes are high: huge financial losses for NHPC; sacrifices in voter numbers; and powerful political interests at national level. Moreover, this being the first successful large-scale anti-dam protest of downstream communities in India, the decision will set an important precedent for hydropower conflicts in the country. Thus, instead of taking a clear stand on the issue, the state government has delegated expert committee after expert committee, each time engaging more senior and "qualified" experts, in hope of obtaining unambiguous evidence on the safety of the dam.

The effect has been the opposite: each round of expert scrutiny has not only revalidated the technical concerns already raised, but also revealed others. It has also made visible that science is not isolated from political pressures. The penultimate expert committee produced four Assamese votes against the dam and four votes by Central Government-appointed non-Assamese experts in favor [103]. Likewise, the latest expert-panel appointed by the MoEF stands accused of a possible conflict of interest, since its three members belong to institutions backing NHPC's position on the project in the past [104].

6. Discussion: Challenging the Production of Unequal Risk through Knowledge Co-Production

With this paper I have sought to explain why considerations of hazard potential and dam safety occupy a marginal position in the governance of Himalayan hydropower projects and their contestation, and how this contributes to the material production of uneven risks and vulnerabilities in hydropower-producing areas. My main analytical focus has been on two arenas of environmental governance: institutional mechanisms and knowledge politics.

While research on uneven geographies of risk has focused on the role of techno-scientific representations in shaping knowledge about risk [33–35], the findings from Sikkim point to a different type of knowledge politics: to legitimize complacency in the face of heightened hazard potential, the question of risk is discursively and institutionally erased and invisibilized. This "environmental denial" [24] (p. 15) by members of Sikkim's hydropower community (and beyond Sikkim [28,32,79]) is facilitated by various types of uncertainty, related for example to the invisibility of large parts of the project infrastructure, the difficulty to expose below-ground hydrogeological processes, or the impossibility to predict future hydro-climatic hazards. Expert discourses and political decision-makers selectively mobilize these uncertainties, but also actively reinforce them, by discrediting alternative sources of knowledge and by influencing what type of knowledge is produced.

These insights speak to a community of research on the deliberate production, perpetuation and institutionalization of ignorance and scientific ambiguity—also referred to as agnotology, antiepistemology, or the strategic unknowns, among others [32,54,105]. Ignorance studies explore the use of ignorance as a resource to deflect, obscure, conceal or magnify knowledge that is dangerous or unpalatable to powerful interest groups [32,55], and fill an important theoretical void: they account for absences in knowledge production—knowledge, which gets lost or never gets made [39,106,107]—and counter the common assumption that knowledge is necessarily more powerful than ignorance [32,55]. As McGoey [32] (p. 5) argues,

> "We need less attention to the politics of knowledge and more to the politics of ignorance, to the mobilization of ambiguity, the denial of unsettling facts, the realization that knowing the least amount possible is often the most indispensable tool for managing risks and exonerating oneself from blame in the aftermath of catastrophic events."

In Sikkim, institutionalized ignorance about certain hydropower risks and impacts has enabled the unchecked construction of hazardous hydropower projects. Further, by undermining the assessment, prevention and mitigation of hydropower risks, it has helped to maintain uncertainty, obscure liabilities, and thus prevent policy amendments and other changes in the official "rules of the game", which could jeopardize the viability of hydropower investments and investment decisions. For example, establishing liabilities in the EIA or adjusting environmental clearance requirements would imply lengthier, more elaborate, and more costly clearance procedures. Obliging developers to acquire all land and rehabilitate all households physically impacted during the entire project cycle would make project costs spiral. Such measures could also increase the likelihood of (organized) resistance by affected communities, implying further delays and expenses.

Admittedly, the strategicness of ignorance about hydropower risks in Sikkim is debatable. As McGoey [55] argues, the power and success of ignorance is underpinned by the difficulty "of proving whether someone is actually ignorant or simply feigning ignorance" (p. 559). In the Himalayas, geological and seismic uncertainty and the unpredictability of certain earth system dynamics complicate the question of strategic ignorance. Rather than a "deliberately wielded tool", could ignorance here also simply be "an unavoidable handicap" (ibid.)? I would argue that deliberation becomes evident when one looks beyond what experts, hydropower professionals and decision-makers claim not to know, and at how they deal with risk and uncertainty. Rather than improving the state of knowledge and applying double precautionary measures, Sikkim's hydropower community has used ignorance as an excuse not to act. Whether this "blissful disinterest" is strategy or merely convenience, the lack of preparation it imposes on the state and its population is hard to justify. As the editor of a prominent local newspaper argues about Sikkim's disaster management culture:

> "Unprepared, is inexcusable for a population that has lived with slides and shakes since forever and specially not at this juncture when there is (. . .) a full-fledged department for Disaster Risk Management and earmarked funds as well. How can any area in Sikkim not have a risk reduction plan to limit the damage caused by landslides and earthquakes? (. . .) Unless one understands the causes, how can safeguards against such disasters in the future be devised? Preparation requires accepting that a threat exists and then understanding the reasons causing (it)" [108].

Further, this paper has shown institutionalized ignorance to be a central mechanism in relational processes of marginalization and facilitation, allowing for the material production and shifting of risk [33–35,52]. Investment incentives have encouraged a hydropower development model, which diverts public resources for private profit and consumption elsewhere, while externalizing environmental risks and costs, thus compromising on factors of socio-environmental impact and structural safety. The exclusion of certain risks from impact assessment and rehabilitation processes has jeopardized households and entire communities, up to the point of unforeseen displacement. The failure to investigate and communicate the causes of incurred damages has obscured the question of liability. Depriving affected communities of systematic, adequate and timely rehabilitation, and undermining the ability of state institutions, communities and households to anticipate and adapt to risk situations has exacerbated vulnerabilities and deepened social marginalization.

Clearly, marginalization and facilitation go hand in hand here: while the risks produced through strategic ignorance threaten the lives, livelihoods and living environments of people in hydropower-producing areas, they represent economic opportunities for powerful corporate and political interest groups. Not surprisingly, the latter are actors usually far removed from these local realities, indicating how the production of geographically and racially uneven hazardscapes perpetuates historical patterns of dependency and exploitation between the Indian state and its margins, but also within frontier spaces like Sikkim [25].

Finally, this paper has drawn on two different case studies to understand what determines compliance and resistance to the production and discursive erasure of risk. The Sikkim case study illustrates that institutionalized ignorance, which instrumentalizes scientific uncertainty is a challenge to public resistance. As Butler and Rest observed in Nepal, laypeople often "cannot marshal an opposition to dam construction (. . .) beyond supposition [and] having little evidence for their position, (. . .) can only warn about the danger of dam breach as a possibility rather than a likelihood" [24] (p. 23). The Assamese resistance against the Lower Subansiri dam is instructive because it has managed to overcome this challenge. While the impetus for public concern came from embodied experiential knowledge, based on the lived memories of flood disasters, the movement has successfully mobilized the epistemic authority of scientific experts to expose the malleability of science and to stake alternative claims about the dam's risks. This public contestation of the risk question has been so powerful that it has caught political decision-makers at state and national level in a bind for nearly a decade.

But how useful is the Assamese experience for those affected by risky dam construction in Sikkim? No doubt, troubled state–society relations complicate the question of resistance in Sikkim, where political pressure and authoritarian government tendencies have long curtailed opportunities for affected communities and civil society groups to challenge development decisions [11]. Interviews in project-affected areas produced multiple accounts of repressive tactics used by project developers and state authorities in response to individual complaints and collective protests, creating widespread fear of dissent. Likewise, several respondents from NGOs explained that they must take a politically neutral stand and cannot work "against the government" for fear of political retaliation. As a result, hardly any environmental NGOs or research organizations in Sikkim are willing to openly engage with hydropower development.

Moreover, questions of ethnic identity and minority politics have historically fractured social mobilization in the state. Like other states of Northeast India, Sikkim's population comprises of numerous ethnic groups, who compete for access to public resources through reservations [109]. While Assam is no less tormented by ethnic politics, the activists have not framed their struggle in ethnic-cultural terms. Unlike in Sikkim, Assamese activists could draw on a powerful culture of class-based resistance and a long history of peasant and student mobilization to build strength in numbers [110,111]. Thus, resistance, which initially emerged in response to individual dam projects eventually turned into a state-wide movement against dam construction on the Brahmaputra River.

It is doubtful whether such a mass movement, articulated in terms of environmental justice could ever materialize in Sikkim. And in fact, this is not to downplay the remarkable successes Sikkimese anti-dam movements have achieved by using ethnic and cultural particularism as mobilizing concepts because they do carry political clout. These achievements deserve greater recognition, as I have argued elsewhere [112]. Nevertheless, to resist "unavoidable externalities" like heightened ecological precarity, disaster risk and social marginalization by projects yet to be built/completed, the Subansiri case does provide a useful lesson: the power to define risks and their acceptability is not entirely monopolized by experts. Hegemonic expert discourses can be punctured and subverted, particularly when they bank on contested knowledges [62]. The activists in Assam have achieved this by entering the field of knowledge production and bridging the lay-expert divide [101].

More generally, the successful challenge this citizen–science alliance has mounted against institutionalized ignorance testifies to the need for a democratization of knowledge production: reconsidering the exclusive authority awarded to experts, and recognizing the value and particularity of the embodied, experiential knowledge, which comes from having to live with ecological precarity [25]. As experiences of human-made, capital-driven disasters—many dam accidents included—indicate, scientific controversies often come to light only in the aftermath of catastrophe, while public resistances based on experiential knowledge of the local climate, geology, hydrology, and disaster historiography can issue effective early warnings [6]. Such contestations therefore deserve more credence by environmental decision-makers, especially for complex governance problems with high political stakes and high uncertainty.

There are already multiple proposals for the co-production of knowledge out there, including post-normal science [113], citizen science [114], street science [115], or activism mobilizing science [116], among others. These approaches center on inclusive strategies, seeing citizens not "as passive receivers [but] as active partners in the production of knowledge" [59] (p. 71), and have proven particularly useful for environmental justice conflicts with uncertain facts, disputed values, high stakes and urgently required decisions [113]. By establishing new linkages between experts/science and citizens/activism, and by promoting alternative perspectives, knowledge co-production frameworks counter the technocratization and scientization of knowledge production [59], and the instrumentalization of scientific complexity and uncertainty for political purposes.

Despite the constrained political space for hydropower activism, Sikkim has potential for creating such citizen–science linkages. First, as mentioned earlier, local knowledge is already consulted and put to use in different development and environmental research initiatives [29,95–97], demonstrating a general appreciation of vernacular perspectives. Second, as this paper has demonstrated, not all experts are complicit in invisibilizing risk and in ridiculing vernacular perspectives. Many government and NGO-experts work closely with local communities, and given the limited influence regional state agencies have on hydropower governance decisions, some experts have openly professed their antagonistic relationship with hydropower developers and their disdain for corporate negligence in environmental management, siding with locals instead [25]. Finally, with Indian environmental activism shifting increasingly to the legal level, civil society groups may be able to push for improvements in the implementation of participatory tools, such as the EIA process or the veto power of the gram sabhas (village assemblies), which are explicitly designed to broaden the knowledge base for governance decisions by including local perspectives.

Nevertheless, to create powerful citizen–science linkages and mobilize Sikkim's extensive environmental research community, concerned scientists and other experts in the state must leave the comfort of alleged objectivity and challenge the power structures, which influence what environmental knowledge may be produced or not. Sikkim's contested hydropower future and its environmental ramifications are after all not technical, but fundamentally political questions, and engaging with or abstaining from these is a fundamentally political choice.

In closing, I want to point to three specific policy implications of strategic and institutionalized ignorance in the hydropower sector, particularly in a context of climate change. First, there clearly is an urgent need to reevaluate the viability and hazard vulnerability of Himalayan hydropower infrastructure. The latter has a significant bearing on the economic potential of new hydropower ventures, too, and for decision-making on future development pathways of Himalayan constituencies [24,74].

Second, as this paper has illustrated, hydropower infrastructure and its neglectful governance can exacerbate climate vulnerability and complicate adequate climate action. Framing hydropower as a climate mitigating energy technology without due attention to associated risks is misleading and leads to maladaptation [117].

Finally, as Lord [57] suggests, with more powerful natural hazards to be expected in mountain regions around the world, and given the significant uncertainties over the looming impacts of climate change, the time seems ripe to take seriously "technologies of humility", such as micro-hydropower and other decentralized renewable solutions, and to dismiss the infrastructural hubris which has long driven risky hydropower investments.

Funding: This research was funded by the European Union's Seventh Framework Programme for research, technological development and demonstration (Marie Curie Actions) grant number 289374 ("ENTITLE").

Acknowledgments: The author would like to thank Begüm Özkaynak, Giorgos Kallis, Mabel Gergan, Austin Lord, Santiago Gorostiza, Manish Kumar, the editors of this special issue, and two anonymous reviewers for their constructive comments on earlier versions of this article. My gratitude also goes to Samuel Thomas, Partha J. Das and K. J. Joy for their invaluable guidance and support while undertaking fieldwork, as well as to Anuradha Bhutia, Sweekriti Pradhan, Prerana Nair and Shubhangi Shukla for research assistance.

Conflicts of Interest: The author declares no conflict of interest. The funders had no role in the design of the study; in the collection, analyses, or interpretation of data; in the writing of the manuscript, and in the decision to publish the results.

References

1. Ives, M. A Day Before Laos Dam Failed, Builders Saw Trouble. 2018. Available online: https://www.nytimes.com/2018/07/26/world/asia/laos-dam-collapse.html (accessed on 31 July 2018).
2. Thakkar, H. Role of dams in Kerala's flood disaster. *Econ. Polit. Wkly.* **2018**, *LIII*, 20–23.
3. McCully, P. *Silenced Rivers: The Ecology and Politics of Large Dams*; Zed Books: London, UK, 2001; ISBN 978-1-85649-901-9.
4. Pearce, F. *The Dammed: Rivers, Dams, and the Coming World Water Crisis*; Bodley Head: London, UK, 1992; ISBN 978-0-370-31609-3.
5. Armiero, M. *A Rugged Nation: Mountains and the Making of Modern Italy*; The Whitehorse Press: Cambridge, UK, 2011.
6. Huber, A.; Gorostiza, S.; Kotsila, P.; Beltrán, M.J.; Armiero, M. Beyond "socially constructed" disasters: Re-politicizing the debate on large dams through a political ecology of risk. *Capital. Nat. Social.* **2017**, *28*, 48–68. [CrossRef]
7. Ahlers, R.; Budds, J.; Joshi, D.; Merme, V.; Zwarteveen, M. Framing hydropower as green energy: Assessing drivers, risks and tensions in the Eastern Himalayas. *Earth Syst. Dyn.* **2015**, *6*, 195–204. [CrossRef]
8. Ahlers, R.; Zwarteveen, M.; Bakker, K.; Flyvbjerg, B. Large Dam Development: From Trojan Horse to Pandora's Box. In *The Oxford Handbook of Megaproject Management*; Flyvbjerg, B., Ed.; OUP: Oxford, UK, 2017.
9. Lord, A. Citizens of a hydropower nation: Territory and agency at the frontiers of hydropower development in Nepal: Nepalese hydropower development. *Econ. Anthropol.* **2016**, *3*, 145–160. [CrossRef]
10. Schneider, H. World Bank Turns to Hydropower to Square Development with Climate Change. 2013. Available online: https://www.washingtonpost.com/business/economy/world-bank-turns-to-hydropower-to-square-development-with-climate-change/2013/05/08/b9d60332-b1bd-11e2-9a98-4be1688d7d84_story.html (accessed on 31 July 2018).
11. Huber, A.; Joshi, D. Hydropower, anti-politics, and the opening of new political spaces in the eastern Himalayas. *World Dev.* **2015**, *76*, 13–25. [CrossRef]
12. Zarfl, C.; Lumsdon, A.E.; Berlekamp, J.; Tydecks, L.; Tockner, K. A global boom in hydropower dam construction. *Aquat. Sci.* **2015**, *77*, 161–170. [CrossRef]
13. Mukerjee, M. The Impending Dam Disaster in the Himalayas. 2015. Available online: https://www.scientificamerican.com/article/the-impending-dam-disaster-in-the-himalayas/ (accessed on 1 August 2018).
14. Merme, V.; Ahlers, R.; Gupta, J. Private equity, public affair: Hydropower financing in the Mekong Basin. *Glob. Environ. Chang.* **2014**, *24*, 20–29. [CrossRef]

15. Parry, M.L.; Canziani, O.F.; Palutikof, J.P.; van der Linden, P.J.; Hanson, C.E. *IPCC 2007. Climate Change 2007: Impacts, Adaptation and Vulnerability*; Cambridge University Press: Cambridge, UK; New York, NY, USA, 2007.
16. Tse-ring, K.; Sharma, E.; Chettri, N.; Shrestha, A. *Climate Change Vulnerability of Mountain Ecosystems in the Eastern Himalayas*; ICIMOD Books; ICIMOD: Kathmandu, Nepal, 2010; ISBN 978-92-9115-142-4.
17. Kelly-Richards, S.; Silber-Coats, N.; Crootof, A.; Tecklin, D.; Bauer, C. Governing the transition to renewable energy: A review of impacts and policy issues in the small hydropower boom. *Energy Policy* **2017**, *101*, 251–264. [CrossRef]
18. Grumbine, R.E.; Pandit, M.K. Threats from India's Himalaya dams. *Science* **2013**, *339*, 36–37. [CrossRef] [PubMed]
19. Winemiller, K.O.; McIntyre, P.B.; Castello, L.; Fluet-Chouinard, E.; Giarrizzo, T.; Nam, S.; Baird, I.G.; Darwall, W.; Lujan, N.K.; Harrison, I.; et al. Balancing hydropower and biodiversity in the Amazon, Congo, and Mekong. *Science* **2016**, *351*, 128–129. [CrossRef] [PubMed]
20. Chettri, M. Ethnic environmentalism in the eastern Himalaya. *Econ. Polit. Wkly.* **2017**, *LII*, 34–40.
21. Gergan, M.D. Living with earthquakes and angry deities at the Himalayan borderlands. *Ann. Am. Assoc. Geogr.* **2017**, *107*, 490–498. [CrossRef]
22. Kohli, K. Inducing vulnerabilities in a fragile landscape. *Econ. Polit. Wkly.* **2011**, 19–22.
23. Rampini, C. Impacts of Hydropower Development along the Brahmaputra River in Northeast India on the Resilience of Downstream Communities to Climate Change Impacts. Ph.D. Thesis, University of California, Santa Cruz, CA, USA, 2016.
24. Butler, C.; Rest, M. Calculating risk, denying uncertainty: Seismicity and hydropower development in Nepal. *HIMALAYA J. Assoc. Nepal Himal. Stud.* **2017**, *37*, 6.
25. Gergan, M.D. Precarity and Possibility at the Margins: Hazards, Infrastructure, and Indigenous Politics in Sikkim, India. Ph.D. Thesis, The University of North Carolina at Chapel Hill, Chapel Hill, NC, USA, 2016.
26. Kumar, D.; Katoch, S.S. Dams turning devils: An insight into the public safety aspects in operational run of the river hydropower projects in western Himalayas. *Renew. Sustain. Energy Rev.* **2017**, *67*, 173–183. [CrossRef]
27. Vagholikar, N.; Das, P.J. *Damming Northeast India*; Kalpavriksh, Aaranyak, and Action Aid India: Pune, India, 2010.
28. Lord, A. Speculation and seismicity: Reconfiguring the hydropower future in post-earthquake Nepal. In *Water, Technology, and the Nation-State*; Menga, F., Swyngedouw, E., Eds.; Routledge: Oxon, UK, 2018.
29. Government of Sikkim. *Sikkim State Action Plan on Climate Change (2012–2030)*; Government of Sikkim: Gangtok, India, 2014.
30. State Disaster Management Authority. *Sikkim State Disaster Management Plan*; State Disaster Management Authority, Government of Sikkim: Gangtok, India, 2015.
31. Government of Sikkim Forests, Environment & Wildlife Management Department | Environment. Available online: http://www.sikkimforest.gov.in/environment.htm#eia (accessed on 1 August 2018).
32. McGoey, L. Strategic unknowns: Towards a sociology of ignorance. *Econ. Soc.* **2012**, *41*, 1–16. [CrossRef]
33. Collins, T.W. The production of unequal risk in hazardscapes: An explanatory frame applied to disaster at the US–Mexico border. *Geoforum* **2009**, *40*, 589–601. [CrossRef]
34. Fraser, A. The missing politics of urban vulnerability: The state and the co-production of climate risk. *Environ. Plan. A* **2017**, *49*, 2835–2852. [CrossRef]
35. Mustafa, D. The production of an urban hazardscape in Pakistan: Modernity, vulnerability, and the range of choice. *Ann. Assoc. Am. Geogr.* **2005**, *95*, 566–586. [CrossRef]
36. Pelling, M. The political ecology of flood hazard in urban Guyana. *Geoforum* **1999**, *30*, 249–261. [CrossRef]
37. Davis, M. *Ecology of Fear: Los Angeles and the Imagination of Disaster*; Metropolitan Books: New York, NY, USA, 1998.
38. Boelens, R.; Shah, E.; Bruins, B. Contested Knowledges: Large Dams and Mega-Hydraulic Development. *Water* **2019**, *11*, 416. [CrossRef]
39. Frickel, S. On missing New Orleans: Lost knowledge and knowledge gaps in an urban hazardscape. *Environ. Hist.* **2008**, *13*, 643–650.
40. Michaels, D.; Monforton, C. Manufacturing uncertainty: Contested science and the protection of the public's health and environment. *Am. J. Public Health* **2005**, *95*, 39–48. [CrossRef] [PubMed]

41. Hewitt, K. (Ed.) *Interpretations of Calamity from the Viewpoint of Human Ecology*; Allen & Unwin: Boston, MA, USA, 1983.
42. O'Keefe, P.; Westgate, K.; Wisner, B. Taking the naturalness out of disasters. *Nature* **1976**, *260*, 566–567. [CrossRef]
43. Waddell, E. The hazards of scientism: A review article. *Hum. Ecol.* **1977**, *5*, 69–76. [CrossRef]
44. Wisner, B.; Blaikie, P.; Cannon, T.; Davis, I. *At Risk: Natural Hazards, Peoples Vulnerability*; Routledge: London, UK, 2004.
45. Beck, U. *World at Risk*, 2nd ed.; Polity Press: Cambridge, UK, 2009.
46. Baldwin, A.; Stanley, A. Risky natures, natures of risk. *Geoforum* **2013**, *45*, 2–4. [CrossRef]
47. Wisner, B. Business-as-usual disaster relief. *Capital. Nat. Social.* **2012**, *23*, 123–128. [CrossRef]
48. Freudenburg, W.R.; Gramling, R.; Laksa, S.; Erikson, K.T. Organizing hazards, engineering disasters? Improving the recognition of political-economic factors in the creation of disasters. *Soc. Forces* **2008**, *87*, 1015–1038. [CrossRef]
49. Clarke, L. Postscript: Considering katrina. In *The Sociology of Katrina: Perspectives on a Modern Catastrophe*; Brunsma, D.L., Overfelt, D., Picou, J.S., Eds.; Rowman & Littlefield: Lanham, MD, USA, 2007; pp. 235–242, ISBN 978-0-7425-5930-1.
50. Davis, H.; Todd, Z. On the importance of a date, or decolonizing the anthropocene. *ACME Int. J. Crit. Geogr.* **2017**, *16*, 761–780.
51. Tognetti, S.S. Revisiting Post-Normal Science in Post-Normal Times & Identifying Cranks. 2013. Available online: http://www.postnormaltimes.net/wpblog/revisiting-post-normal-science-in-post-normal-times-identifying-cranks/ (accessed on 1 August 2018).
52. Stanley, A. Natures of risk: Capital, rule, and production of difference. *Geoforum* **2013**, *45*, 5–16. [CrossRef]
53. Woodhouse, E.; Hess, D.; Breyman, S.; Martin, B. Science studies and activism: Possibilities and problems for reconstructivist agendas. *Soc. Stud. Sci.* **2002**, *32*, 297–319. [CrossRef]
54. Proctor, R.N.; Schiebinger, L. (Eds.) *Agnotology: The Making and Unmaking of Ignorance*; Stanford University Press: Stanford, CA, USA, 2008. [CrossRef]
55. McGoey, L. The logic of strategic ignorance. *Br. J. Sociol.* **2012**, *63*, 533–576. [CrossRef] [PubMed]
56. Oreskes, N.; Conway, E.M. *Merchants of Doubt*; Bloomsbury: London, UK, 2010.
57. Lord, A. Humility and hubris in hydropower. *Limn* **2017**, *9*. Available online: https://limn.it/articles/humility-and-hubris-in-hydropower/ (accessed on 1 August 2018).
58. Gramsci, A. *Selections from the Prison Notebooks*; International Publishers: New York, NY, USA, 1971.
59. Lidskog, R. Scientised citizens and democratised science. Re-assessing the expert-lay divide. *J. Risk Res.* **2008**, *11*, 69–86. [CrossRef]
60. Beck, U. *Risk Society: Towards a New Modernity*; Theory, culture & society; Sage Publications: London, UK, 1992; ISBN 978-0-8039-8345-8.
61. Finger, Y.; Jebri, L.; Kühne, F.; Scheffel, L.; Schnippe, M. Politicisation of science in the process of dealing with manufactured risk. *MaRBLe* **2016**, *4*. [CrossRef]
62. Cooper, T.; Bulmer, S. Refuse and the "risk society": The political ecology of risk in inter-war Britain. *Soc. Hist. Med.* **2013**, *26*, 246–266. [CrossRef] [PubMed]
63. Maeseele, P. On neo-luddites led by ayatollahs: The frame matrix of the GM food debate in northern Belgium. *Environ. Commun.* **2010**, *4*, 277–300. [CrossRef]
64. Whatmore, S.J. Mapping knowledge controversies: Science, democracy and the redistribution of expertise. *Prog. Hum. Geogr.* **2009**, *33*, 587–598. [CrossRef]
65. Saitta, P.; Lazzerini, I. Environment and the citizens: Popular struggles, popular epidemiology, and other forms of resistance "from below" in areas at risk worldwide—An introduction. *Capital. Nat. Social.* **2015**, *26*, 35–38. [CrossRef]
66. Hewitt, K.; Mehta, M. Rethinking risk and disasters in mountain areas. *J. Alp. Res. Rev. Géogr. Alp.* **2012**. [CrossRef]
67. Pandit, M.K.; Manish, K.; Koh, L.P. Dancing on the roof of the world: Ecological transformation of the Himalayan landscape. *BioScience* **2014**, *64*, 980–992. [CrossRef]
68. Lutz, A.F.; Immerzeel, W.W.; Shrestha, A.B.; Bierkens, M.F.P. Consistent increase in high Asia's runoff due to increasing glacier melt and precipitation. *Nat. Clim. Chang.* **2014**, *4*, 587. [CrossRef]

69. Nepal, S.; Shrestha, A.B. Impact of climate change on the hydrological regime of the Indus, Ganges and Brahmaputra river basins: A review of the literature. *Int. J. Water Resour. Dev.* **2015**, *31*, 201–218. [CrossRef]
70. Shrestha, U.B.; Gautam, S.; Bawa, K.S. Widespread climate change in the Himalayas and associated changes in local ecosystems. *PLoS ONE* **2012**, *7*, e36741. [CrossRef] [PubMed]
71. Shrestha, B.; Mool, P.K.; Bajracharya, S.R. *Impact of Climate Change on Himalayan Glaciers and Glacial Lakes: Case Studies on GLOF and Associated Hazards in Nepal and Bhutan*; International Centre for Integrated Mountain Development (ICIMOD); United Nations Environment Programme (UNEP): Kathmandu, Nepal, 2007; ISBN 978-92-9115-032-8.
72. Chopra, R.; Das, B.P.; Dhyani, H.; Verma, A.; Venkatesh, H.S.; Vasistha, H.B.; Dobhal, D.P.; Juyal, N.; Sathyakumar, S.; Pathak, S.; et al. *Assessment of Environmental Degradation and Impact of Hydroelectric Projects during the June 2013 Disaster in Uttarakhand*; The Ministry of Environment and Forests, Government of India: New Delhi, India, 2014.
73. Poudel, R.R. Sunkoshi Dam Weakened by August Flood, Landslide. Available online: http://kathmandupost.ekantipur.com/news/2014-09-18/sunkoshi-dam-weakened-by-august-flood-landslide.html (accessed on 2 August 2018).
74. Schwanghart, W.; Ryan, M.; Korup, O. Topographic and seismic constraints on the vulnerability of Himalayan hydropower. *Geophys. Res. Lett.* **2018**, *45*, 8985–8992. [CrossRef]
75. Schwanghart, W.; Worni, R.; Huggel, C.; Stoffel, M.; Korup, O. Uncertainty in the Himalayan energy–water nexus: Estimating regional exposure to glacial lake outburst floods. *Environ. Res. Lett.* **2016**, *11*, 074005. [CrossRef]
76. Nixon, R. *Slow Violence and the Environmentalism of the Poor*; Harvard University Press: Cambridge, CA, USA, 2011; ISBN 978-0-674-04930-7.
77. Drew, G. Mountain women, dams, and the gendered dimensions of environmental protest in the Garhwal Himalaya. *Mt. Res. Dev.* **2014**, *34*, 235–242. [CrossRef]
78. Erlewein, A.; Nüsser, M. Offsetting greenhouse gas emissions in the Himalaya? Clean development dams in Himachal Pradesh, India. *Mt. Res. Dev.* **2011**, *31*, 293–304. [CrossRef]
79. Buechler, S.; Sen, D.; Khandekar, N.; Scott, C. Re-linking governance of energy with livelihoods and irrigation in Uttarakhand, India. *Water* **2016**, *8*, 437. [CrossRef]
80. Manish, S. A Paradise Dammed. 2011. Available online: http://archive.tehelka.com/story_main50.asp?filename=Ne081011PARADISE.asp (accessed on 2 August 2018).
81. Lepcha, T. Chungthang—The Kalapani of the 21st Century. Available online: http://www.actsikkim.com/docs/Press_Release_Chungthang_EQ.pdf (accessed on 2 August 2018).
82. Mazoomdaar, J. In fact: And the Rivers be Dammed. Indian Express 2016. Available online: https://indianexpress.com/article/explained/kedarnath-uttarakhand-floods-garhwal-earthquake-2867742/ (accessed on 1 August 2018).
83. Dharmadikary, S. *Mountains of Concrete: Dam Building in the Himalayas*; International Rivers: Berkeley, CA, USA, 2008.
84. WCD. *Dams and Development: A New Framework for Decision-Making: The Report of the World Commission on Dams*; Earthscan: London, UK, 2000; ISBN 978-1-85383-798-2.
85. Moore, D.; Dore, J.; Gyawali, D. The world commission on dams + 10: Revisiting the large dam Controversy. *Water Altern.* **2010**, *3*, 11.
86. Wangchuk, P.D. Sikkim's Hydel Story: The Journey from 50 KW in 1927, 30 MW in 1994, to 2200 MWs in 2018. Available online: https://www.summittimes.com/single-post/2018/04/04/Sikkim%E2%80%99s-Hydel-Story-The-journey-from-50KW-in-1927-30MW-in-1994-to-2200-MWs-in-2018 (accessed on 2 August 2018).
87. Bhattacharjee, U. *Dam Planning Under the Spotlight: A Guide to Dam Sanctioning in India*; International Rivers: Berkeley, CA, USA, 2013.
88. Energy and Power Department. *Energy and Power Sector Vision 2015*; Government of Sikkim: Gangtok, India, 2010.
89. Joshi, D. Like water for justice. *Geoforum* **2015**, *61*, 111–121. [CrossRef]
90. Hill, D.P. Where Hawks Dwell on water and bankers build power poles: Transboundary waters, environmental security and the frontiers of neo-liberalism. *Strateg. Anal.* **2015**, *39*, 729–743. [CrossRef]

91. Rahman, M.Z. *Territory, Tribes, Turbines: Local Community Perceptions and Responses to Infrastructure Development along the Sino-Indian Border in Arunachal Pradesh*; Institute of Chinese Studies: New Delhi, India, 2014; Volume 7.

92. Alley, K.D.; Hile, R.; Mitra, C. Visualizing hydropower across the Himalayas: Mapping in a time of regulatory decline. *HIMALAYA J. Assoc. Nepal Himal. Stud.* **2014**, *34*, 17.

93. Vagholikar, N. Risks without enough gain. *Telegr. India* **2007**. Available online: http://weepingsikkim.blogspot.it/2007/12/risks-without-enough-gain.html (accessed on 27 August 2013).

94. Sharma, A.; Sherpa, N.; Lepcha, G.T.; Luitel, K.K. *Report on Damages Caused Due to Tunnel Excavation and Other Activities under Teesta-Hydro-Electric Project Stage V*; Department of Mines, Minerals & Geology, Government of Sikkim: Gangtok, India, 2010.

95. Chaudhary, P.; Bawa, K.S. Local perceptions of climate change validated by scientific evidence in the Himalayas. *Biol. Lett.* **2011**, *7*, 767–770. [CrossRef] [PubMed]

96. Tambe, S.; Arrawatia, M.L.; Bhutia, N.T.; Swaroop, B. Rapid, cost-effective and high resolution assessment of climate-related vulnerability of rural communities of Sikkim Himalaya, India. *Curr. Sci.* **2011**, *101*, 9.

97. Tambe, S.; Kharel, G.; Arrawatia, M.L.; Kulkarni, H.; Mahamuni, K.; Ganeriwala, A.K. Reviving dying springs: Climate change adaptation experiments from the Sikkim Himalaya. *Mt. Res. Dev.* **2012**, *32*, 62–72. [CrossRef]

98. Anderson, P.N. The GE Debate: What is at risk when risk is defined for us? *Capital. Nat. Social.* **2001**, *12*, 39–44. [CrossRef]

99. Swyngedouw, E. Impossible "sustainability" and the postpolitical condition. In *The Sustainable Development Paradox: Urban Political Economy in the United States and Europe*; Krueger, R., Gibbs, D., Eds.; Guilford Press: New York, NY, USA, 2007; pp. 13–40. ISBN 978-1-59385-498-0.

100. Maeseele, P. The risk conflicts perspective: Mediating environmental change we can believe in. *Bull. Sci. Technol. Soc.* **2015**, *35*, 44–53. [CrossRef]

101. Baruah, S. Whose river is it anyway? Political economy of hydropower in the Eastern Himalayas. *Econ. Polit. Wkly.* **2012**, *XLVII*, 41–52.

102. Chakravartty, A. Thousands Marooned in a "Dam-Induced" Flood in Golaghat. 2018. Available online: https://www.downtoearth.org.in/news/thousands-marooned-in-a-dam-induced-flood-in-golaghat-61326 (accessed on 2 August 2018).

103. Sengupta, A. Assam experts wary of dam renewal—Call solve seismic, geological issues. *Telegraph*, 2016. Available online: https://www.telegraphindia.com/1160312/jsp/frontpage/story_74100.jsp (accessed on 2 August 2018).

104. Mukul, J.; Jai, S. NHPC's "Biggest" Subansiri Hydropower Project Stalled on Panel Names. 2018. Available online: https://www.business-standard.com/article/economy-policy/nhpc-s-subansiri-project-stalled-over-composition-of-expert-committee-118080201217_1.html (accessed on 2 August 2018).

105. Galison, P. Removing knowledge. *Crit. Inq.* **2004**, *31*, 229–243. [CrossRef]

106. Frickel, S.; Vincent, M.B. Hurricane katrina, contamination, and the unintended organization of ignorance. *Technol. Soc.* **2007**, *29*, 181–188. [CrossRef]

107. Frickel, S. Absences: Methodological note about nothing, in particular. *Soc. Epistemol.* **2014**, *28*, 86–95. [CrossRef]

108. Summit TIMES. Editorial: Sensitive to Disasters. 2018. Available online: https://www.summittimes.com/single-post/2018/10/04/Editorial-Sensitive-to-Disasters (accessed on 10 October 2018).

109. Chettri, M. *Ethnicity and Democracy in the Eastern Himalayan Borderland: Constructing Democracy*; Amsterdam University Press: Amsterdam, The Netherlands, 2017; ISBN 978-90-485-2750-2.

110. Saikia, A. *A Century of Protests: Peasant Politics in Assam Since 1900*; Routledge: New Delhi, India, 2014; ISBN 978-1-317-32559-8.

111. Baruah, M. Suffering for Land: Environmental Hazards and Popular Struggles in the Brahmaputra Valley (Assam), India. Ph.D. Thesis, Syracuse University, Syracuse, NY, USA, 2016.

112. Huber, A.; Joshi, D. Hydropower conflicts in Sikkim: Recognizing the power of citizen initiatives for socio-environmental justice. In *Water Conflicts in Northeast India*; Joy, K.J., Das, P.J., Chakraborty, G., Mahanta, C., Paranjape, S., Vispute, S., Eds.; Routledge: Oxon, UK, 2018; pp. 71–91, ISBN 978-1-351-68594-8.

113. Funtowicz, S.O.; Ravetz, J.R. Uncertainty, complexity and post-normal science. *Environ. Toxicol. Chem.* **1994**, *13*, 1881–1885. [CrossRef]

114. Irwin, A. *Citizen Science: A Study of People, Expertise and Sustainable Development*; Routledge: Oxon, UK, 1995.
115. Corburn, J. *Street Science: Community Knowledge and Environmental Health Justice*; MIT Press: Cambridge, MA, USA, 2005; ISBN 978-0-262-53272-3.
116. Conde, M. Activism mobilising science. *Ecol. Econ.* **2014**, *105*, 67–77. [CrossRef]
117. Barnett, J.; O'Neill, S. Maladaptation. *Glob. Environ. Chang.* **2010**, *20*, 211–213. [CrossRef]

water

MDPI

Article

Negotiating Water and Technology—Competing Expectations and Confronting Knowledges in the Case of the Coca Codo Sinclair in Ecuador

Tuula Teräväinen

Department of Geographical and Historical Studies, Environmental Policy, University of Eastern Finland, P.O. Box 111, 80101 Joensuu, Finland; tuula.teravainen@uef.fi

Received: 8 July 2018; Accepted: 19 November 2018; Published: 26 February 2019

check for updates

Abstract: Recent and on-going mega-hydraulic development in the global South implies profound socio-technical, ecological, territorial and cultural transformations at different levels and spaces of society. The transformations often involve conflicts and also new governance arrangements between different knowledge regimes, local practices and national and global frameworks of climate mitigation, water resources management and the green economy. Significantly, they also entail varying expectations concerning the meaning of water and the political promises of technology in advancing more sustainable futures. Drawing on sociological science and technology studies, particularly the sociology of expectations, this article analyses competing, parallel and confronting expectations regarding water and technology that different actors produce, negotiate and contest in the context of the recently launched 1500 MW hydropower megaproject Coca Codo Sinclair in Ecuador. It takes expectations as performative as they may shape and challenge policies, discourses, social interactions, institutions and power relations. By analysing and comparing these expectations, the article scrutinises the socio-technical imaginaries and related knowledge regimes they represent, derive from and support, and what kinds of repercussions these have in terms of water resources management in particular and sustainability governance in general.

Keywords: expectations; hydroelectric megaprojects; socio-technical imaginaries; Ecuador; energy policy

1. Background

Recently, academic scholars, national governments and international organisations such as the OECD and the UN have increasingly called for system-level sustainable transitions driven by new policy concepts like green growth and the green economy [1–6]. These concepts have highlighted the urgency of climate change while seeking to simultaneously ensure sustained economic growth and responsible environmental governance. National responses to these challenges have entailed new policy visions accompanied by increased investment in clean technologies and renewable energy sources in order to enable and accelerate substantial changes in national energy matrices [3,4,6–8]. This article addresses such efforts in the context of the Andean country Ecuador, which exemplifies recent developmental challenges and the ways in which a new hydroelectricity-driven energy policy approach has been negotiated, contested and legitimised. In this context, the role of expectations is crucial: the ways in which the new hydraulic megaprojects such as the 1500 MW Coca Codo Sinclair (CCS) are motivated and imagined have important repercussions for how energy futures, values, industrial transformations, social organization and governance are understood, rationalised and managed [9–13].

Hydropower can be understood not only as a form of electricity production but also as a socio-technical, cultural-symbolic, discursive and political phenomenon. This article focuses on expectations and confronting knowledge regimes (see the introduction in this volume) concerning water and technology, understood as hydropower infrastructure, in the CCS as it exemplifies current challenges in sustainability governance and green transitions. Drawing on sociological science and technology studies, in particular the sociology of expectations [10,12–14] and transitions studies literature [1,2,9,12], the article takes an actor-oriented approach and examines recent expectations and future visions regarding the CCS among various actors at different levels and spheres of Ecuadorian society. We show how the project is understood by diverse actors, ranging from instrumental and functionalist-pragmatic interpretations to addressing an intrinsic value of water and technology or generating symbolic understandings. Thus, the meanings of water and technology vary from resource-based views to energy supply and security-oriented interpretations, and further to drawing on esthetical and cultural values or seeing them as mediators of techno-economic advancement and systemic transitions. These meanings have important repercussions not only in terms of understanding the CCS and the Ecuadorian context but also, more broadly, in planning and governing pathways to more sustainable energy futures. Moreover, they show how hydro-social realities (see the introduction in this volume) are formed and negotiated by confronting knowledge regimes based on different grounds for claiming the truth.

The sociology of expectations has been applied, for example, to biomedical research [10,15] but less often in relation to natural resources governance or hydropower (see, however, [9,12]). This article emphasises the performativity of expectations [10,13], i.e., how expectations shape politics, the future, technology, stakeholder interactions and concrete policies. Examining socio-political and cultural-symbolic dimensions of hydropower infrastructures opens up insights into understanding how perceptions concerning water and technology shape levels and forms of knowledge and governance. This highlights how expectations shape and are shaped by both the everyday lives in riverside communities and national energy policy visions. The article also generates knowledge about the challenges of new hydropower in the global South by examining the degree to which different understandings and knowledge regimes resemble or depart from each other. It does not take the benefits or problems often associated with hydraulic mega-projects as a given, but instead considers them as social constructs and as subjects of continuous political (re)negotiation (see also [16] in this volume).

The analysis is based on interview data and written material collected in Ecuador in 2016–2017. The interview data consists of 39 open-ended, semi-structured interviews with representatives from relevant ministries (two), regulatory authorities (five), regional administration (two) and Coca Codo Sinclair S.A. (three) as well as non-governmental organizations (NGOs) (four), project workers (three) and local inhabitants (twenty) in the riverside villages of the Rio Coca. All interviews were conducted with and transcribed by a local research assistant (we would like to thank Ricardo Andrade and Paolo Aranda for their valuable contribution in the data collection and the transcriptions of the interview data). The interview data was collected using snowball sampling for the expert and authority interviews to allow finding key actors relating to the CCS. In the riverside villages, snowball sampling was complemented by purposeful sampling [17] to illustrate what kinds of perspectives are shared by not only the politically active or knowledgeable actors but also among local residents in the CCS's impact area. The purposeful sampling strategy was based on identifying and interviewing informants who lived in the impact area and collecting data until the point of theoretical saturation, i.e., the point at which the collected data did not provide additional major insights. The data also includes participatory observation in the project site and surrounding areas (2016–2017). This part of the data was used as background and contextual information in the analyses. The interviews and notes from the participatory observation were complemented by written material, especially regarding the official views and the NGO perspectives. This material consists of official documents, strategies, evaluation reports, project documents and other written communications produced by the Ecuadorian government, key ministries, national and regional regulatory authorities and provincial actors as well as NGOs and the Coca Codo S.A. (2005–2018).

The main method employed is data-driven qualitative content analysis [18] which allows a data-oriented and inductive approach. Rather than using pre-selected categories or theory-driven themes, the thematic codes, categories and broader themes were identified and elaborated directly from the data through several readings of the material. The analysis was then conducted on the basis of the identified categories and themes.

The first two sections discuss the theoretical framework and context of the study including a brief introduction to Ecuadorian energy policy in general and the case of the CCS in particular. The following section entails an analysis of expectations that various actors have in relation to the CCS and what kinds of socio-technical imaginaries and knowledge regimes they draw on, shape and represent. The concluding section summarises the key findings and discusses them in relation to how the different imaginaries are reflected in the dynamics of various levels and forums of negotiation and contestation.

2. Expectations Shaping Energy Futures, Technological Choices and Policy

The meanings and roles given to water and technology shape the ways in which water rights and futures are constructed, negotiated and contested. Parallel, competing and confronting expectations play a crucial role in imagining energy futures and making technological and political choices—and in turning policy visions into concrete policies and practices [13]. The sociology of expectations literature suggests that an important aspect of expectations is their future orientation [10,13]. National and supranational policy documents [3,4,19,20] imply that increasing electricity demand, together with climate mitigation and the ideas of the green economy, necessitate a particular set of politico-economic, technological and institutional changes. In this way, new hydropower projects contain a promise of providing solutions to broad societal challenges and can be seen as a kind of an 'imagined world' ([14], Anderson's term imagined community´ [21]), wherein hydraulic technology serves to prevent human and ecological disasters in a sustainable and profitable way (cf. [12,13,15]). Hydropower is, therefore, not only an economic or techno-scientific project but, importantly, a political one.

This article analyses the expectations different actors from the state to grassroots level regarding water and technology. It approaches expectations and visions concerning hydropower as historically and culturally constituted socio-technical imaginaries. This concept refers to nationally or locally produced expectations and visions related to techno-economic and socio-political possibilities [14,21,22]. They are constitutive because they generate expectations, but at the same time, they are based on local practices, history and public reasoning [14]. This interpretative flexibility [23] implies that water and technology are subject to varying forms of reasoning in particular contexts, which necessitates going beyond technical aspects and focusing on their social and political dimensions [12–14].

Another key aspect is that expectations and socio-technical imaginaries are actively produced [13,23] and performative [10] because they create actions, define roles and responsibilities and shape political agendas [2,12,13]. This ties the sociology of expectations to transition studies [1,2,9,10,12–14] wherein socio-technical transitions are seen as enabled and/or hindered by prevalent or changing politico-economic conditions and political activities at different levels of a society. As Sovacool and Brossman [12] (p. 839) have summarised, literature on technology and future visions, or "fantasies" (e.g., [22,24]) suggests that successful socio-technical imaginaries often entail four common characteristics; they (1) are concrete enough to be applied in the real world, (2) are critical towards the present situation, (3) provide convincing arguments for socio-technical transition, and (4) suggest that the socio-technical vision in question is powerful enough to make previous changes irrelevant. This implies that expectations are important not only in generating policy visions but also in accelerating or hindering sustainability transitions.

Expectations are also crucial in defining cooperation and common goals and brokering relationships between different actors and stakeholders who often draw on different knowledges and modes of social interaction [9,10,12–15]. They are thus important in understanding how and to what extent different rhetorical strategies and rationales resemble or depart from each other, how they take shape in politics, and how they represent and shape different knowledge regimes [9,11–13]. Rather than neutral or universally agreed-upon entities, this points to hydropower infrastructures as negotiated and co-produced constructs [14] that have contextually embedded meanings and manifestations. Socio-technical imaginaries, with expectations and counter-expectations concerning water and technology as delivering progress, modernity, hope and prosperity, thus shape the ways in which society perceives new hydropower and its potential or challenges (see also [25] in this volume).

As will be discussed below, expectations concerning the CCS reflect divergent socio-technical imaginaries (Table 1). They tend to frame the Coca River not only *instrumentally*, as a resource for electricity or a service for the nation, but also *functionalist-pragmatically*, as an ecosystem with rich biodiversity, a source of local income and regional development, and as having an important recreational value; or by *addressing an intrinsic value of water*, as an historically and culturally situated place with esthetical and cultural value. In addition, *symbolic understandings* have also attached new meanings to water. The river has been understood as a mediator of techno-economic advancement and a key element in enabling and accelerating socio-technical transitions. These four ways of constructing meanings for water imply divergent knowledge regimes, resulting in rather different socio-technical, cultural and environmental imaginaries and policy alternatives.

Table 1. Expectations towards water and technology in the context of the Coca Codo Sinclair.

	Instrumental	Functionalist-Pragmatic	Intrinsic Value	Symbolic
WATER	• resource for electricity • service for the nation	• ecosystem with rich biodiversity • a source of local income and regional development • recreational value	• historically and culturally situated place • esthetical and cultural value	• mediator in techno-economic advancement • enabling and accelerating sociotechnical transitions
TECHNOLOGY	• contribution to national economy • potential for regional development • a means for social and environmental compensations	• improved access to and supply of electricity • provision of lightning in households • control and regulatory infrastructures	• development of the national knowledge base and technological know-how • modernization and progress	• future hopes and national pride • mediator in environmental, social and economic policy goals • symbolizing techno-economic and industrial competence with contributions to *buen vivir*

The four ways of reasoning also imply different understandings concerning technology that draw on divergent knowledge regimes. Instrumental constructions tend to emphasise the contribution of hydraulic technology to the national economic and regional development and consider it a means through which various social and environmental compensations can be attained. The functionalist-pragmatic perspectives focus on expectations regarding enhanced direct influences such as improved access to and supply of electricity and more secure provision of lighting in households. Since both the technology and know-how used by the CCS are largely imported, these perceptions highlight national competence and technological infrastructures particularly within the systems of control and regulation. Constructions that provide technology an intrinsic value are closely related to imaginaries of the importance of technological advancement as such and for the broader development of the national knowledge base. These conceptualisations tend to value technological advancement as signifying modernisation and progress. Finally, symbolic views attach future hopes and national pride to hydraulic technology. In these perceptions, technology is viewed as mediating environmental, social and economic policy goals and as symbolising techno-economic and industrial competence with contributions to good living (*Buen vivir*, see below). Moreover, hydraulic technology is represented in the symbolic understandings as a means to control natural resources and as an enabler of system-level sustainable transitions. These conceptualisations highlight the performativity of expectations as they may have important political repercussions and shape broader understandings regarding sustainability and technology.

3. Ecuadorian Energy Policy and the Case of the Coca Codo Sinclair (CCS)

3.1. 'Buen Vivir' and the Grand Energy Transition

The CCS was constructed in the context of former (2007–2017) President Rafael Correa's political program of 'Citizen Revolution' (*Revolución Ciudadana*) adopted in 2006 as a response to the previous market-based regime, or, as Correa put it, the 30-year 'long and dark night of neo-liberalism' [26]. The program has been based on the notion of 'good living' (*Buen Vivir*), a concept borrowed from the Kitcha term *Sumak Kawsay*. It has to be noted, however, that somewhat similar ideas have been visible for instance in Bolivia as well. Under the presidency of Evo Morales (2006–), the ideas of "vivir bien" have included an attempt to defend 'Mother Earth' (*Pachamama*) through sustainable use of natural resources [27]. Correa's program has been characterised as a mixture of a statist and neoliberal models [28] due to its strong, state-led orientation in governing the 'strategic sectors' defined by the 2008 Constitution, i.e., energy, telecommunications, non-renewable natural resources and water, combined with neoliberal and market-based mechanisms. The National Plan for Good Living 2013–2017 (*Plan Nacional de Buen Vivir*) [8] outlines key aspects of *Buen vivir* and suggests a shift towards long-term planning and a holistic view of (state-led) governance. It entails several ambitious objectives to improve education, health care and infrastructure while protecting nature and managing natural resources in a sustainable way. A focal aspect in this effort has been strengthening the state's role in resource management and re-nationalising natural resources as well as re-enforcing state power in the strategic sectors and increasing public expenditure in fields such as transport infrastructure, public health care and hydropower [26,29–31].

A key element in the new policy programme is an ambitious energy transition aimed at diversifying the energy matrix, contributing to climate change mitigation and improving national energy security and sovereignty [26,31,32]. The Electricity Master Plan 2007–2016 [20] emphasises generating a substantial change in the national energy matrix largely by accelerating greater use of renewable energy sources. The CCS is one of the eight new hydropower projects included in the energy transition that seeks to increase the share of hydropower from 58% of electricity generation in 2015 to 90% in upcoming years (see also [25] in this volume, [29]). The policy priorities also include accelerating sustained economic growth, e.g., via electricity export and reducing the public deficit that has resulted from the recent oil price fall [33].

Generally, the Ecuadorian economy has grown steadily over past few decades (excluding the recessions in 1999–2000 and 2015–2016) and especially since the early 2000s, while the growth in population has been moderate (Table 2). The gross domestic product (GDP) grew from 74,111 billion USD in 1990 to 160,097 billion USD in 2015, while the total population increased from 10.22 million to 16.14 million during the same period with average annual growth of 1.4% [33–35]. The rise in total greenhouse gas (GHG) emissions has been explained by the growing energy sector, increasing demand and transportation [30]. However, this development has been relatively moderate in relation to the GDP or per capita (Table 2).

Table 2. Ecuador's GDP, population growth, GHG emissions * and oil production and consumption in 1990–2015.

	1990	2000	2005	2010	2015
Real GDP (constant prices, million USD 2011)	74,111	93,842	118,922	140,492	160,097 **
Real GDP per capita (constant prices, USD 2010)	3721	3679	4287	4657	5353
Population (million)	10.22	12.63	13.74	14.93	16.14
Total GHG emissions (kton CO_{2eq})	45,300	42,210	53,240	65,970	67,940 **
GHG emissions (kton CO_{2eq}) per GDP	0.59	0.45	0.45	0.47	0.43 **
GHG emissions (kton CO_{2eq}) per capita	4.44	3.34	3.87	4.42	4.41 **
Oil consumption (t of barrels/day)	97	134	159	243	251
Oil production (t of barrels/day)	285	395	532	486	504

* Excluding the land-use, land-use change and forestry sector ** 2012. Sources: [33,35–37].

In 2012, Ecuador's total GHG emissions amounted to 67,940 kton CO_{2eq}, which accounted for a 0.15% share of the world's GHG emissions [36]. The country's energy sector is still highly reliant on fossil fuels. For instance in 2014, oil accounted for 88.50% of the total production of primary energy. Ecuador is a member of OPEC and a net exporter of oil; nonetheless, its dependency on price fluctuations in the world market has had significant implications for the country's economy, and current plans entail exporting the oil surplus or refining it to a higher value. In terms of GHG emissions, the energy sector is responsible for almost half (44.63% in 2012) of national emissions [38]. While the new policy approach has been driven by development, climate mitigation, conservation and resource nationalism, the rhetoric concerning the sustainable transition in the national energy matrix has been somewhat inconsistent with recent policy actions such as new oil fields concessions [30]. The large Chinese loans and investments, e.g., to finance the CCS, have also limited resource nationalism, and the recent economic situation has partly subjected conservation demands to developmental imperatives [30,39]. It can thus be questioned to what extent these policies are in line with the 2008 Constitution, which gives nature rights of its own and seeks to preserve the environment. Related to this point, concerns have also emerged regarding stakeholder engagement, integrative environmental planning and transparency and accountability in the new mega-projects (interview data, [40]).

3.2. The Coca Codo Sinclair

The CCS is located in the Napo and Sucumbios provinces, approximately 100 km east of Ecuador's capital Quito, in an area where the River Coca is formed by the waters of the Quijos and Salado Rivers (Figure 1). The CCS diverts water just below the confluence, piping flows to a power plant about 25 km downstream of the diversion dam. The total drop at the powerhouse is 620 m [41].

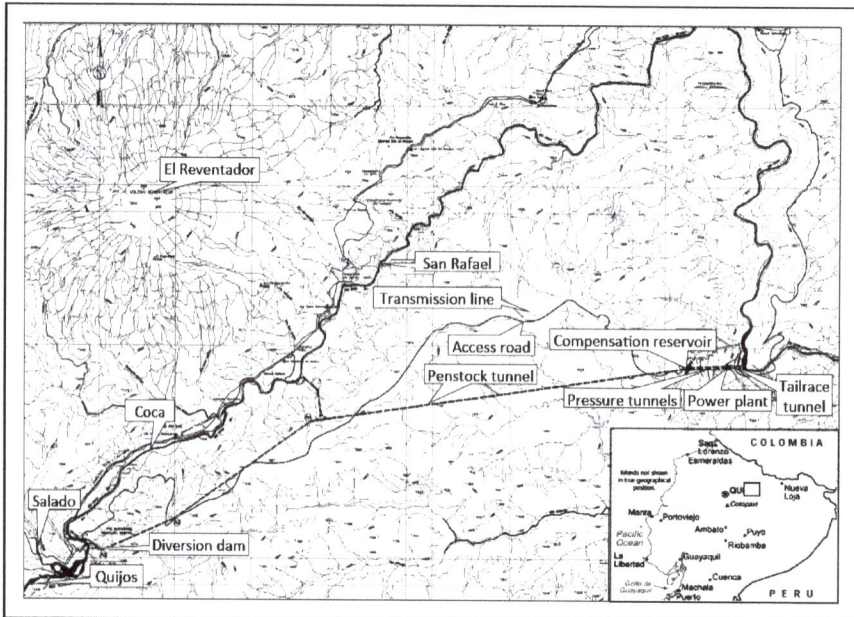

Figure 1. The Coca Codo Sinclair project and the surrounding area. Source: modified from [42].

Initial studies concerning the project began in the 1970s. A pre-feasibility study in 1976 and a feasibility study in 1992 were conducted on the basis of registered flow calculations from 1972–1990. It suggested an 859 MW project with two units (432 MW and 427 MW) using a total flow of 127 m^3/s. Because of financial constraints and the eruption of the Reventador volcano close to the project area in

1987, the plans were halted for almost two decades. The 1992 plan was updated in 2007 to include a potential of up to 1500 MW and a maximum usage of 222 m^3/s flow out of the estimated average annual flow of 287 m^3/s. However, the new feasibility studies were, according to a representative of the Ecuadorian Rivers Institute (interview data), based on historical hydrological data of questionable validity, which led to overestimations concerning water availability [43,44]. The average annual flow figures also are not very informative because of large differences in wet and dry season flows [43,45,46].

The preliminary Environmental Impact Assessment (EIA) studies were conducted quite rapidly, between September 2007 and March 2008, by the consultant Entrix, a contractor of the Ecuadorian generator Termopichincha [46]. The national electricity coordinator, Coordinador Eléctrico Nacional (CONELEC), approved these in record time, within one week, in the beginning of April 2008. At the time, the project had already been included in the 2007 Electrification Master Plan, and the Ministry of Environment had given the project a certificate concerning protected areas in February 2008. These did not, however, include official environmental permissions or licenses for the project [46].

In April 2008, Rafael Correa and former Argentine head of state Cristina Fernandéz de Kirchner broke ground on the project by establishing a joint venture between Termopichincha and the Argentinean state energy company Enarsa. At the same time, the Ministry of Electricity and Renewable Energy (Ministerio de Electricidad y Energía Renovable, MEER) promoted the project locally and sought to reduce opposition through special agreements with municipal governments entailing aims to educate local residents, build a health center and improve inter-institutional cooperation. Entrix was hired again to conduct public consultations, together with Coca Codo Sinclair S.A., in the area in May and June 2008. These entailed informing local residents about the project through a public hearing and an advertising campaign in local newspapers as well as establishing two public information centers in the impact area. However, little detailed information is available regarding the community consultations, environmental audits or the EIA processes (see also [40,45]).

In 2009, Argentina sold its share to Ecuador's state power generation holding company, CELEC EP (*La Corporación Eléctrica del Ecuador*), and in 2010, the state-owned special-purpose company Coca Codo Sinclair EP was established for the project's development. The project was constructed by a Chinese company, Sinohydro, which won the engineering, construction and procurement contract for the project in 2009. Initially, the Sinohydro-Andes JV consortium consisted of the Chinese Sinohydro (89%); an Ecuadorian company, Coandes (8%); and the Chinese Yellow River and Italian Geodato consultant companies (3%). In September 2009, however, Coandes withdrew from the consortium, and the contract was awarded to Sinohydro alone. The project is highly dependent on foreign debt: the Export-Import Bank of China financed 85% of it with a 1.68 billion USD loan, while the Ecuadorian government was responsible for the rest of the funding. The total costs of the project rose to 2.25 billion USD.

The water intake consists of a concrete-face rockfill dam, a concrete spillway and an intake between them (Figure 2). The water diverted from the intake runs via the sedimentation basin through the 24.85 km tunnel into the compensating reservoir (with 800,000 m^3 usable volume), and via the almost 2 km-long penstocks to the eight Pelton-type turbines, each with a capacity of 187.5 MW. The CCS's run-of-river intake has a maximum capacity of 7500 m^3/s. The project has been in operation since 2016; the first commercial phase entailed taking four units to operation in August 2016, and the second included the remaining units in December 2016. According to the MEER [40], the project contributed to the National Interconnected System (NIS) by 11,603.76 GWh by May 2018. Nevertheless, the project has also suffered from technical difficulties. In 2012, a tunnel collapse during the construction caused the deaths of 13 people. In October 2018, CELEC EP announced that three out of eight generation units were not operating because of major disconnections in the system. The power cuts in the transmission lines affected many Ecuadorian cities, and two newly launched hydroelectric projects (Minas San Francisco and Delsitanisagua) were used to replace the energy shortage [47].

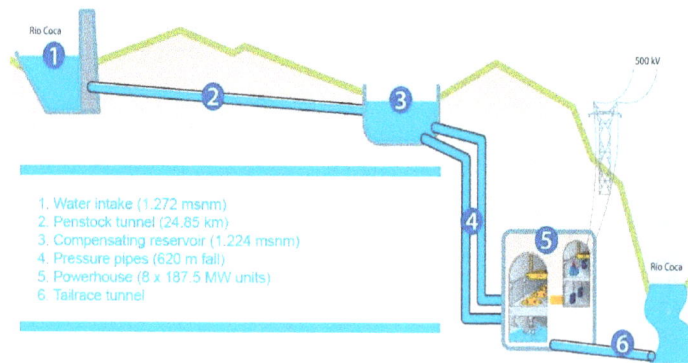

Figure 2. The structure of the Coca Codo Sinclair. Source: modified from [41].

The CCS's impact area covers almost 40,000 hectares, and it has been estimated to directly influence about 2000 people [46,48]. The population in the area mostly consists of small-scale farmers and entrepreneurs. As a run-of-river project situated in a highly active seismic area, the CCS does not include a massive compensation reservoir and thus it is not comparable with typical mega-dams in terms of carbon emissions. Nor has it entailed displacement of people or violations of indigenous rights as has been the case in many other mega-dams. Concerns have emerged, however, concerning the project's other impacts, including increased sedimentation upstream and significant lowering of the water flow below the dam, affecting fish supplies. It also is claimed to practically dewater Ecuador's tallest (146 m) waterfall, San Rafael [49], which is a major attraction of the United Nations Educational, Scientific and Cultural Organization (UNESCO) Sumaco Biosphere and is located about 20 km downstream of the water diversion dam (see Figure 1).

The available water flows vary considerably in the dry and wet seasons, which makes the CCS's commitment to the minimum flow requirement of 20 m^3/s (originally 56 m^3/s)—the amount of water being left in the river to maintain the waterfall—challenging [48–50]. In particular during dry seasons, the prioritisation of different water uses (e.g., power generation vs. recreational use) thus is an important question. Yet, this matter is difficult to accurately estimate beyond observations, given the problems of accessing hydrologic and operational data of the CCS (see also [45]). Other concerns entail the CCS's impacts on the flora and fauna of the Sumaco Reserve and the Cayambe-Coca National Park, the high seismic risk, as well as deforestation caused by the construction of the transmission lines. Social concerns relate to employment opportunities and working conditions in the construction sites, infrastructural transformations, healthcare and sanitation as well as broader regional development and the living conditions of the local residents (interview data, [40,45]). One criticism also concerns the partiality of the EIAs, which were conducted separately for the dam and the transmission lines and, thus, allegedly lacked a comprehensive view of the impact of the project as a whole.

At the same time, high expectations are visible at the national level regarding the ability of the project to deliver a pathway to a more sustainable and economically viable energy future. Described as an 'emblematic' project of the national government, the government expects the CCS to provide approximately 30–44% of the supply of national energy demand, contribute to 3.45 tons of CO^2 emission reductions annually, generate annual savings worth 617 million USD by reducing the import and consumption of fossil fuels, create 7739 new jobs (mostly in the construction work) and directly benefit over 16,000 inhabitants through its compensation programs such as public infrastructure and improved access to electricity. Whether and to what extent these estimations are met remains yet to be seen because of the lack of accurate, updated hydrologic and operational data and pending compensation activities [45,46]. In any case, the CCS can work at full capacity only about five or six months per year because of the changes in the availability of water. The following section discusses in more detail the different expectations stakeholders have expressed concerning the project and some of the key rationalities and diverging knowledge regimes that shape them.

4. Competing, Parallel and Confronting Expectations around the CCS

4.1. The Government's Logic: Economic Growth, Energy Security and Buen Vivir

Hydropower has been an attractive electricity generation option in Ecuador not because the technology as such is superior to other renewable energy technologies but rather because it significantly contributes to national policy priorities such as energy security, self-sufficiency and the reduction of CO_2 emissions (interview data, [19,20,31,32,41]). The official rhetoric has emphasised how the CCS is a crucial part of the transition of the national energy matrix towards more reliable, self-sufficient, cost-efficient and cleaner energy production [38,41]. Thus, it is portrayed as a kind of a systemic innovation and an important part of a society-wide transformation that also provides export opportunities. As a representative of the CELEC EP stated,

> "We, rather than being suppliers of infrastructure, say that to produce electricity, the intention is to be an exporter of energy. A vision of 20 years from now is for Ecuador to be a regional supplier of electricity, that is, not to develop infrastructure or equipment technologies but selling energy."

Behind these arguments is also the international climate debate vis-à-vis national concerns concerning the reliance on fossil fuels and the increasing need for electricity related to economic development. Moreover, the Ecuadorian government's hype around hydraulic mega-projects aligns with the broader rhetoric of modernity and progress (see also [26,29,30,41]), which has partly justified the official energy policy. Ecuador is thus presented in the government's arguments as a kind of energy policy pioneer with advanced policy visions, enabling the modernisation of the national energy system while boosting economic growth and social well-being (interview data).

The official rhetoric also resembles the notions of green growth and the green economy. Advocated by the Organization for Economic Cooperation and Development (OECD) [3] and the United Nations Environment Program (UNEP) [4] in the 2000s, these concepts build on previous eco-modernist ideas of decoupling economic growth from environmental depletion and emphasise the economic potential of a system-level transition [1,2,9,12] to cleaner energy production, green technologies and resource efficiency. This has entailed considerations concerning energy security and climate change. The OECD [3] (p. 9) defines green growth as "fostering economic growth and development while ensuring that natural assets continue to provide the resources and environmental services on which our well-being relies." This formulation calls for synchronising economic and environmental policy goals into a coherent, cross-sectoral policy approach in which the role of public policies is crucial.

The Ecuadorian government's rhetoric similarly emphasises clean electricity production and a change in the national energy matrix towards a greater use of renewable energy sources while enabling improvements in the country's economic performance and climate action [19,31,38,41]. Here, water is constructed instrumentally as a resource for development and a service for the nation (Table 1, interview data).

> "Advantageously, for the size of our country, we have sufficient and extensive resources that logically allow us to make a strong investment to transform the energy matrix. [. . .] We consider that the [local] impact is quite low compared to the benefits that Coca Sinclair brings" (a representative of MEER).

The relatively strong role attributed to the state in the vision of *buen vivir* also highlights the importance of public policies in achieving sustainable transitions and infrastructure improvements to enable long-term economic growth and societal well-being [8]. In this context, technology is viewed in a functionalist-pragmatic way as contributing to infrastructural development and energy supply. Unlike in other countries where energy policy objectives are largely motivated and rationalised by climate policy arguments, Ecuadorian energy policy rhetoric particularly emphasises national prosperity and energy security and conditions climate change concern in terms of energy sovereignty and self-sufficiency (interview data). The strong focus on hydropower also raises a question concerning the future potential of new green technologies (e.g., solar, wind, new biomass applications) that have received increasing global emphasis. The technological choices the government makes today have important repercussions for the country's energy future, and the potential technological lock-ins and path dependencies [1,2] also crucially shape longer-term socio-technical transformations and the country's innovative potential.

So far, there has been relatively little discussion concerning the potential of other renewable energy sources because the government's high expectations of hydropower (interview data) have dominated the discussion. As Sovacool and Brossman [12] (p. 839) have noted, "Those who advance a rhetorical vision naturally shape and limit the scope of how the vision is discussed." In the Ecuadorian case, the public debate on energy policy has indeed been quite limited. The Correa regime had a high level of control over national media and non-state actors such as NGOs, and the terms, conditions and content of all kinds of public debates were, thus, strongly determined by official authorities (interview data). Representatives of NGOs and inhabitants saw that one reason for the lack of public resistance and protests against the CCS has been the Correa government's repressive policy towards political action and organisation (interview data). In addition, the government policy has been secretive, and little information concerning public consultations, project implementation details, China–Ecuador relations or alternative energy policy choices has been publicly available. Currently, with the new government, the situation may change in this respect, and according to many interviewees, one welcome change under the new president Lenín Moreno (2017–) is at least some degree of liberalisation in terms of the freedom of speech. Yet, it needs to be noted that Moreno also was actively involved with the CCS negotiations and acted as a mediator between the Chinese Eximbank and the Ecuadorian government in the preparatory phases of the project. Despite new hopes and expectations towards enhanced possibilities for a more open political dialogue, few thus consider broader political changes attainable in the near future (interview data).

4.2. Technological Optimism Underpinning Regulatory Practices and Regional Development

Whereas the government's official rhetoric demonstrates a broader strategic policy approach and reflects high expectations of economic development and prosperity through cleaner electricity production, energy security and self-sufficiency, the expectations of regulatory authorities and national energy policy agencies highlight infrastructural innovations and technology within the regulatory systems (interview data).

These exemplify how the CCS is perceived symbolically in this rhetoric, not only as an electricity generation project but also as a showpiece of national competence and pride (Table 1). This, together with the government's rhetoric of the CCS as an 'emblematic' project, implies its construction as a kind of a monument, a signifier of how the government transforms and modernises the whole society through its massive infrastructural and energy projects [26,30,32,41]. While the technology utilised by the CCS is not new per se, the advanced technology used within the infrastructure for the regulation and control of its operations is interpreted as representing modernity and technological know-how, thus symbolising hopes for national technological advancement (interview data, [51]).

"It is important to continuously enhance the knowledge of technology and professional development that will be even more relevant in the future. [. . .] There is a very clear policy of the government's electricity system." (representative of CENACE).

In this context, water is perceived instrumentally as a resource for increased electricity demand without any (explicit) socio-political or culturally embedded value. This apoliticisation of water frames electricity generation in general and the CCS in particular not only as a necessity but also a possibility for the nation because it seems to offer a potential for the development and utilisation of technologies especially within the established structures of regulatory infrastructure (interview data).

> "Here, within Ecuador, the opportunities for technological development entail all the maintenance that we have to take care of. [...] We have to be technically sustainable to be able to repair all components. [...] All this is our responsibility that we are going to develop, and it also is an opportunity to understand the latest-generation hydrodynamic profiles" (a representative of ARCONEL).

While the politics of water and questions concerning the social and environmental impacts and public and political acceptability of the CCS are largely hidden by in this rhetoric, infrastructural innovations and supportive technologies is a source of national competence. The pride over technological advancement is exemplified by the regulatory authorities' references to the project's technology-driven and allegedly well-functioning systems of regulation and control (interview data, [51]), in a way demonstrating a faith in national technological expertise within the regulatory systems. These expectations are thus largely shaped by hopes and future visions for technological and economic modernisation, a mastery over natural resources and an enhanced techno-scientific knowledge base. This kind of technological optimism, or legitimation, has also been visible in other contexts such as the early 20th-century hydropower projects in the U.S. [11], nuclear new-build in Finland, France and the UK in the 2010s [52], and in Peruvian water resources management [53–55]. At the same time, regulatory authorities and energy policy agencies perceive their role as more or less politically neutral (interview data). This rhetoric thus implies a pragmatic, apoliticised and techno-economic rationalism fueled by a sense of national pride and a technology-driven imaginary of the future energy system (cf. [6,16] in this volume).

A certain type of technological optimism is also visible among regional authorities. Here future expectations are related to the compensation schemes and the potential of introducing new solutions for water treatment because one major problem in the area is the contamination of water by municipal waste (interview data). The project's presence in the area symbolises future potential in the form of techno-economic cooperation and investment in innovation, and expectations favour its ability to contribute to regional development through technological upgrading (interview data). In a sense, these expectations reflect technocratic visions of modernisation (see also [53–55]) with imaginaries concerning the exploitation of clean technology solutions that would improve the overall living conditions within the area. In this rhetoric, water is also perceived instrumentally as a resource, not only for electricity production for national needs or economic purposes, but also as a source of local income and well-being for inhabitants (Table 1). In general, the regional authorities' expectations centre on balanced regional development, sustainable socio-technical transitions, environmental protection and improvements in the everyday living conditions of the local residents. While many CCS-related initiatives have remained at the level of political promises (interview data), strong expectations prevail concerning the future developments and local support that the project might offer. The lack of resources, unequal distribution of benefits and inadequate attention from the project's side towards local developmental needs are considered a major hindrance to technological and social improvements (interview data).

4.3. Expertise Outside and Beyond the State: The Value of the Multiple Uses of Water

In contrast to economically or technologically oriented argumentation or symbolic framings (Table 1), many NGOs emphasize the multiple uses of water in the case of the CCS. Instead of appealing to culturally embedded or historically constructed understandings, however, their argumentation tends to draw primarily from research and statistical data. This professionalisation of political activism is increasingly used by NGOs in many other countries as well to gain a voice in political debates and legitimise knowledge claims that partly emerge outside and beyond the state structures, as in the case of nuclear power opposition in Europe [52]. It is a kind of activism-expertise that draws its legitimacy from evidence and sophisticated techno-economic calculations rather than from arguments appealing to normative or emotional rationalities (as visible for instance in the government's politically appealing promises of *buen vivir*).

In the case of the CCS, environmental organisations have been concerned about the local effects the project may have on the flow of the river, its biodiversity, deforestation and the impacts of the transmission lines on the vulnerable Amazonian area, especially the UNESCO Sumaco Biosphere Reserve (interview data, [43,44,48]). They also have criticised the environmental and social impact assessments and consultations for their inadequate implementation, hurried schedule, insufficient local participation, and flawed or outdated background data (interview data). According to some interviewees, however, many NGOs have considered the project's overall impacts relatively limited in comparison to some other mega-dams (interview data). Another factor in this respect has been the lack of indigenous population in the area, and as, for instance, a representative of Acción Ecológica Ecuador mentioned, many NGOs have focused on other hydraulic projects with more direct and excessive impacts on local livelihoods in areas with more indigenous population. In these cases, the influence of the NGOs has been seen as potentially more concrete and effective (interview data).

Currently prevailing concerns relate to the operational capacity of the CCS. NGO leaders criticise the government for overestimating the availability of water especially in the dry season, and the flow estimations are claimed to have been made with inaccurate and historical data (interview data, [43–45,48]). Related to this matter, the future of the San Rafael watershed has raised severe concerns among environmentalists, perhaps most visibly in the calculations of the International Rivers and Save America's Forests (interview data, see also [43,44]). Another issue the NGOs emphasise is the recreational value of the river Coca, which in many official communications has gained little attention.

The characteristics of the environmentalist rhetoric, however, are not only a focus on concrete questions and impacts of the project, but also on the processes of governance, policy negotiations and democracy at a more general level. The government's oppressive policies and the lack of freedom of speech are referred to as delimiting the possibilities for open public dialogue and political influence [interview data]. Many interviewees mentioned that multiple voices and understandings concerning water and technology are often not sufficiently heard because the official communication practices have remained closed, and public consultations have been organised with short notice (interview data). Moreover, the environmentalists criticise the established evaluation practices for their lack of case-sensitivity and sufficient contextual understanding. They highlight the local contexts and the need to develop and refine the assessment procedures in a more detailed, case-sensitive and contextually embedded direction (interview data). Representatives of NGOs have also mentioned the insufficient monitoring practices and the lack of transparency and accountability of the CCS operations as key problems of its governance. Recent studies regarding the impacts of the CCS also support these claims (e.g., [40,45]).

An interesting finding in the case of the CCS is that the visibility of NGOs in the project's area of influence is nearly non-existent. This raises a question concerning the modes, levels and forums in which expectations and policy visions are articulated, interpreted and contested. Indeed, it seems that many environmental NGOs' operations are largely nationally or internationally oriented, and the debates take place in forums such as informal policy networks, internet-based forums, blogs and other official or unofficial communications. In this sense, the different levels and forums of policy negotiation do not necessarily meet and, therefore, the policy-making and governance processes lack mutual dialogue and multi-level stakeholder interaction. Arguably, the Ecuadorian state-centrist political context is also behind this (see also [28,56]); there has been only limited participatory processes, public debates and political inclusion. This has also been visible in the minimal critical discussion concerning energy policy issues or the CCS in the national media. An important exception is the studies and debates steered by an international science-activist network, Foro Recursos Humanos (RRHH) (see [57]).

4.4. Mundane Expectations and Hopes for Enhanced Local Development

At the local level, expectations concerning water and technology are largely framed by the mundane concerns of everyday life. These perspectives highlight the concrete implications that the CCS may have in the riverside villages and the lives of the local inhabitants (Table 1). In general, the local inhabitants perceive the direct environmental and socio-political impacts on their livelihoods as relatively small (interview data). Most agricultural activities focus on products that are not highly dependent on the river and, thus, the impact of the CCS in this respect is commonly seen as limited (interview data). In terms of environmental impacts, the local residents expect the government and the CCS to take care of necessary environmental protection measures such as reforestation programs and preventing the contamination of the available water resources (interview data).

Perhaps partly affected by the government's strong policy rhetoric and the promises of the 2008 Constitution, many local residents also believe that the level of environmental protection is sufficient, and the CCS causes no major harm. One issue, however, is the project's impact on the biodiversity and species in the river, particularly in terms of substantially reduced possibilities for fisheries. The residents perceive this as having a direct negative effect on fishermen's income opportunities and the recreational value of Coca (interview data), which confirms the concerns some NGOs have anticipated. The impact on the river's natural flow has in turn raised local concerns about the watershed San Rafael and its implications for eco-tourism in the area. According to a guide of the San Rafael, the dewatering (estimated to amount to about 60% of its flow) is further accompanied by the deforestation of some parts of the UNESCO natural reserve, and uncertainty regarding the project's degree of commitment remains in terms of mitigating these impacts (interview data). At the same time, however, some interviewees saw the project itself as a potential tourist attraction, thereby indirectly benefitting the whole area because of its promotional value and the expected international visibility as a flagship project of the new national energy policy.

Some degree of change has taken place since the inauguration of the project in 2016. While, in the beginning, the hegemony of the government's promising rhetoric was locally quite visible, more recently (as of October 2017), the perceptions among the residents have also entailed more critical aspects. The expectations they had towards the benefits of the project faded to some extent, especially in terms of the distribution of benefits and compensation between different parts of the area (interview data). Some communities are seen to have benefitted from the project more than others, and in some areas improvements in schooling and health care in particular are still pending (interview data, [45]). A representative of local administration in Gonzalo Pizarro also shared this viewpoint (interview data). Another issue is the CCS's local employment effects. According to a recent study, the high expectations generated during the public consultation process were not fulfilled because the project hired approximately only 40% instead of the promised 70% of its labour force from the area of impact, mainly including jobs for unskilled labour (45). Yet, there have been varying estimations on this matter and, thus, it remains somewhat unclear who, and to what extent, has benefitted from the added employment.

However, many local residents and administrative staff have regarded the increased availability of potable water, which had previously been a substantial problem in some parts of the area, as a major improvement. In addition, an enhanced system of and access to the electricity supply are generally seen as positive developmental impacts of the CCS in the area (interview data). Some local inhabitants, however, argue that a more secure supply of electricity has been accompanied by increased electricity prices and, therefore, their expectations concerning the benefits of the CCS have not been fully met. The government's dominant rhetoric and the state-led governing rationale have also been experienced as delimiting opportunities for political influence. In this respect, the historically and culturally constituted skepticism and lack of trust towards public authorities still seems to prevail in local contexts.

At the same time, the major concerns at the local level have remained largely unchanged; they include everyday matters such as sustained income, the availability of jobs, the functioning of the electricity system and overall living conditions (interview data). One issue has been the loss of jobs after the completion of the construction work, which has directly impacted the lives of many local villagers (interview data). Related to this matter is that a majority of the Chinese and Ecuadorian project workers have left the area, which has had multiple direct and indirect impacts on the local economy and society. A socio-cultural impact has also been the intercultural interactions and cultural diffusion of the Chinese and Ecuadorians; many new multicultural families have formed in the area (interview data).

Unlike some other more confrontational projects (see [7,58–63] in this volume) that may have substantial, direct impacts on local livelihoods, however, municipal authorities and local residents have claimed that the CCS has not generated organised opposition or major criticism in the area (interview data). There seems to be neither major interest in nor substantial concerns regarding the project beyond mundane expectations related to practical matters (interview data). One reason for this might be that the project has only limited direct influence on everyday lives in the surrounding villages as their livelihoods are typically not directly dependent on the river. Another issue is that the government's rhetoric has been largely hegemonised through mechanisms of necessitation and naturalisation, i.e., de-politicising arguments visible in several government documents claiming that the project is 'emblematic', 'neutral' and a 'necessity' for national economic development and energy policy. Moreover, symbolic actions, such as Correa's personal visit to the area and the municipalities during the construction phase, may have contributed to local acceptance of the project (interview data).

5. Conclusions: Expectations Embedded in Diverging Socio-Technical Imaginaries

Socio-technical imaginaries are related to the notion of imagined communities [21], which, in this case, are largely defined by sustainable energy futures and hopes for technology's ability to generate economic prosperity and social well-being. Official imaginaries of national energy policy in general and the CCS in particular are constructed around the rhetoric of progress, economic development and national competence. In this context of state paternalism combined with elements of neo-liberal rationality, the official rhetoric constructs a particular national imagined community wherein the context and rules of policy negotiation are strongly defined by the state. As the findings indicate, this form of public reasoning implies that the modes and spheres of political participation and influence are essentially conditioned by the dominant policy vision, which determines not only the desired energy future of the country but also the rationales for citizenship, identity and participation.

The rather hopeful political visions evident in the government's rhetoric concerning water and technology have indeed been influential in legitimising the CCS. Expectations emerging around substantial national economic benefits accompanied by visions of enhanced energy security and self-sufficiency, climate-friendliness and improved local well-being have been appealing arguments also at the local level. In this way, the dominant rhetoric has been strategically used to provide political acceptability and public legitimacy for the government's policy program. At some points, however, the dominant socio-technical imaginaries have been confronted by local perceptions of failed political promises, NGOs' views regarding misleading information and secretive policymaking practices, as well as regional authorities' and residents' concerns over future regional development.

The four ways of constructing meanings for water and technology (instrumental, functionalist-pragmatic, with intrinsic value and symbolic) discussed in this article illustrate how various expectations and counter-expectations generate competing, and at some points confronting, visions concerning energy futures. They also entail varying understandings concerning *buen vivir* and good society. The instrumental views emphasise economic growth, GDP development and electricity export as key elements of *buen vivir* as they are seen to not only contribute to economic goals but also enable and support national well-being and prosperity. The functionalist-pragmatic perceptions highlight climate benefits, biodiversity and social progress through enhanced employment and energy supply as well as technological regulation and multiple uses of water, thereby pointing to social and environmental aspects and control over technology. The expectations addressing an intrinsic value to water and technology instead highlight the importance of recognising cultural aspects and local knowledge on the one hand, and including ideas of modernisation and progress into the core of *buen vivir* on the other hand. Finally, the symbolic understandings point to *buen vivir* as inherently mediated by technology. They emphasise the importance of techno-economic and industrial competence in boosting system-level transitions, controlling natural resources and integrating various policy goals.

At a more general level, the findings illustrate the performativity and the contextual embeddedness of expectations. The results show how the current political climate emphasising the urgency of climate policy actions, the replacement of fossil fuels by renewable energy sources, and concerns over energy supply and security have provided strong justifications for the national policy approach. Moreover, the rhetoric of sustainable transitions and green growth has provided further legitimisation for these arguments, as socio-technical transitions and renewable energy production are seen as needed in order to achieve more sustainable futures and economic prosperity. The official rhetoric has also presented the CCS as a matter of national security and sovereignty, a 'necessity' for broader developmental and economic objectives, in this way de-politicising and neutralising the national policy approach. The generation of economic abundance, environmental protection and social well-being have been politically appealing promises in an age of uncertainty, politico-economic turbulence and new global fears [64,65]. This shows how the official rhetoric has strategically employed elements from both, national (and local) policy concerns and international climate policy, in justifying the government's policy program. It also has been nearly hegemonised, given the lack of opposition and the very limited opportunities for counter-arguments and alternative rationalities to gain credibility or recognition.

At some point, however, locally embedded concerns have raised critical questions with repercussions at various levels and across sectoral policies. These include, for instance, the connections between hydraulic development and extractive industries in relation to regional socio-economic development and impacts; the reliance of Ecuador on foreign (Chinese) investment and its relation to new governance arrangements; and the potential of sustainable socio-technical transitions in a context in which some people's basic needs are yet not fulfilled. These concerns highlight policy challenges that have so far continued to gain somewhat little attention but might provide important entries for future research, also in other countries. In addition, it can be questioned how and to what extent divergent expectations, knowledges and future imaginaries could be effectively integrated into national decision-making, and what kinds of policymaking structures and forums would allow the generation of democratic innovations for more transparent policy-making and governance that might emerge from more open public debates and co-construction of knowledge. These findings provide a basis for reflection for further analyses concerning the politics, legitimacy and acceptability of energy transitions in other contexts as well. Theoretically, they illustrate the importance of expectations not only in constructing meanings but also in shaping politics and energy futures.

Funding: The founding sponsors had no role in the design of the study; in the collection, analyses, or interpretation of data; in the writing of the manuscript; and in the decision to publish the results. This research was funded by the University of Eastern Finland (UEF) as part of the author's post-doctoral research project at the UEF (2016-2018).

Conflicts of Interest: The author declares no conflict of interest.

References

1. Geels, F.W.; Smith, W.A. Failed technology futures: Pitfalls and lessons from a historical survey. *Futures* **2000**, *32*, 867–885. [CrossRef]
2. Falcone, P.M. Sustainability transitions: A survey of an emerging field of research. *Environ. Manag. Sustain. Dev.* **2014**, *3*, 61–83. [CrossRef]
3. OECD. *Towards Green Growth*; OECD: Paris, France, 2011.
4. UNEP. Towards a Green Economy: Pathways to Sustainable Development and Poverty Eradication, 2011. Available online: www.unep.org/greeneconomy (accessed on 20 February 2018).

5.   ECLAC (Comisión Económica para América Latina y el Caribe). *Recursos naturales: Situación y tendencias para una agenda de desarrollo regional en América Latina y el Caribe*; UN: Santiago, Chile, 2014.

6.   Teräväinen, T. *The Politics of Energy Technologies—Debating Climate Change, Energy Policy and Technology in Finland, the UK and France*; Into Publications: Helsinki, Finland, 2012.

7.   Susskind, L.; Kausel, T.; Aylwin, J.; Fierman, E. The Future of Hydropower in Chile. *J. Energy Nat. Resour. Law* **2014**, *32*, 425–481. [CrossRef]

8.   The Government of Ecuador. *The National Plan for Good Living 2013–2017*; The Government of Ecuador: Quito, Ecuador, 2013.

9.   Steen, M. Reconsidering path creation in economic geography: Aspects of agency, temporality and methods. *Eur. Plan. Stud.* **2016**, *24*, 1605–1622. [CrossRef]

10. Brown, N.; Michael, M. A sociology of expectations: Retrospecting prospects and prospecting retrospects. *Technol. Anal. Strat. Manag.* **2003**, *15*, 3–18. [CrossRef]

11. Segal, H.P. *Technological Utopianism in American Culture*; Syracuse University Press: Syracuse, NY, USA, 2005.

12. Sovacool, B.K.; Brossnam, B. The rhetorical fantasy of energy transitions: Implications for energy policy and analysis. *Technol. Anal. Strat. Manag.* **2014**, *26*, 837–854. [CrossRef]

13. Borup, M.; Brown, N.; Konrad, K.; van Lente, H. The sociology of expectations in science and technology. *Technol. Anal. Strat. Manag.* **2006**, *18*, 285–298. [CrossRef]

14. Jasanoff, S.; Kim, S.-H. Sociotechnical imaginaries and national energy policies. *Sci. Cult.* **2013**, *22*, 189–196. [CrossRef]

15. Goven, J.; Pavone, V. The Bioeconomy as Political Project. A Polanyian Analysis. *Sci. Technol. Hum. Values* **2015**, *40*, 302–337. [CrossRef]

16. Boelens, R.; Shah, E.; Bruins, B. Contested Knowledges: Large Dams and Mega-Hydraulic Development. *Water* **2019**, *11*, 416. [CrossRef]

17. Patton, M.Q. *Qualitative Research & Evaluation Methods. Integrating Theory and Practice*, 4th ed.; Sage: London, UK, 2015.

18. Hsiesh, H.-F.; Shannon, S.E. Three Approaches to Qualitative Content Analysis. *Qual. Health Res.* **2005**, *15*, 1277–1288. [CrossRef] [PubMed]

19. The Inter-American Development Bank (IDB). The IDB Country Strategy 2012-2017: Ecuador, 2012. Available online: http://idbdocs.iadb.org/wsdocs/getdocument.aspx?docnum=37893787 (accessed on 20 February 2018).

20. CONELEC. Plan Maestro de Electrificación 2007–2016. Quito, 2017. Available online: http://www.regulacionelectrica.gob.ec/plan-maestro-de-electrificacion-2007-2016/ (accessed on 20 February 2018).

21. Anderson, B. *Imagined Communities: Reflections on the Origin and Spread of Nationalism*; Verso: London, UK, 1991.

22. Olson, R.L. Sustainability as a social vision. *J. Soc. Issues* **1973**, *51*, 15–35. [CrossRef]

23. Bijker, W.E. *Of Bicycles, Bakelites, and Bulbs: Toward a Theory of Sociotechnical Change*; MIT Press: Cambridge, MA, USA, 1995.

24. Segal, H.P. *Future Imperfect: The Mixed Blessings of Technology in America*; University of Massachusetts: Boston, MA, USA, 1994.

25. Hidalgo-Bastidas, J.P.; Boelens, R. Hydraulic Order and the Politics of the Governed: The Baba Dam in Coastal Ecuador. *Water* **2019**, *11*, 409. [CrossRef]

26. Correa Delgado, R. Ecuador: El desarrollo como proceso político. Conferencia magistral del economista Rafael Correa Delgado en la catedra Raúl Prebisch. Ginebra, 24 de Octubre de 2014. Available online: http://unctad.org/en/PublicationsLibrary/prebisch15th_RCorrea_es.pdf (accessed on 20 February 2018).

27. Ranta, E. In the Name of Vivir Bien. Indigeneity, State Formation, and Politics in Evo Morales' Bolivia. Ph.D. Thesis, Dept of Political and Economic Studies (Development Studies), Faculty of Social Sciences, University of Helsinki, Helsinki, Finland, 2014.

28. Boelens, R. *Water, Power and Identity: The Cultural Politics of Water in the Andes*; Routledge: London, UK, 2015.

29. Boelens, R.; Hoogesteger, J.; Baud, M. Water reform governmentality in Ecuador: Neoliberalism, centralization, and the restraining of polycentric authority and community rule-making. *Geoforum* **2013**, *64*, 281–291. [CrossRef]

30. Escribamo, G. Ecuador's energy policy mix: Development versus conservation and nationalism with Chinese loans. *Energy Policy* **2013**, *57*, 152–159. [CrossRef]

31. Ministerio Coordinador de Sectores Estratégicos (MCSE). *Agenda Nacional de Energía 2016-20140*; MCSE: Quito, Ecuador, 2016.

32. Correa Delgado, R. Discurso Presidencial en la colocación de la primera piedra del Proyecto Coca Codo Sinclair. Simón Bolívar, 29.4.2008. Available online: www.presidencia.gov.ec (accessed on 10 January 2016).

33. EIA. Country Analysis: Ecuador, 2017. Available online: https://www.eia.gov/beta/international/analysis.php?iso=ECU (accessed on 10 October 2016).

34. IMF. Country Report: Ecuador, 2015. Available online: https://www.imf.org/en/countries/ECU (accessed on 18 September 2018).

35. United Nations. WPP, 2015 Revision, 2016. Available online: https://population.un.org/WPP/ (accessed on 18 September 2018).

36. EC. Fossil CO2 & GHG Emissions of All World Countries 2017. Edgar—Emissions Database for Global Atmospheric Research. Available online: https://edgar.jrc.europa.eu (accessed on 18 September 2018).

37. World Bank, Constant GDP per Capita for Ecuador, Retrieved from FRED, Federal Reserve Bank of St. Louis. Available online: https://fred.stlouisfed.org/series/NYGDPPCAPKDECU (accessed on 1 November 2018).

38. Ministry of Environment (MAE). *Tercera Comunicación Nacional del Ecuador a la Convención Marco de las Naciones Unidas Sobre el Cambio Climático*; Ministry of Environment: Quito, Ecuador, 2017.

39. Zambrano-Barragán, P. The Role of the State in Large-Scale Hydropower Development. Perspectives from Chile, Ecuador, and Peru. Master's Thesis, Dept of Urban Studies and Planning, Massachusetts Institute of Technology, Cambridge, MA, USA, 2012.

40. Ray, R.; Callagher, K.P.; Sanborn, C. *Standardizing Sustainable Development? Development Banks in the Andean Amazon*; Boston University and University del Pacífico: Boston, MA, USA; Lima, Peru, 2018.

41. MEER. Coca Codo Sinclair, 2018. Available online: http://www.energia.gob.ec/coca-codo-sinclair/ (accessed on 20 February 2016).

42. Sector Electricidad, 2011. Map of the Coca Codo Sinclair. Available online: http://www.sectorelectricidad.com/1675/ecuador-proyecto-hidroelectrico-coca-codo-sinclair-1500mw/ubicacion-coca-codo-sinclair/ (accessed on 1 November 2018).

43. International Rivers. Ecuador's Most Spectacular Waterfall Threatened by Chinese-Funded Hydroelectric Project. Press Release, 2016. Available online: https://www.internationalrivers.org/resources/ecuador-s-most-spectacular-waterfall-threatened-by-chinese-funded-hydroelectric-project (accessed on 20 February 2016).

44. Finer, M.; Jenkins, C.N. Proliferation of hydroelectric dams in the Andean Amazon and implications for Andes-Amazon connectivity. *PLoS ONE* **2012**, *7*, e35126. [CrossRef] [PubMed]

45. Vallejo, M.C.; Espinosa, B.; Venes, F.; López, V.; Anda, S. *Evading Sustainable Development Standards: Case Studies on Hydroelectric Projects in Ecuador*; GEGI Working Paper 19; Boston University: Boston, MA, USA, 2018.

46. Lopez, V.A. Implicaciones del Proyecto C.C.S. para la Amazonia Ecuatoriana. Ecociencia, 2008. Available online: http://ecociencia.org/implicaciones-del-proyecto-coca-codo-sinclair-para-la-amazonia-ecuatoriana/ (accessed on 20 February 2016).

47. El Universo. Cinco de ocho turbinas operan en hidroeléctrica Coca Codo Sinclair. 10.10.2018. Available online: https://www.eluniverso.com/noticias/2018/10/10/nota/6992942/cinco-ocho-turbinas-operan-hidroelectrica (accessed on 10 October 2016).

48. Nathanson, M. Damming or Damning the Amazon: Assessing Ecuador/China Cooperation. Mongabay, 22.11.2017. Available online: https://news.mongabay.com/2017/11/damming-or-damning-the-amazon-assessing-ecuador-china-cooperation/ (accessed on 10 October 2016).

49. Finer, M.; Terry, M. La Cascada San Rafael Amenazada por Proyecto Hidroeléctrico Coca Codo Sinclair. Press reléase, 2010. Available online: http://saveamericasforests.org/SanRafaelFalls/San%20Rafael%20Press%20Release%20-%20Espanol.pdf (accessed on 20 February 2016).

50. Ramos De Mora, J.P. Análisis del proceso gubernamental de toma de decisiones para la implementación del proyecto hidroeléctrico Coca Codo Sinclair y sus repercusiones en la hidrología del área en el periodo 2008-2016. Disertación licenciado. Quito, 2017. Available online: http://repositorio.puce.edu.ec/bitstream/handle/22000/12755/Tesis%20Juan%20P.%20Ramos.pdf?sequence=1 (accessed on 20 February 2016).

51. Notes from a personal visit to CENACE. *The Centro de Control de Transmisión*, Quito, Ecuador, 5 January 2017.

52. Teräväinen, T.; Lehtonen, M.; Martiskainen, M. Climate change, energy security and risk—Debating nuclear new build in Finland, France and the UK. *Energy Policy* **2011**, *39*, 3434–3442. [CrossRef]
53. Oré, M.T. *Agua. Bien común y usos privados. Riego, estado y conflictos en la Achirana de Inca*; Pontifícia Universidad Católica del Perú: Lima, Peru, 2005.
54. Damonte, G.; Gonzales, I.; Lahud, J. La construcción del poder hídrico: Agroexportadores y escasez de agua subterránea en el valle de Ica y las pampas de Villacurí. *Anthropologica* **2016**, *34*, 87–114. [CrossRef]
55. Damonte, G.; Óre, M. (Eds.) *¿Escasez de agua? Retos para la gestión de la cuenca del río Ica*; PUCP: Lima, Peru, 2014.
56. van Teijlingen, K.; Hogenboom, B. Debating alternative development at the mining frontier: Buen vivir and the conflict around El Mirador mine in Ecuador. *J. Dev. Soc.* **2016**, *32*, 1–39. [CrossRef]
57. Foro Recursos Humanos (RRHH), 2018. Available online: https://https://www.fororecursoshumanos.com/ (accessed on 3 July 2018).
58. Alhassan, H.S. Butterflies vs. hydropower: Reflections on large dams in contemporary Africa. *Water Altern.* **2009**, *2*, 148–160.
59. Hensengerth, O. Chinese hydropower companies and environmental norms in countries of the global South: The involvement of Sinohydro in Ghana's Bui Dam. *Environ. Dev. Sustain.* **2013**, *15*, 285–300. [CrossRef]
60. Kliot, N.; Shmueli, D.; Shamir, U. Institutions for management of transboundary water resources: Their nature, characteristics and shortcomings. *Water Policy* **2001**, *3*, 229–255. [CrossRef]
61. Duarte-Abadía, B.; Boelens, R.; du Pre, L. Mobilizing water actors and bodies of knowledge. The multi-scalar movement against the Río Grande Dam in Málaga, Spain. *Water* **2019**, *11*, 410. [CrossRef]
62. Dukpa, R.; Joshi, D.; Boelens, R. Contesting Hydropower Dams in the Eastern Himalaya: The Cultural Politics of Identity, Territory and Self-Governance Institutions in Sikkim, India. *Water* **2019**, *11*, 412. [CrossRef]
63. Fox, C.; Sneddon, C. Political Borders, Epistemological Boundaries, and Contested Knowledges: Constructing Dams and Narratives in the Mekong River Basin. *Water* **2019**, *11*, 413. [CrossRef]
64. De Cauter, L. *The Capsular Civilization: On the City in the Age of Fear*; NAI Publishers: Rotterdam, The Netherlands, 2004.
65. Dogan, E.; Stupar, A. The limits of growth: A case study of three mega-projects in Istanbul. *Cities* **2017**, *60*, 281–288. [CrossRef]

water

MDPI

Article

The Fantasy of the Grand Inga Hydroelectric Project on the River Congo

Jeroen Warner *, Sarunas Jomantas, Eliot Jones, Md. Sazzad Ansari and Lotje de Vries

Sociology of Development and Change, Social Sciences Group, Wageningen University, 6706KN Wageningen, The Netherlands; sha.jomantas@gmail.com (S.J.); eliotandpipe@gmail.com (E.J.); sazzad.ans@gmail.com (M.S.A.); lotje.devries@wur.nl (L.d.V.)
* Correspondence: jeroen.warner@wur.nl or jeroenwarner@gmail.com

Received: 5 May 2018; Accepted: 3 January 2019; Published: 26 February 2019

check for updates

Abstract: The Congo River is the deepest in the world and second-longest in Africa. Harnessing its full hydropower potential has been an ongoing development dream of the Democratic Republic of Congo (DRC) and its more powerful regional allies. If completed, the Grand Inga complex near Kinshasa, the capital of the DRC, will be the largest dam project in the world. Its eight separate dams (Inga 1–8) are envisioned to be "lighting up and powering Africa". Opponents claim, however, that the rewards will be outsourced to corporate mining interests rather than meeting the needs of the local population, and that the project is flawed economically, socially and environmentally. The planned construction of the Inga dams and associated infrastructure has been stuck in limbo since it was mooted in the 1960s; a fantasy rather than a reality. This article attempts to analyse the rivalry underlying the Grand Inga scheme beyond the "pro" and "contra" reports. Embracing Lacanian psychoanalysis and triangulating multiple sources, we seek to unmask Grand Inga as a potent fantasy. Whilst exhibiting its purpose to serve as a screen to protect both proponents of and opponents to the dam from encountering their own self-deception, we conclude the scheme to be at its most powerful whilst the dream remains unfulfilled.

Keywords: Jacques Lacan; psychoanalysis; fantasy; mega-dam; Inga; DR Congo; hydropolitics

1. Introduction

The Grand Inga Project is the largest, most powerful and possibly most controversial prospective hydroelectric dam development project ever imagined. It is the flagship of the Democratic Republic of the Congo's (DRC) development strategy, a proposal of overwhelming prestige which has captured global imagination since the 1960s [1]. Despite receiving considerable support from leading financial institutions, however, it has, so far, not been built. In 2017, the Grand Inga project made the headlines when the World Bank (WB) declared it was going withdraw its support from the project, citing a lack of transparency and failure to observe international good practices as major causes [2]. Despite the failure of the original Inga dams 1 and 2, a World Commission on Dams (WCD) report deeming mega-dams economically and environmentally unviable [3] and left without its major investor, the Inga dream pushes on. In October of 2018, the DRC signed off on a deal with two international private companies to outsource the construction in return for mining rights [4]. Bruno Kapandji, DRC's Minister of Electricity and head of the Inga development and promotion agency, ADEPI, even states that "as Congolese we have no choice but to build Inga 3" [5]. In this article we venture to question why the dream perseveres and has done so for such a long time in the face of considerable setbacks.

In the early 1990s, the influential field of post-development emerged, represented by authors such as Escobar, Ferguson and Crush, seeking to 'decolonise the imagination' of development from its (Western) preconceptions. It was preoccupied with asking why projects of this nature proceed regardless of their persistent failure. It attempted to reconcile the often inherently contradictory realities of international development and political projects [6]. In this field of endeavour an emerging strand of scholars within this post-development domain draw on psychoanalysis to explain why many continue to harbour desires for development. In this line of thought, we postulate that in the case of the dam, it may actually be of more benefit to the DRC as a fantasy than as a possible reality. In doing so, we employ an interpretation of Jacques Lacan's theories on psychoanalysis, an emerging theoretical grounding that looks deeply into the more human, emotional aspects of development [7] and constitutes a linguistic re-interpretation of Sigmund Freud's teachings. Lacan has been adopted in multiple disciplines of social sciences, in response to frameworks that failed to account for the irrational dimension of capitalism and development [8].

Interpreting Lacan's psychoanalytical theory in development studies, as done by scholars such as Kapoor, Sato, de Vries and Fletcher, desire is the ability to 'produce dreams and utopias that are both evoked and betrayed by actual development projects' [7–10]. De Vries' interpretation, we claim, sheds new light on the question of why the Inga 3 dam persists in its state of limbo, and reflects on the relevance and merit of this framework within the context of (capitalist) development. Using Lacanian terminology, we showcase the Grand Inga project as a fantasy, not in the typical definition of the word, but specifically as an imagined entity, formed by an amalgamation of unconscious desires [11]. These desires belong to a variety of actors composed of those "in favour of" the dam in the form of the DRC government and its financial investors, and those "against" including environmental NGOs and academia. The force behind the Grand Inga is proclaimed as the pinnacle of progress, the greatest and best project yet in terms development and sustainability. It departs from traditional contemporary development practice of deliverable small-scale projects [12] and builds upon an enormous drive to achieve major goals. Our argument is, however, that these ideas are so grandiose that they actually work to hinder the project by their own ambition.

2. Lacan and Development

The Parisian psychiatrist Jacques Lacan transformed psychoanalysis from an approach to curing neuroses to an analytical instrument to analyse developments in society, a (still controversial) idea that has since widely been taken up in the social sciences (In addition to those already mentioned, these include geography [13], geopolitics [14], ecology [15,16] critique of capitalism [17] and spatial planning [18]). When used in the analysis of post-development, Lacan's logic of the individual is applied to whole populations and even institutions and international organisations.

Here, we will zoom in on the Lacanian *Triad*, which Lacan, in his final years of writing, came to see as his most applicable analytical tool [19]. He posits that there are three equal orders of consciousness within this *Triad*; the Real, the Symbolic and the Imaginary. The latter two realms are where we, as humans, generally find our 'reality'; the Symbolic, most evidently expressed in language, is the means by which we apply meaning to the world, the basis of knowledge defining how we come to know ourselves and one another; the Imaginary creates the ego, that which offers the conception of a self as opposed to the other, an individual identity acting apart from the broader collective. This allows people to act in accordance with their own personalised rationale and feelings. The Real is all that resists substantiation in the other two orders forever unattainable and inaccessible and beyond grasp, always escaping signification in the Symbolic [9,11,19]. This *Triad* will be our entry point for discussing the dam.

As humans, we are born into the Real, with no understanding of the world beyond our instinctual needs, which are met by our parents. Lacan recognises two essential stages in a child's development for forming adult consciousness. The first, the mirror phase, occurs when a child is first met with their own reflection and begins to recognise itself as a separate entity from their mother. The second stage is when that child begins to understand and use language as a tool of representation and signification. It is the language that the parent uses to describe the child that defines how it understands itself and comes to structure their entire existence. These phases are the child's introduction to the

Symbolic and Imaginary orders. They are its most significant influences on becoming an individual but, simultaneously, they create a void, an emptiness within the child's psyche. The separation from the maternal body and the subjection to the 'abstract structure' of words forms in the adult individual, or the 'Subject', an insatiable yearning to find satisfaction in that which cannot be achieved [9,11].

The Subject represents a tension between the lack within and the defined societal order outside. Structured and constituted by society and its rules, the Subject then reproduces such an order by entering into it and being 'filled' with its meaning. This can be any form of ideology, from communism to sustainability [20]. Even if agency for change to the Subject is granted, the Symbolic order in itself never ceases to exist, evolving only to represent new values, practices and rules of engagement. Moving past the libidinal fixations of Lacan's original writings, the same theory can be then employed to critique economic development [7–9,11].

The work of French philosopher Michel Foucault has offered the grounding theory in development critique for the past two decades [8]. Despite great leaps in the understanding of power within the field, this does not extend to the Real order in developmental discourse. Scholars analysing those offering and receiving development can appeal to the Imaginary and Symbolically defined needs of their subjects, but this risks only reifying the existing rhetoric and repeating actions that have been shown to fail. Recognising the Real order is essential to determine an antagonistic dimension that reproduces and challenges the other two orders.

The void that is created by the tension between the Real and Symbolic, produces in the Subject two fundamental unconscious forces: drive and desire. Post-development scholars, including Ilan Kapoor, have adopted these concepts to explain the apparent satisfaction in failure of development ([7] p. 67). Kapoor differentiates the two processes as follows: 'desire targets a lost object to cover up our lack—an object that is never found' as opposed to drive, the process of finding enjoyment in that lack ([7] p. 69). Within capitalism, the apparent drive is to consume, finding solace from the void in our psyche. Kapoor argues that development practice pursues the very same logic of consumption, preceding capitalism's tendency to reproduce inequalities. Development functions as a trade-off, exchanging the potential for progress for actual access to resources, e.g., mining rights and hydropower. Therefore, we see that while the movement of capital may on the surface appear to be leading toward a more 'developed' world (which is the desire), the drive unconsciously promotes the same structures of hegemonic capital accumulation. This has potentially disastrous future consequences as, rather than pursuing the goals of equality that development purports to promote, it is actually empowering the forces that produced this disparity of wealth in the first place [7].

What Kapoor proposes is the return of human passions in the analysis of development. Rather than looking at the rationale behind what is evidently an irrational phenomenon, pursuing a goal that only exacerbates the problem first addressed, he sees development as a result of an unconscious libidinal force, something innately human and beyond the scope of traditional analytical tools. We argue that the Inga project, the most illustrious vision of international development and its proponents in and around the DRC, is an amalgamation of coinhabiting desires. It is an imagined object promised to alleviate all of the Republic's problems, both Symbolic and Real. Emanating from the work of psychoanalyst Darian Leader, objects of desire have both Real and Symbolic elements [19]. As Lacan writes 'it is the Real which creates . . . desire by reproducing in it the relationship between the Subject and the lost object' ([19] p. 724). The 'lost object' in this sense, forces the Subject to cling onto the Imaginary order, providing an identity. Yet the resurfacing of the Real, reminds the Subject of the void within; this unbearable realisation ultimately drives it to seek what Lacan calls *jouissance*.

Roughly translated as 'enjoyment', *jouissance* is understood as the insatiable pursuit of a certain goal (desire) despite experiencing both great pleasure (presence) and pain (absence) [9]. Drawing on *jouissance*, Robert Fletcher maintains that the scourge of overpopulation remains a popular motivation for pursuing development among its propagators, including the WB. The desire in this case, for the world to become 'developed', is pursued (presence/pleasure) even though this premise has been widely shown as false (absence/pain) and, therefore, contradictory of its goal [9,11]. Demonstrating the irrationality behind development discourse and its presentation in the Symbolic, he shows how the desire for development is deeply rooted in constructing ones *jouissance* in relation to an 'Other', in this instance, the undeveloped. Developed nations perceive themselves as domesticated and civil in relation to the animalistic and primitive sexual tendencies of so called 'Third-World' nations. They feel threatened by the Others' apparent lack of control in breeding and, therefore, invest much time and energy into solving this issue [9].

This leads to the final concept relevant to this study: fantasy. As Lacan writes 'desire is propped up by a fantasy, at least one foot . . . of which is in the Other, and precisely the one that counts, even and above all if it happens to limp' ([11] p. 658). Here Lacan refers to the desires imposed upon the Subject within the Symbolic order; he explains that to engage socially in the world, the subject must conform to its demands. The 'fantasy' Lacan describes is produced by this compulsion to fulfil or to counter the desires that are drawn from the Other. Whereas, when he writes 'if it happens to limp' he confirms that this desire only reminds the Subject of the lack, reproducing the void even in an attempt to hide it. Translating this to capitalist development rhetoric, Kapoor states that there is a need to be 'the best, biggest, tallest, richest, most original . . . [covering] everything from coffee and art to national monuments, airports and dams' ([7] p. 71). To be a significant voice in the world of international NGOs, governments and development institutions, each party must in some way play toward ideals embedded within the Symbolic order, both to gain recognition and to feed their drive.

Rather than focusing on those institutions that push top-down development, as do Fletcher and Kapoor, Pieter de Vries addresses how 'to be developed' also has become a fantasy [10]. He shows how the desire of external development agencies, representing here the Lacanian Other, instilled onto a Peruvian indigenous group in the Andes a desire to become 'developed'. By adopting and embodying western bureaucratic elements, the tribe, while bizarre to them, then assumes to be on their right path of 'progress'. The concept of development, argues de Vries [10], produces nothing but a 'desiring machine'—imposing via the Symbolic order undeliverable aspirations upon already impoverished peoples.

This article identifies the Grand Inga project as such a fantasy or 'desiring machine' [10]. Drawing on a broad range of materials, extensive review and triangulation of over 90 journal articles and 40 online sources, our analysis combines three specific readings of the Grand Inga case, each exploring how the dam's persistence in the Symbolic maintains the deadlock. In the first, we present specific groups and individuals and the discursive actions of each, to consider what is produced and what kinds of rhetoric is drawn upon and envisioned. The next step shows, what possible *jouissance* is pursued, and the role played by the language, presented in the first reading. Finally, in the third step, we investigate the actual repercussions of said fantasy. We investigate what tensions are maintained, and how these both undermine and reproduce the dam as a dream. Before we delve into this in-depth analysis, however, we offer a short review of the history of the Inga dams.

3. Grand Inga in Context

Water resources in Africa have bedazzled prospective developers since colonial times, promising extremely inexpensive energy for mining and the export of raw materials. With its 42,000 m^3/s water flow at the Inga falls (rapids forming part of the Livingstone Falls, see Figure 1 below), the Congo fuelled fantasies of a megaproject as early as the 1930s when technologies of this nature first became available on the African continent [21].

The DRC gained its independence in 1960, five years after the Belgian colonial powers first announced an Inga dam scheme. The then President Mobutu took the project forward and the current President, Kabila, evidently continues to champion it [22]. The country is known for its rich mineral wealth, yet the Fund for Peace's annual Fragile States Index ranked DRC as the 7th most fragile state in 2017 [23]. Its political instability, harsh terrain, poor infrastructure, largely dysfunctional institutions and ongoing violent conflicts in the central and eastern part of the country, have repeatedly caused the development process to falter [24]. The DRC faces a serious political crisis; some 42 million Congolese are undernourished [25] and despite the country's plentiful water resources, 75% of the Congolese population has no access to clean drinking water [26]. The shrinking administration has left public servants and local authorities with hardly any means to operate and salaries are often left unpaid. The relation between public authority and its citizens remains characterized by deep mutual mistrust and democratic deficits [22].

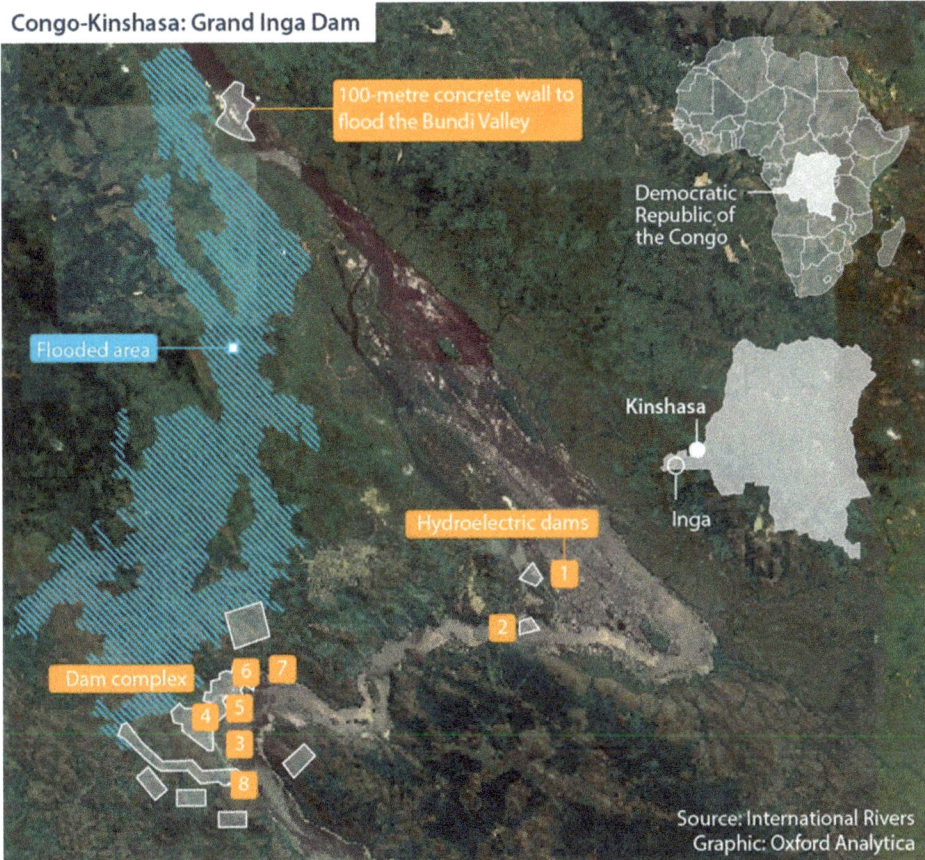

Figure 1. The Inga Valley, the two existing and the six planned dams forming part of the Grand Inga Complex (https://dailybrief.oxan.com/Analysis/GI239798/Congo-Kinshasa-Grand-Inga-dam).

The existing Inga dams 1 and 2 have done very little to alleviate any of this disparity. Constructed in 1972, the first of the two was relatively successful [27]. Completed in only four years, it initially provided a reliable source of energy to Kinshasa and served as a strong success story for promoting Mobutu Sese Soko's emerging dictatorship. The second dam, completed in 1982, however, has not proven so fruitful. It was intended to supply mining endeavours in the east of the country but was built five years before any power grid had been established and was thus useless until this project was completed. During this hiatus the dam fell into disrepair and, due to this poor planning, only 20% of Inga 2's potential power has ever been utilised. Furthermore, the construction costs of both dams went way over budget, driving the DRC into substantial debt. Albeit, the plan to build a third dam was first conceived in the late 1980s but stalled due to violent conflict in the region [1,27].

After a slew of controversies over large infrastructural projects in the 1990s (Narmada, Arun, Pergau, Lesotho Highlands and many more) a tripartite World Commission on Dams was instated. In 2000, it concluded that mega-dams, i.e., those over 15 m high, are unsustainable and cost-ineffective. Yet dams are still popular as purported engines of economic growth and development, albeit under new guises such as climate buffers [28]. According to the Commission [3], 40–80 million people worldwide have been forced off their settlements, agricultural lands, forests and other resources due to dam-related flooding. Over 45,000 large dams have been constructed worldwide and reservoirs inundate approximately 500,000 km of land surface [3,29]. The adverse impacts of large dams also fall disproportionately on subsistence farmers, indigenous peoples and ethnic minorities, who often rely on common property regimes of resource utilization [30]. Despite this, the pursuit of mega-dams persists.

While the World Bank did not reject the WCD's recommendations, it largely continued along the earlier path [31]. In 2004 the Bank started a two-year project to refurbish the dilapidated Ingas 1 and 2, with a significant investment of $200 million USD [1]. With this new interest in the Congo River, sparked by the UN's Millennium Development Goals, plans concerning Inga 3 began to re-emerge. Some five years later, the WB and WESTCOR (Western Power Corridor) feasibility study concluded that damming the Congo River via a single dam, Inga 3, would inundate and flood the capitals of DRC (Kinshasa) and the Republic of the Congo (Brazzaville). Considering this, a further new project was proposed; the Grand Inga dam Complex, consisting of six separate dams, rather than one [32]. Eleven teams expressed an interest in the construction of the first dam, Inga 3 (GCR1). British aluminum smelter BHP Billiton, one of the frontrunners for the construction of the new Inga, pulled out a year later, without any clear reasons cited [32,33]. In 2012, negotiations were then escalated to presidential level between the DRC and South Africa (SA) [33]; neighbouring countries grew eager to reap a portion of the dam's potential energy.

The resulting Grand Inga Complex is envisioned to be twice the size of the Three Gorges dam in China, currently the biggest hydropower project in the world. It will surmount the great Congo River, the deepest in the world and second-longest in Africa. The African Development Bank (ADB) claims its eight separate dams and associated power stations and transmission network could lead to 'Lighting up and powering Africa' [34].

In 2014, however, the US Congress passed a law forbidding "any loan, grant, strategy or policy ... to support the construction of any large hydroelectric dams" [35] withdrawing support for the WB Technical Assistance (TA) Package of Inga 3. It cited the disappointingly slow rehabilitation of Ingas 1 and 2 and the unsuccessful reformation of the DRC state electrical utility as main causes. Lack of evidence of direct benefits to the local Congolese people, of social safeguards concerning the construction, or of environmental and social impact assessments are also mentioned.

Despite these setbacks, in March of 2014 the WB approved a further $73 million USD to finance environmental, social and technical studies to proceed with Inga 3. These never took off, however, due to irregularities in the procurement process [36]. The same year, South Korea's Posco and Daewoo, in partnership with Canada's SNC-Lavalin withdrew, leaving Chinese company Sinohydro and Grupos ACS from Spain as the only serious bidders to construct the dams remaining from the initial 11 [37].

By 2016, Inga 1 and 2 rehabilitation programmes were still not finished. 10 years overdue, the projected costs had risen to a staggering $1 billion USD, five times the initially predicted figure [38]. Then, to the dismay of the DRC government and the surprise of others, the WB suspended its support for the Inga 3 dam preparations [36]. The Congolese Project Director then, to please his last remaining financiers, proposed to proceed with the dam's construction without environmental, social or technical studies [36]. In June of 2017, the DRC government asked two last bidders to join and submit a single bid to start developing the now long-delayed Inga 3 [39,40]. This bid apparently has been accepted and the Congolese Energy Minister Kapandji bullishly announced construction is to commence this year and to be completed in 2025 [41]. Analysts, however, claim 2030s–2040s is the best possible envisioned date for the remaining five stages of Grand Inga [33]. A grid capable of transferring energy to the Congolese population is yet to be built.

4. The Inga Fantasy

The following analysis is based on Lacan's concepts of drive, desire and his definition of a fantasy [11]. Drive, the process of finding *jouissance* in the pursuit of an object rather than in the object itself, divides the Subject from their initial intention of obtaining said object. The desire is only sustained by imagining a false relationship between the drive and the object, disavowing the satisfaction found in failure. Lacan writes: 'fantasy is the means by which the Subject maintains himself at the level of his vanishing desire, vanishing inasmuch as the very satisfaction of demand deprives him of his object' ([11] p. 532). Scholars have recognized this phenomenon within development. Japhy Wilson states that large-scale projects such as Grand Inga represent 'development' as a machine, perpetually aiming to improve the lives of those living in the Global South, however, in order to maintain this machine, more and more people need to be considered 'undeveloped' and, therefore, must take its own failure to accomplish its objectives into account in advance [42]. We place Grand Inga within this formulation of failure, showing how Inga as a fantasy feeds the desires of the DRC, the WB and its private investors, on the one hand, and International Rivers (IR) and their scientific backing on the other.

4.1. The Triad

Central to the current position of the Grand Inga Project is its Symbolic efficiency [19]. How, why and by whom Inga 3 is discussed allows us a glimpse not only into the various groups and their stakes in the project, but also the intertwined character of different realms pertaining to the Lacanian *Triad*; the Real, the Symbolic and the Imaginary. The scheme's name alone pulls no punches: 'grand' evokes its impressive size and appearance. It is characterised as stately, majestic, dignified, ambitious, magnificent and noble. The action of describing the dam in such a way feeds its Symbolic potency, furthering its appreciation of the Other, how far it is removed from the Real. As its only signification in the Real is its absence, the dam project can remain unchallenged in this regard. Compared with other, smaller dams, the exaggerated size, the river it aims to tame and the required investment, ranging between $12 billion and $14 billion USD, Inga puts all other projects to shame [37,38]. In brief, the scheme is in a league of its own [43]. These claims bear little relation to reality beyond the actual size and scope of the Congo River. A brief look at previous attempts to achieve major projects in the region provides significant reason not to believe in the efficacy of such claims. Within the worlds of international development, finance and political relations, however, these may carry a lot of weight. The prestige, excitement and promotion brought to the Congolese government by these plans, a renowned corrupted dictatorship, is extraordinary.

Claims that the hydropower of Grand Inga could supply Egypt via the Sahara, stretching to Europe [44] are very potent. Presented in such Symbolic light, it resurrects hope and support for Grand Inga after two previous failures. This works perfectly for the benefit of both the WB and the ADB, who in lending money to the project also prop up Inga as a driver for regional cooperation [45]—another symbolically positive connotation. The dam, however, is merely an absent proposition, promising to turn the DRC from a third-world economy into a global exporter of 'clean' energy, with little to no evidence.

Through Inga, the DRC likewise projects itself as the new hub for Pan-Africanism, an age-old rallying cry focused on uniting the African continent under one banner [46]. Emerging as a response to the oppression and exploitation African people suffered under colonialism and the slave trade, this movement is a highly potent ideology among continental statesmen. With the DRC geographically at its center, it is already in a great position to take on this role. Cooperation by delivering renewable energy to all its neighbours could also promote the DRC as an emerging authority amongst other regional nation states. As such, Inga has attracted much attention from SA, the current regional leader. The DRC has proclaimed itself to the Western world at large as a progressive and environmentally conscious new African power [47]. This has been validated by the huge support coming from the Congo's regional competitors as they have prepared to invest heavily in the prospective project [48].

It seems paradoxical, however, that a country deemed the world's 7th most fragile state [23], which has been in the same perilous political situation for the last 70 years, could become the new face of African solidarity. The Symbolic efficiency attributed to the dam within the discursive sphere of International Relations and development (big, clean, sustainable) has obfuscated its presence in the Real to such an extent that the DRC's poor reputation can be condoned. Using the *Triad*, we can see how Imaginary identities and social collectives (DRC, SA, WB) cling to their Symbolic rhetoric to remain relevant and important. They project themselves in congruence with the desire of the Other, maintaining a cognitive shield between themselves and their grounding in the Real.

A similar knowledge dispute is reflected by those opposing the dam, where groups use the same terminology to represent contradictory points of view. Many NGOs standing against its construction claim to represent the interest of the environment. They posit that the natural and human cost of building the dam far outweighs its prospective benefits, flooding vast amounts of land where people currently reside [27]. They are supported by much of the scientific community and the WCD report which, as noted, agreed that such projects tend to be neither sustainable nor cost-effective. This same rhetoric, however, is used to promote the dam. The DRC maintains that the dam will provide 'clean' energy and lead to a more environmentally sustainable future for Africa. Bruno Kapandji has insisted that environmentalists are "fabricating" stories in regards to the damaging impacts of the dam [1]. The WB and other parties are working in line with UN sustainable development goals, putting the DRC at the forefront of climate adaptation.

Further contradictions in the Symbolic battle of Inga 3 undermine many of the DRCs promises. While claiming humanitarian goals, bringing energy to the poor and deprived, most of the energy that Inga 1 and 2 produce, is absorbed by mining companies in the eastern Katanga province [1]. In addition, SA, as a key contributor, is assured to be the first and largest recipient of Inga's energy [49].

One would expect that the WB's financial withdrawal might have endangered the whole Inga project. In response, however, the DRC has promised to end their commitment to proper environmental testing, seeking that the private financial interest in pursuit of constructing the dams remains [38]. The actions of those for and against the dam, however, exhibits the power behind the development discourse as the project can survive such a blow. The WB has invested at the very least $273 million USD into Inga 1, 2 and 3 over the past ten years and has seen almost zero return [2]. The DRC has managed to keep the dream alive despite losing the support of the largest and most respected development financier on the planet due to its own poor planning. These actions can be explained further when considering how these desires function, but first we shall delve into the enjoyment gained, linking the Symbolic value of each group through *jouissance*.

4.2. Jouissance

Jouissance is born of the Symbolic order; it structures desires and future goals, feeding the drive to continue. The semblance of *jouissance* expected, however, is not the substance of its actualisation (*jouissance* obtained) [11,19]. It would compel the Subject to acknowledge the ugliness and insufficient nature of the object of their desire. Furthermore, such transcendence would make that Subject question the nature of their belief system and self-esteem. The Subject will repeatedly act, consciously or unconsciously, to abide by and conform to the Symbolic order, to reap the rewards of '*jouissance* expected'. The deadlock between contested knowledges around Inga 3 manifests itself in buttresses and threats, which each party exerts as hindrances in the quest of the other to obtain *jouissance*.

Among the most vocal opponents to the Grand Inga project is IR, an activist non-profit non-governmental, environmental and human rights organization. IR questions the development claims attached to dam-building projects and aims to protect rivers while defending rights of dependent communities. IR continuously reaffirms its opposition to the Grand Inga project [50]. As a values-based organisation, its figures are often disputed but IR's critiques resonate in policy circles. Other vocal opponents include Friends of the Earth and African human rights organisations.

While their discourses reject dam construction, the case for their *jouissance*, their financial support, media attention and international success requires that other actors and discourses—the financial and governmental stakeholders it sets out to oppose—to pursue the Inga fantasy [31]. Lacan writes: 'my formula for fantasy allows us to bring out the fact that the subject here makes himself the instrument of the Other's *jouissance*' ([11] p. 697). We may see, by referring to the previously laid out debate over the sustainability of the dam, that each party, by contesting their opponent's Symbolic value, defines their struggle. Each creates the demands the other side must pursue to find *jouissance*. Its existence, therefore, is premised on this rivalry and is equally culpable for the maintenance of the Inga fantasy. What unites every party within this dispute, aside from the collective fantasy, is the obedience to the symbolic order and the fruits provided thereby.

Here we see different parties, the government of the DRC, the WB, the WCD and IR, claim the same discourse to represent their cause and garner funding and support. While both proponents and opponents present evidence and truth claims, the underlying Symbolic function means that neither can be wholly substantiated in the Real, and the argument is mooted. Both cases can cast doubt upon the other and, therefore, both can manipulate language to best present themselves. In opposition to the previous case, figures regarding the DRCs poor governance record can be ignored because it is assumed that Inga is a solution. Acknowledging this inefficacy, however, would undermine the cause for development and is shunned. This rivalry retains precedence as it serves to increase each party's Symbolic potency, which we will return to in the following reading.

4.3. Desire and the Fruits of Fantasy

The object of *jouissance*, the desire within the DRC's leadership, is to build the dam and reap its Symbolic rewards. The Grand Inga is perceived by the Congolese government and purveyed to its people as a saviour to end all problems, from guerrilla fighting to economic and political upheaval. It will project the country into an important role in political relations and classify it as a 'developed' nation [5]. The exact notion that this is the object of their desire, however, contradicts its potential.

Desire functions by finding a lack in the Other. Therefore, while desire belongs to the Subject, it is vitally socially constructed within the Other. It can only be fulfilled by utterly appeasing the Other, which, according to Lacan, is not possible for extended periods of time, as he or she also has their own insatiable desires [11,19].

In this manifestation, then, the WB recognises a lack of wealth and poverty in the DRC. It is the Bank's desire that the country become 'developed' according to their conceptualisation and agenda. It is fundamental, however, that to regulate their *jouissance*, and to continue to maintain their relevance and Symbolic power, they must not succeed in this goal. Therefore, they unconsciously disavow their inadequacy by continuing to hold the fantasy of Grand Inga in such high regard. In doing so, the broader development drive may remain intact. After the WB has withdrawn support, the DRC can continue to push for the project to satisfy their own destitution and the WB can drop the project, having exhausted huge resources without repercussion since this may be blamed on the Congolese. In addition, the DRC can realign and push forward their Symbolic cause as an oppressed nation seeking development despite the rejection of worldly institutions [5,38]. In this rendering, the WB is the doting financier pushing for development but not succeeding because of corrupt politicians in donor countries [2]. While these stories are somewhat contradictory, their desire does not only survive but is perhaps even strengthened by failure.

It is noteworthy here to remind the reader that this is not the first time that the WB has undergone such a reversal. In 1993, after years of support for the Narmada dam project in India, the WB pulled out due to mounting environmental controversy [51]. This piece of evidence sheds light upon the decades of contestation the WB has undergone regarding mega-dams and its repeated support despite consistent, complimentary problems. An almost identical loan to that offered to the DRC was withdrawn over 20 years previously for the same reasons as Inga, yet the fantasy obfuscates this Real connection. The desire to harness the promises of the Congolese dam shield its unfeasibility right up until the moment that the WB must come to terms with the hopelessness of its goal, finally letting go in order to maintain its *jouissance*.

The DRC has played toward this 'lack' in the international push for development. If a fantasy is that which moves further as one tries to capture it, the DRC in its pursuit of the Inga fantasy has fed the desires of first the WB and SA, and now China, but severely contradicted itself in the process [1,38]. SA has become a leader of sorts in the region. Its involvement across the continent is unparalleled to any other African nation, from railway to airport and gas pipeline projects, up to involvement in transboundary river basin governance [48,52]. Driven to challenge SA's hegemony in the region, the DRC is seeking international prestige of its own. Inga could potentially grant this shift in power, promising to extend energy supply all the way to Europe [53]. The DRC, however, has also made promises to SA in the form of energy prospects and construction contracts which contradict these assurances. As a leading financier, the DRC must comply with their demands in order to achieve its goals. SA on the other hand continues to exercise its power over DRC, disavowing side deals, expecting up to half of the energy from Inga 3 alone. By assisting the DRC, it is showing its gracious generosity to its developing neighbour. In the trade balance across the Southern African regional development community, SADC, the balance of payments favours SA ten-to-one [33].

The void left by the Bank was soon filled. When the WB pulled out, the Chinese, the next big player on the continent, particularly in the case of dam projects, stepped in [53]. China started to promote overseas investments, exports and contracts to engineering projects outside of China from 2001 onward [54]. Pursuing their resources-for-infrastructure "development" policies, which are often based on build-own-produce norms, free of taxation, China not only builds dozens of overseas dams, but also uses the energy to mine tax-free local resources as part of the package. China's investments in Africa between 2003 and 2015 were $27 billion USD [55], whereas loans between 2000 up to 2017 totaled $143 billion USD. In the DRC alone, its financial injections and commitments have added up to $9 billion USD [56]. In return, the Congo will cede the majority rights in a joint venture to develop copper and cobalt concessions in the region of Katanga, to which energy is crucial.

Here we see an old form of imperial domination and colonialism across Africa giving way to a new form of fiscal hegemony, as described by Kapoor [7]. African states continuously look out for investment and aid, they encourage corporate invasion, to materialize and capitalize on resources to their benefit. These rich resources have then been noted to be 'cursing' DRC, laying the foundations for conflicts and continuing violence across its eastern regions [57]. Therefore, by appealing to development, the DRC has invited violent forces into the country, impeding the development process. The desire for international prestige, a reputation for sustainability, peace and prosperity in the DRC, to be found in the construction of Grand Inga, in actuality seem to be escalating problems. Attempts to gain independence might lead to greater economic subordination, particularly in the case of SA. By turning to the Chinese, the DRC must promise much of its resources away, potentially leading to further conflict.

In addition, by dissuading the WB from supporting the project, dam opponents turned it into an even greater environmental threat. Until recently, Sinohydro was on the World Bank's blacklist because of poor compliance with environmental norms [58]. In 2016, the company claimed it could achieve the landmark project in only four years, if they were "free to do whatever they want to" [38], a demand President Kabila has appeased. In their agreement, the DRC has pledged more of its mining reserves. As a consequence, the Chinese will use the dam's clean energy to engage in environmentally degrading practices.

5. Conclusions

In each of the cases outlined above, we have shown how desire for an object is sustained, first, by *jouissance* and, second, that object being projected as a fantasy. *Jouissance* is found in the pursuit of unattainable goals, reaping Symbolic rewards without actually having to achieve them. Fantasy then asserts that those desires and goals have a potential manifestation and can realistically be achieved.

This belief, however, is imaginary and mistaken; as Lacan writes, a 'misrecognition' between the drive and the object ([11] p. 724).

The potency of the Grand Inga fantasy is then revealed when considering its ability to support existing regimes of Symbolic power. In other words, the fantasy of Inga 3 instills agency and legitimacy to various groups working both for and against it. Its powerful image projects the possibility of solving the persisting problems for the DRC and the African continent. On the other hand, engaging with this Congolese dream, IR contests the Grand Inga project as a means to promote environmental superiority over economic development. We have shown how, by embracing symbolic interpretation of the dam, IR reproduces the fantasy and secures its position to contest it. The tension between the two sides could be then mediated by private development and construction companies. It seems, however, that they hinge predominantly upon ulterior capitalist interest, rather than the ethical goals they refute.

The Imaginary that precedes and exists within the social construction of the yet immaterial Grand Inga Complex, that permeates between the government, its corporate backers and international onlookers, continues to profess Grand Inga as a development saviour. As such, it ignores and surpasses the Real social, environmental, and economic drawbacks of such a dream. The Inga fantasy, an amalgamation of a variety of contesting desires and knowledges, disavows that which opposes it, suspending the dam in a limbo-like state, constantly on the cusp of 'vanishing desires' ([11] p. 724).

Acting as what Lacan calls a 'narcissistic mirror', the Inga fantasy reflects only positive elements of the Subject [19,42]. The WB can justify its support for corporations that continue to exploit and conduct violence in the Congo as they have attempted to rectify these evils by funding the dam, if ultimately failing. IR won a victory over the WB but now have an even greater opponent in Sinohydro. The prospective Inga 3 and its surrounding projects, therefore, seem to be more useful to its proponents and opponents on the horizon of possibility than as a reality.

In the case of DRC, their desire for the dam is sustained by a drive for development, yet their efforts to pursue it only seem to contradict their intentions, inviting violent corporations into their country. For IR, pushing against the dam has only discouraged the more legitimate institutions, such as the WB, allowing less environmentally conscious organisations to appear in their wake. We argued, therefore, that the Grand Inga Complex is more powerful as a fantasy than it would be if it became a reality. The rhetoric that pervades the project has far greater potential to satiate each party's enjoyment as a mere idea and has done so for various groups for over 50 years. Given the evidence above, each time the dam inches closer to construction, the desire of each group is impeded, as evidenced by the recent manoeuvres from Sinohydro. When the dam seems further from reach, as in the WB's withdrawal, however, each group's Symbolic power seems inflated.

We do not predict that the dam will not be built, although there is plenty of evidence to suggest so. Rather, we have shown that, even if the dam would be built, the goals and achievements it claims to encourage will not be realised. Furthermore, the current efforts to build the dam, and their undergirding motivations are contradictory, leading to intensification of the problems they first sought to alleviate.

Borrowing a novel approach from Lacanian psychoanalysis, this study has examined and deconstructed the ongoing Grand Inga dam polemic beyond the classical for-and-against assertions. By recognising the irrational, human construction of the controversy that surrounds the project, maintaining its Symbolic status, we may see how each party works in accordance with the next, how, by standing in opposition, each subject regulates the position of his enemy and how, in every victory, forgoing the Other's desire, the subject only damages his own struggle for enjoyment.

The potential application of Lacan's work could be highly productive in conducting research into the representation of dams and other environmentally related development projects. By showing in the case of Grand Inga, that knowledge contestation is of greater significance, we hope to assist those approaching such deadlocks to gain a critical outlook of the underlying force field.

Author Contributions: The conceptualization and methodology were constructed by authors E.J. and S.J. They were responsible for devising the frame of analysis, employing Lacanian Psychoanalysis. M.S.A. then validated this framework based on his experience with water management. The combined efforts of these three authors culminated in the original draft. Having seen the novelty in the argument, J.W. and L.d.V. proceeded to review and radically edit the draft for publication in close collaboration with E.J. and S.J.

Funding: This research received no external funding.

Acknowledgments: We thank Robert Fletcher and Esha Shah for their feedback to an earlier version of the present article. Any error of fact or interpretation remains the authors' responsibility.

Conflicts of Interest: The authors declare no conflict of interest.

References

1. Klemm, J. World Bank Pitches Mining to Drive Energy Investment in Africa. *Bretton Woods Project*. 31 March 2015. Available online: http://www.brettonwoodsproject.org/2015/03/world-bank-pitches-mining-to-drive-energy-investment-in-africa/ (accessed on 14 June 2017).
2. Fabricius, P. The World Bank has Suspended Funding for the DRC's Inga 3 Hydropower Scheme. Where Does this Leave the Project? Institute for Security Studies, 2016. Available online: https://issafrica.org/iss-today/inga-dream-again-deferred (accessed on 14 June 2017).
3. World Commission on Dams. *Dams and Development: A New Framework for Decision-Making*; Earthscan Publications Ltd.: London, UK; Sterling, VA, USA, 2000.
4. Gandhi, D. Africa in the News: DRC Inga Dam Agreement and Somaliland Port Expansion. 2018. Available online: https://www.brookings.edu/blog/africa-in-focus/2018/10/20/africa-in-the-news-drc-inga-dam-agreement-and-somaliland-port-expansion/ (accessed on 25 November 2018).
5. Vidal, J. Construction of World's Largest Dams in DR Congo Could Begin within Months. *The Guardian*. 28 May 2016. Available online: https://www.theguardian.com/environment/2016/may/28/construction-of-worlds-largest-dam-in-dr-congo-could-begin-within-months (accessed on 14 June 2017).
6. Boelens, R.; Shah, E.; Bruins, B. Contested Knowledges: Large Dams and Mega-Hydraulic Development. *Water* **2019**, *11*, 416, doi:10.3390/w11030416.
7. Kapoor, I. What "DRIVES" capitalist development? *Hum. Geogr.* **2015**, *8*, 66–78.
8. Sato, C. Subjectivity, enjoyment, and development: Preliminary thoughts on a new approach to postdevelopment. *Rethink. Marx.* **2006**, *18*, 273–288. [CrossRef]
9. Fletcher, R.; Breitling, J.; Puleo, V. Barbarian Hordes: The overpopulation scapegoat in international development discourse. *Third World Q.* **2014**, *35*, 79–99. [CrossRef]
10. De Vries, P. Don't compromise your desire for development! A Lacanian/Deleuzian rethinking of the anti-politics machine. *Third World Q.* **2007**, *28*, 25–43. [CrossRef]
11. Lacan, J.; Fink, B. *Écrits*; W.W. Norton: New York, NY, USA, 2007.
12. PNUD. Republique Democratique du Congo: Rapport national, Energie Durable pour tous a L'horizon 2030. 2013. Available online: http://www.cd.undp.org/content/dam/dem_rep_congo/docs/eenv/UNDP-CD-RAPPORT-ENERGIE-DURBALE-POUR-TOUS-HORIZON-2030.pdf (accessed on 14 June 2017).
13. Pile, S. *The Body and the City: Psychoanalysis, Space and Subjectivity*; Routledge: London, UK, 2013.
14. Bjelic, D.I. *Normalizing the Balkans: Geopolitics of Psychoanalysis and Psychiatry*; Routledge: London, UK, 2016.
15. Dodds, J. *Psychoanalysis and Ecology at the Edge of Chaos: Complexity Theory, Deleuze, Guattari and Psychoanalysis for a Climate in Crisis*; Routledge: London, UK, 2012.
16. Robbins, P.; Moore, S.A. Ecological anxiety disorder: Diagnosing the politics of the Anthropocene. *Cult. Geogr.* **2013**, *20*, 3–19. [CrossRef]
17. Tomšic, S. *The Capitalist Unconscious: Marx and Lacan*; Verso: London, UK, 2015.
18. Gunder, M.; Hillier, J. *Planning in Ten Words or Less: A Lacanian Entanglement with Spatial Planning*; Routledge: London, UK, 2016.
19. Leader, D.; Groves, J. *Introducing Lacan*, 2nd ed.; Icon Books Ltd.: London, UK, 2014.
20. Žižek, S. *The Sublime Object of Ideology*; Verso: London, UK, 1989.
21. Jullien, M. Can DR Congo's Inga Dam Project Power Africa? 2013. Available online: https://www.bbc.com/news/world-africa-24856000 (accessed on 9 October 2018).

22. Rosen, A. The Origins of War in the DRC. 2013. Available online: https://www.theatlantic.com/international/archive/2013/06/the-origins-of-war-in-the-drc/277131/ (accessed on 9 October 2018).

23. *The Fragile States Index 2017*; Messner, J.J. (Ed.) The Fund for Peace: Washington, DC, USA, 2017; Available online: http://fundforpeace.org/fsi/wp-content/uploads/2017/05/951171705-Fragile-States-Index-Annual-Report-2017.pdf (accessed on 11 October 2018).

24. Nguh, A. *Corruption and Infrastructure Megaprojects in the DR Congo*; International Rivers: Oakland, CA, USA, 2013.

25. Sanyanga, R. *Right Priorities for Africa's Power Sector: An Evaluation of Dams under the Programme for Infrastructure Development in Africa (PIDA)*; International Rivers: Washington, DC, USA, 2015; Available online: https://www.internationalrivers.org/sites/default/files/attached-files/pida_report_21_oct_for_web.pdf (accessed on 4 April 2018).

26. Maupin, A. Energy and Regional Integration: The Grand Inga Project in the DR Congo. In *A New Scramble for Africa?: The Rush for Energy Resources in Sub-Saharan Africa*; Scholvin, S., Ed.; Routledge: London, UK, 2016.

27. Sanyanga, R. Inga 1 and Inga 2 Dams. 2017. Available online: https://www.internationalrivers.org/resources/inga-1-and-inga-2-dams-3616 (accessed on 9 October 2018).

28. Warner, J.F.; Mirumachi, N.; Farnum, R.L.; Grandi, M.; Menga, F.; Zeitoun, M. Hydrohegemony 10 years later. *WIREs Water* **2017**, *4*, 1042e. [CrossRef]

29. International Commission on Large Dams (ICOLD). *World Register of Dams*; ICOLD: Paris, France, 1998.

30. Nüsser, M. Political ecology of large dams: A critical review. *Petermanns Geogr. Mitt.* **2003**, *147*, 20–27.

31. Eeten, M.V. Sprookjes in rivierenland: Beleidsverhalen over wateroverlast en dijkversterking. *Beleid en Maatschappij* **1997**, *24*, 32–43.

32. Showers, K.B. Beyond mega on a mega continent: Grand Inga on Central Africa's Congo River. In *Engineering the Earth: The Impacts of Mega-Engineering Projects*; Brunn, S.D., Ed.; Springer: London, UK, 2011; pp. 1651–1679.

33. Gottschalk, K. Hydro-politics and hydro-power: The century-long saga of the Inga project. *Can. J. Afr. Stud./Revue Canadienne des Études Africaines* **2016**, *50*, 279–294. [CrossRef]

34. Donati, M. Lighting up and powering Africa. A Priority for African Development Bank. Supply Management, Chartered Institute for Procurement and Supply. 2016. Available online: https://www.cips.org/en/supply-management/news/2016/march/lighting-up-and-powering-africa-a-priority-for-african-development-bank/ (accessed on 3 December 2018).

35. Bretton Woods Project. US Congressional Legislation Threatens IMF and World Bank Plans. 2014. Available online: https://www.brettonwoodsproject.org/2014/01/us-congress-threatens-imf-world-bank-plans/ (accessed on 9 October 2018).

36. EJAtlas. Inga 3 and Gran Inga Complex Hydropower Project on Congo River, DRC. 2018. Available online: https://ejatlas.org/conflict/gran-inga-hydropower-project-on-congo-river-drc (accessed on 9 October 2018).

37. Rogers, D. Chinese Bidder Praised for Speed in DR Congo's Massive Dam Contest. 2016. Available online: http://www.globalconstructionreview.com/news/chinese-bidder-praised-speed-d7r-co7ngos-mas7sive/ (accessed on 9 October 2018).

38. Nevin, T. Congo's Grand Inga Plan Faces a Watershed. *Business Live*. 13 May 2016. Available online: https://www.businesslive.co.za/bd/opinion/2016-05-13-congos-grand-inga-plan-faces-a-watershed/ (accessed on 14 June 2017).

39. Reuters. BHP Pull-Out a Problem for Inga 3 Project. 2012. Available online: https://www.reuters.com/article/bhp-congo-inga-idUSL5E8DF2AZ20120215 (accessed on 9 October 2018).

40. Bruno, S.E. Inga 3 Project: The Two Remaining Candidates Invited to Submit a Common Offer. *BusinessWire*. 13 June 2017. Available online: https://www.businesswire.com/news/home/20170613006491/en/ (accessed on 15 June 2017).

41. Clowes, W. Congo Plans to Start $13.9 Billion Hydropower Project This Year. *Bloomberg*. 2018. Available online: https://www.bloomberg.com/news/articles/2018-06-13/congo-plans-to-start-13-9-billion-hydropower-project-this-year (accessed on 17 Ocotber 2018).

42. Wilson, J. Fantasy machine: Philanthrocapitalism as an ideological formation. *Third World Q.* **2014**, *35*, 1144–1161. [CrossRef]

43. Van der Zaag, P.; Robinson, P.; Groverman, V. Inga Basse Chute and Mid Size Hydropower Development TA Project Advice on the Adequacy of the Information Underlying Decision Making. 2014. Available online: www.eia.nl/dsu (accessed on 11 October 2018).

44. Mathiason, N. Fury at Plan to Power EU Homes from Congo dam. *The Observer/The Guardian*. 23 August 2009. Available online: http://www.theguardian.com/world/2009/aug/23/power-eu-congo-dam (accessed on 28 June 2017).

45. AfDB. Multinational Project Appraisal Report for Inga Site Development and Electricity Access Support Project (PASEL). 2013. Available online: https://www.afdb.org/fileadmin/uploads/afdb/Documents/Project-and-Operations/Multinational_Inga_Site_Development_and_Electricity_Access_Support_Project_PASEL__-_Appraisal_Report1.pdf (accessed on 30 June 2017).

46. James, C.; Kelley, R. *A History of Pan-African Revolt*; PM Press: Oakland, CA, USA, 2012.

47. Söderbaum, F. Modes of Regional Governance in Africa: Neoliberalism, Sovereignty-Boosting and Shadow Networks. 2004. Available online: Papers.ssrn.com/sol3/papers.cfm?abstract_id=2398949 (accessed on 10 June 2017).

48. Turton, A.R.; Earle, A. Post-apartheid institutional development in selected Southern African international river basins. In *Water Institutions: Policies, Performance and Prospects*; Gopalakrishnan, C., Tortajada, C., Biswas, A.K., Eds.; Springer: Berlin, Germany, 2005; pp. 154–168.

49. McEwan, C. Spatial processes and politics of renewable energy transition: Land, zones and frictions in South Africa. *Political Geogr.* **2017**, *56*, 1–12. [CrossRef]

50. Bosshard, P. *The World Bank's Inga 3 Project Goes from Bad to Worse*; International Rivers: Oakland, CA, USA, 2014; Available online: https://www.internationalrivers.org/blogs/227/the-world-bank%E2%80%99s-inga-3-project-goes-from-bad-to-worse (accessed on 14 June 2017).

51. Wood, J.R. India's Narmada River Dams: Sardar Sarovar under Siege. *Asian Surv.* **1993**, *33*, 968–984. [CrossRef]

52. Meissner, R.; Warner, J. Hydro-hegemony or Water Security Community? Collective Action, Cooperation, and Conflict in a Transboundary Security Complex in the SADC region. In *Water Governance and Collective Action: Multiscale Challenges*; Suhardiman, D., Nicol, A., Mapedza, E., Eds.; Routledge: London, UK, 2018.

53. Hensengerth, O. Chinese hydropower companies and environmental norms in countries of the global South: The involvement of Sinohydro in Ghana's Bui Dam. *Environ. Dev. Sustain.* **2013**, *15*, 285–300. [CrossRef]

54. Pamlin, D.; Long, B.J. *Rethink: China's Outward Investment Flows*; Worldwide Fund for Wildlife: Grand, Switzerland, 2007; Available online: assets.panda.org/downloads/wwf_re_think_chinese_outward_investment.pdf (accessed on 10 January 2019).

55. SAIS. Chinese FDI Flow to African Countries, China-Africa Research Initiative, School of International and Advanced Studies, Johns Hopkins University. 2018. Available online: http://www.sais-cari.org/s/Upload_LoanData_v11_October2018.xlsx (accessed on 26 November 2018).

56. Atkins, L.; Brautigam, D.; Chen, Y.; Hwang, J. *China-Africa Economic Bulletin #1: Challenges of and Opportunities from the Commodity Price Slump*; China Africa Research Initiative, Johns Hopkins University School of Advanced International Studies: Washington, DC, USA, 2017.

57. Brautigam, D. *The Dragon's Gift: The Real Story of China in Africa*; Oxford University Press: Oxford, UK, 2009.

58. Nordensvard, J.; Urban, F.; Mang, G. Social innovation and Chinese overseas hydropower dams: The nexus of national social policy and corporate social responsibility. *Sustain. Dev.* **2015**, *23*, 245–256. [CrossRef]

water

MDPI

Article

Political Borders, Epistemological Boundaries, and Contested Knowledges: Constructing Dams and Narratives in the Mekong River Basin

Coleen A. Fox * and Christopher S. Sneddon

Department of Geography and Environmental Studies Program, Dartmouth College, Hanover, New Hampshire, NH 03755, USA; christopher.s.sneddon@dartmouth.edu
* Correspondence: coleen.a.fox@dartmouth.edu; Tel.: +1-603-646-0440

Received: 13 July 2018; Accepted: 3 January 2019; Published: 26 February 2019

check for updates

Abstract: The Mekong River Basin of mainland Southeast Asia is confronting a series of intertwined social, political, and biophysical crises. The ongoing construction of major hydroelectric dams on the river's main channel and tributary systems—particularly in the basin's lower and more populated reaches—is leading to significant socioecological changes. Multiple scientific studies have suggested that proceeding with the planned dam construction will disrupt the region's incredibly productive fisheries and threaten the livelihoods of millions of basin residents. These effects will almost certainly be exacerbated by global and regional climate change. Yet increased understanding of the adverse consequences of dams for the Mekong's hydrological and ecological processes is having minimal impact on decision-making around hydropower development. While local communities, non-governmental organizations (NGOs), and certain scientists draw on this knowledge to oppose or question accelerated dam building, state officials and hydropower developers have turned to the expertise of engineering and technological assessments in order to justify dam construction. Drawing on work in political geography, political ecology, and science and technology studies (STS), we ask two primary questions. First, why does engineering/technological knowledge retain so much legitimacy and authority in the face of mounting scientific knowledge about ecological change? Secondly, how are narratives of progress deployed and co-produced in the contested epistemologies of large dams as development? We conclude with some examples of how contestations over dams seem to be shifting epistemological boundaries in meaningful ways, creating new spaces for knowledge production and transfer. To answer these questions, we focus on three contested dams that are at various stages of construction in the basin: the nearly complete Xayaburi Dam, the under-construction Don Sahong Dam, and the planned Pak Beng Dam. The research advances understandings of the politics of contested knowledges as they become manifest in the conceptualization and governance of large dams in transboundary basins.

Keywords: hydropower; Mekong River Basin; political ecology; STS; public knowledge controversies

1. Introduction

"The Don Sahong and other mainstream dams are foolhardy and dangerous, as they threaten to fundamentally change the nature of the river and its resources, which serves as the lifeblood for millions of people in the region." Kumpin Aksorn, from the Thai community-based organization Hug Namkhong (https://www.internationalrivers.org/resources/continued-work-towards-the-don-sahong-dam-threatens-havoc-for-mekong-fisheries-8023 (retrieved on 31 July 2018))

"I visited (the dam). It does not have any impacts."

Cambodia Prime Minister Hun Sen, after visiting the Don Sahong dam site (https://www.voacambodia.com/a/no-impact-on-fish-from-major-hydropower-projects-pm-claims/4045015.html https://www.voacambodia.com/a/no-impact-on-fish-from-major-hydropower-projects-pm-claims/4045015.html (retrieved on 2 August 2018))

On 23 July 2018, the Xe-Pian Xe-Namnoy dam collapsed in southern Laos, sending 481 million cubic meters of water downstream. Thousands of people were left homeless in Laos and Cambodia. At least 27 people died, and more than 100 went missing. Livestock and crops were wiped out. The dam, which was built by a Korean engineering and construction firm, collapsed when water overtopped the structure following torrential rains. International Rivers, a non-governmental organization (NGO) that has long criticized dam building in the region, issued a statement saying that the dam was not designed to deal with extreme weather events, and that locals were never meaningfully consulted before construction commenced (https://www.internationalrivers.org/dam-collapse-in-laos-displaces-thousands-exposes-dam-safety-risks (retrieved on 3 August 2018)). Ian Baird, a geographer with expertise on the impact of hydropower on fisheries in Laos, stated that the collapse was likely from "faulty construction or a decision to store too much water in the dam's reservoir at a time when heavy rains should have been expected", and that the companies were "trying to play this out as a natural disaster that wasn't their fault", which he did not "believe for one second" (https://www.nytimes.com/2018/07/26/world/asia/laos-dam-collapse.html (retrieved on 3 August 2018)).

This event underscores the crux of our inquiry into the Mekong Basin, where the ongoing and planned construction of major hydroelectric dams on the river's main channel and tributary systems—particularly in the basin's lower and more populated reaches—is significantly changing the system's hydrology, fisheries production, and the livelihoods of millions of basin residents. Recent biophysical research has demonstrated that the present suite of dams on the river's main channel (concentrated in the Mekong's upper reaches in China) and tributaries has had significant downstream and cumulative effects on the basin's water flows and sediment transport [1–4]. Environmental research has explored scenarios involving up to 12 major hydroelectric dams on the river's mainstream and concluded these projects will have significant negative impacts on fish production, fish diversity, and food security in the region [5–11]. Recently published hydrologic models show likely significant negative effects on downstream socioecological systems such as floodplains in the Mekong's lower reaches if dam construction proceeds as currently planned [12]. These effects will almost certainly be exacerbated by global and regional climate change [13].

Studies on the potential negative socioecological impacts of large dams, as well as questions about their economic benefit [14], have had little influence thus far on the plans of the region's governments to proceed with hydropower development projects. The Lao government is currently building two large hydroelectric schemes—the Xayaburi and Don Sahong projects—on the Mekong mainstream, and has announced plans to construct a third, the Pak Beng Dam [15]. Given the sheer magnitude of fish production in the basin, the impacts of hydropower development on fish-based livelihoods will be significant—one study indicated a 26–42% reduction in the current two million metric ton annual production, should planned mainstream dams move forward [16]—and almost certainly negative. In the face of criticism from NGOs and academics regarding these projects' likely deleterious impacts, government officials have argued that the dams will have minimal effects, and that displaced communities and negative environmental effects can be effectively mitigated. The Chinese government claims that the cascade of hydroelectric facilities in the upper Mekong will benefit water governance in the river's lower reaches by augmenting dry season flows and exercising flood control during the wetter months. Yet officials in Yunnan province, where the dams are located, have been reluctant to release flow data that might reinforce these claims [17]. In a similar vein, Cambodian Prime Minister Hun Sen disparaged research that predicted that the Lower Se San II dam will negatively affect 70% of fish in the Cambodian part of the basin, asking "Do our country's fish know how to climb trees? Do they know how to climb mountains?" [18]. His comments suggest that even nonsensical claims about the lack of environmental consequences (that seemingly equate dams with other natural features on the landscape) can prevail over ecological assessments if promoted by powerful decision makers.

In nearly all cases of current and planned hydroelectric development in the Mekong—as well as in transboundary basins such as the Nile, Nu/Salween, and Amazon—project proponents and opponents alike muster technological, scientific, and locally-based knowledge claims to defend their positions vis-à-vis dam construction. The public knowledge controversies [19] engendered by large dams underscore the primacy of the politics of knowledge production, circulation, and use. Crucial decisions regarding the siting, design, and operation of hydroelectric facilities throughout the basin ostensibly rely on state-of-the-art scientific understandings of the basin's critical biophysical processes such as water flows, sediment transfer, and fisheries production. Yet, it remains unclear precisely how such knowledge is incorporated into water governance in the Mekong or other large transboundary basins. It is even less clear whether the myriad of types of knowledge produced at local sites throughout the basin, on occasion facilitated by Mekong River Commission (MRC)-sponsored experts, independent scientists, and non-governmental organizations (NGOs), are received and integrated into the decision-making around any given project.

These intertwined conceptual themes—knowledge, borders/boundaries, and technology—are at the core of our inquiry. We argue that interrogating political borders and epistemological barriers as they relate to knowledge production and knowledge flows is an important part of understanding the acceleration of large dam construction in a transboundary basin [see also the related conceptual notions in this special issue's introductory article [20]. To facilitate our inquiry, we ask the following questions. Why does engineering/technological knowledge retain so much legitimacy and authority in the face of mounting scientific knowledge about ecological change? How are narratives of progress deployed and co-produced in the contested epistemologies of large dams as development? A key finding of our research is that technological and engineering knowledge, evolving in conjunction with narratives of development, inequitable power relations, and institutional arrangements, creates epistemological barriers that devalue or de-legitimize local and ecological knowledge. Our research relies on multiple documents focusing on the competing knowledge claims regarding disputed hydropower projects in the Mekong—including governmental and intergovernmental reports and websites, independent assessments of socioecological impacts from the scientific community, newspaper articles, and the statements of activist and NGO groups (both authors have over 20 years of research experience in the Mekong basin, and the analysis draws heavily on hundreds of documents and observations regarding the changes that have occurred during this period). We use constructionist discourse analysis to uncover how "particular lines of argument have become taken as truths, while others are dismissed" [21]. Our perspective is informed by critical theory as well as interpretive policy analysis [22], which means that we are investigating dominant assumptions about society–nature relations, exposing how promotional and oppositional communities form around particular policy issues (e.g., hydropower development), and ultimately gaining a more robust understanding of the social and ecological implications of mainstream, hegemonic definitions of development. Our inquiry draws primarily on developments in the region beginning in 2010, when the Lao government submitted prior consultation documents to the Mekong River Commission. In the same year, the International Centre for Environmental Management submitted a report on mainstream dams, which recommended a 10-year deferment on mainstream dam building [23].

We begin the paper by explaining two crucial levels at which borders and boundaries shape and regulate knowledge flows in transnational river basins. The first level, which is familiar to scholars of political geography, is constituted by the socially constructed yet potent political borders that separate national territories and define a river system as transboundary. The second boundary, which is familiar to science and technology studies (STS) scholars and political ecologists, is constituted by the epistemic barriers that divide different types of 'expert' knowledge (e.g., engineering, technical knowledge, and scientific knowledge) and multiple types of 'non-expert' knowledge, which are often manifest through cultural and linguistic differences. Both types of borders figure prominently in public knowledge controversies surrounding economic development, and particularly the construction of large dams. We use these understandings of borders and boundaries to focus our questions around the Xayaburi, Don Sahong, and Pak Beng dams. An overarching goal is to reveal the power relations that underpin the knowledge claims that have been put forward by multiple actors, and better understand how those claims are shaping the development and governance in this transboundary basin. Addressing these questions allows for more informed explanations about how conflicts over dams might be shifting epistemological boundaries and creating new spaces for knowledge production and transfer.

2. Borders, Boundaries, and Knowledge Controversies

2.1. Constructing Borders

In the Mekong River Basin, contestations over knowledge surrounding hydropower development are heightened by the transboundary nature of the river, which flows for 4800 km from the Tibetan Plateau in China through Myanmar, Laos, Thailand, Cambodia, and Vietnam (see Figure 1). Water, sediment, and migratory fish pay no heed to the political lines on a map of the region, but basin boundaries define the geopolitical relations that shape knowledge flows and hence basin governance [24]. Although the complexity of the political landscape is not reflected fully by the map of sovereign states, borders can enclose projects, demarcate domains of development, and obstruct the flow of environmental knowledge in the Mekong, while also explaining which knowledge is promoted. Mekong development depends on states pursuing their sovereign right to harness the river within their borders [25], but development also benefits from the narrative of a region of "traversed boundaries" [14], whereby electricity can flow freely from Laos to Thailand, or state-focused knowledge can be channeled through the multilateral MRC to align with "sustainable development" agreements (the Mekong River Commission comprises Lao PDR, Vietnam, Cambodia, and Thailand) [26]

Our investigation into the relationship between hydro development, contested knowledge, and sovereign states builds on work in political geography and critical geopolitics that understands borders as co-produced through contingent political, economic, and technological processes [27,28]. We also derive insights from STS scholarship, which reveals the "messy, impure, and political" nature of science [29], and which explores how scientific knowledge evolves in conjunction with representations, institutions, and identities [30,31], thereby helping to give meaning to technological objects such as large dams. Together, these frameworks shed light on processes of "border work" [32] occurring in the region, whereby borders, technological and engineering knowledge, and hydropower projects are co-produced. In other words, states have development agendas, and they have the sovereign right to define and carry out development within their borders. Dams, as material expressions of development, depend on the state embracing and elevating the technological and engineering knowledge that underpins their construction and operation. Backed by powerful state and corporate interests, this knowledge gains legitimacy as expertise shaping the development of natural resources. Once built, dams help ensure that the relationship between hydropower and development becomes even more tightly interwoven in state discourses and policies. In countries such as Laos, where there is little opportunity for opposition to these agendas, the space for pursuing alternative development paths becomes increasingly narrow.

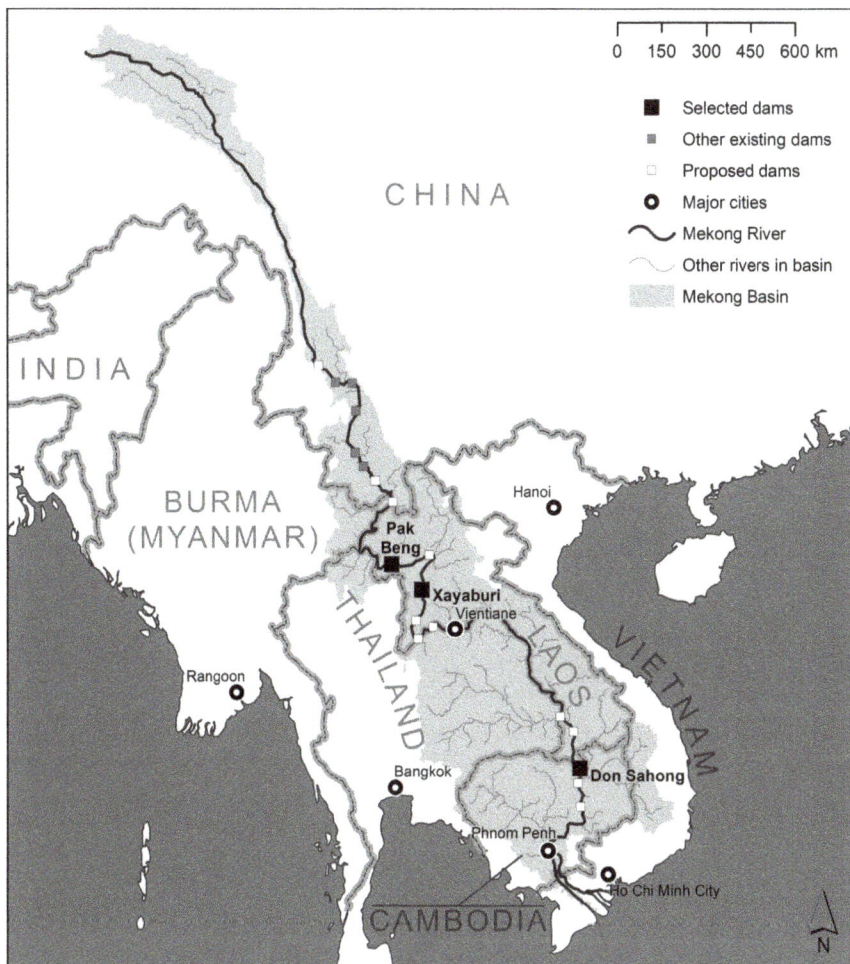

Figure 1. Map of Mekong River basin showing existing dams and ongoing hydropower projects mentioned in the article.

These dynamics are at play in the cases of the Xayaburi, Don Sahong, and Pak Beng dams. The Lao government sees the energy sector (primarily hydropower) as the means to elevate itself to middle-income status and increase its political–economic influence in the region [33]. The Ministry of Energy and Mines, which is the implementing ministry for hydropower projects, notes that "the power sector has been classified as a development strategy of the national economy", and that the sector "has the potential to play a pivotal role in achieving the social and economic development objectives" (http://www.poweringprogress.org/power-sector (retrieved on 3 August 2018)). To support this narrative, the state relies on a process that has been termed "rendering technical"—a set of practices concerned with defining boundaries, assembling information, and representing "the domain to be governed as an intelligible field with specifiable limits and particular characteristics" [34]. In this case, the river becomes intelligible to the state primarily as an underutilized resource. This 'problem' is framed "in terms that are amenable to technical solutions" such as large dams [34]. Threats to wild-capture fisheries and the livelihoods that depend on them become less important than a lack of electricity to power development. Furthermore, when fisheries are degraded, engineering and technical solutions are understood as adequate to address the damage. As knowledge claims and state interests adapt to one another in a process of mutual development [29], the river is transformed in ways that invite further state control and more 'expert' engineering knowledge, while simultaneously rendering the river ecosystem less intelligible to local communities and their traditional knowledges. In this way, the co-production of borders is closely linked to the epistemological boundaries that characterize the landscape of contested knowledge in the basin.

2.2. Maintaining and Challenging Epistemological Boundaries

While state actors are potent forces in knowledge controversies in the Mekong region, a growing network of non-state actors are deeply implicated in the transformation of the river and are increasingly important nodes of knowledge production and transfer [24,35–37]. Non-state actors in the region include multinational corporations, local communities, NGOs, multilateral organizations, scientists, and academics. Woven throughout the relationship between these actors and hydropower development are epistemological boundaries, which become apparent when what is perceived by powerful actors as legitimate knowledge (often associated with technology and engineering knowledge) confronts other forms of knowledge (often lay or local knowledge, but in this case also knowledge produced by ecologists and social scientists). Building on STS scholarship that examines complex environmental debates [38–41], political ecologists have expanded research on the epistemic boundaries between expert and non-expert knowledge, emphasizing how the production and circulation of environmental knowledge regarding particular socioecological interventions (e.g., development projects) are contingent on the power relations of particular places and tend to serve the interests of those who stand to benefit economically and politically from those interventions. Moreover, in the case of Mekong hydro development, it is clear that boundaries exist between different types of expert knowledge. For example, a key boundary lies between the engineering and technological knowledge produced by consultants, states, and hydropower companies, and the knowledge produced and supported by ecologists and social scientists. We make a distinction between these two types of expert knowledge, without claiming that one is inherently more legitimate than the other. Rather, we emphasize how engineering and technical knowledge is elevated by powerful political and economic interests to serve a particular development agenda associated with hydropower in the present moment. For example, this knowledge could just as easily drive a solar power revolution in the basin in the future, with a different set of environmental and social outcomes). Engineering and hydropower technologies are associated with types of expert knowledge that are typically inaccessible to the public and therefore often generate strong boundaries. We contrast this with ecological science, which is also a type of expert knowledge, but one that is often aligned with how locally based, resource-dependent people are experiencing changing Mekong ecosystems. While no knowledge claims are free from biases and agendas [42], and the boundaries among multiple types of expert knowledge are blurred in multiple ways, the distinction between different expert knowledges is critically important in understanding dam building decisions and the subsequent conflicts in the region.

Political ecology has been particularly effective in highlighting the 'non-expert' side of the epistemic boundary, because it focuses on how "local", "traditional ecological", "indigenous", "experiential" or "tacit" knowledge has been used to both counter and abet scientific knowledge in the context of international development and conservation [43]. Years of academic research in the Mekong system have highlighted the value of local knowledge in terms of generating crucial knowledge regarding fish production and migration [44,45]. While place-based knowledge claims regarding human–environment relations are often a welcome palliative to environmental narratives of states and other powerful development actors, local knowledge systems should not be perceived as a universal panacea for environment development dilemmas. Such knowledge is neither foolproof nor immune from a variety of social influences that challenge its veracity and its uses [46]. Evidence suggests that integrating scientific and local forms of knowledge produces hybrid knowledges that present robust, practical insights on how to ameliorate environmental degradation and livelihood vulnerabilities [47,48]. Yet, the awareness and implementation of hybrid or co-produced knowledge in Mekong water governance arrangements remain uncommon.

Scholarship on the boundaries between different types of knowledge can be fruitfully integrated and extended by focusing on what have been termed public knowledge controversies [19]. Such controversies regarding how to best resolve environmental problems might involve a variety of conflict points, including disagreements about: evidence, interpretations of evidence, the nature of expertise, the relative transparency of expertise, or the level of participation of non-experts [19]. When these disputes involve transnational environments and networks of actors (e.g., oil pipelines, international river basins), the issues driving the controversy can remain unresolved for years; they also tend to engage a diversity of interested parties beyond research scientists (e.g., government agencies, corporations, non-governmental organizations, and media outlets), and often mobilize transnational norms of governance (which are often unavailable at national levels) that provide a template around which different actors can support or resist knowledge claims [49]. These conflicts also raise important questions regarding how expertise might be democratized to consider and asses a variety of knowledge domains, particularly ones that stand apart from formal sites of knowledge production in the academy or other professional arenas [50].

Large dams are exemplars of public/transnational knowledge controversies, and existing and future hydroelectric projects on the Mekong's mainstream and its tributaries can be used to rethink the role of political and epistemic borders in shaping the flows of environmental knowledge in transnational river basins. Current hydropolitics in the Mekong basin are dominated by interstate tensions regarding the potential downstream impacts of dams built or underway in China and Laos on critical downstream environments in Cambodia (particularly the productive fisheries of the Tonle Sap region) and the Mekong Delta region of Vietnam [51]. With regard to mainstream dams in Laos, the Chinese government is fully supportive, since a highly regulated Mekong River is aligned with its development vision for the basin. The Thai government is mostly supportive, since it will be a primary consumer of the power generated by the projects. Cambodian and Vietnamese officials have voiced some opposition due to the downstream impacts, but both support the sovereign right of countries to develop the river for national development. At another level, dam projects—past and present—have been criticized by an assortment of local, national, regional, and global scientists and advocacy organizations who insist that government assessments of their likely socioecological impacts have been inadequately scrutinized and rely on tentative and incomplete scientific understandings of the flow dynamics of large rivers, fish ecology, and other biophysical processes [52]. A crucial part of these critiques is that government-sponsored impact assessments almost never account for the local knowledge of observed effects of river alterations, which is generated by the communities that are most affected by the dam projects in the region. Thus, a focus on conflicts regarding large dams in the Mekong basin will shed light on how borders at political and epistemological levels feed into knowledge controversies and affect knowledge flows throughout the basin.

3. Supporting Hydropower through Technical and Engineering-Based Knowledge

The Xayaburi, Don Sahong, and Pak Beng dams are the first three of 11–12 dams planned for the mainstream in the lower basin. Xayaburi is nearly complete, Don Sahong is in the early stages of construction, and Pak Beng is still in the planning phase. Each dam has been characterized by controversy, and all three are situated on the river within Lao territory. Construction of the 1285-megawatt Xayaburi dam in northern Laos began in 2010. With financing coming from six Thai commercial banks, the $3.2 billion project is being built by Xayaburi Power Company Limited (XPCL), which is a subsidiary of the Thai construction company CH Karnchang Public Company Limited. Thailand's electricity utility, the Electricity Generating Authority of Thailand (EGAT), has agreed to purchase 95% of the dam's electricity. Laos and dam developers have made the case that Xayaburi is a run-of-river project, but critics dispute that characterization, given that dam retains water for up to two days. Key concerns include the dam's effect on migratory fish, downstream ecological impacts, the relocation of more than 2000 people from the site, and negative consequences for the livelihoods of more than 200,000 people living near the dam [53,54]. Construction of the 25-meter-high Don Sahong Dam, which is located in southern Laos a few kilometers from the border with Cambodia, began in early 2016. It is expected to generate 260 MW of electricity for domestic use and export to Cambodia or Thailand. The project's developer, Don Sahong Power Company, is a joint venture between the Government of Laos and Mega First Corporation Berhad (MFCB), which is a Malaysian company. The company will build, operate, and transfer the project over a period of 25 years. A crucial worry is that the dam will irreparably harm both commercial and subsistence fisheries, because it will block the Hou Sahong channel, which is of critical importance for year-round fish migration [25,55]. The developers of both Xayaburi and Don Sahong have proposed engineering-based mitigations to address fishery concerns. The Pak Beng Dam would be the northernmost of the dams proposed for construction on the lower Mekong River mainstream. Datang (Lao) Pak Beng Hydropower Company is the project developer (with Thai and Chinese partners), and 90% of the energy produced by the 912-MW project will be sold to Thailand. The remaining 10% will go to Laos' state-owned utility, Électricité du Laos [56]. The dam will displace 25 villages in Laos and two villages in Thailand. There is concern and controversy over the potential ecological and social impacts both at the dam site and downstream [57].

Over the past century, states have largely controlled the generation and sharing of environmental data, making the environment legible to the logics of economic growth and national security [58]. The development of natural resources is an important articulation of sovereignty, reflecting states' desires to improve the condition of their populations [34]. To support these efforts, states, development banks, and their corporate partners produce and promote specific narratives about the sustainable development of resources [59]. Hydropower is assumed to be sustainable, with "sustainable hydropower" generally being used to describe all of the projects. This is reflected in documents such as the Lao government's 2015 Policy on Sustainable Hydropower Development (which actually represents a lowering of human rights and environmental standards in hydropower projects) (https://www.internationalrivers.org/blogs/294/new-policy-proposed-on-hydropower-development-in-lao-pdr-puts-developers-interests-first (retrieved on 3 August 2018)). Similarly, in response to critiques of the Xayaburi Dam design, the Lao government insisted that the project "is a green energy and shall be strongly promoted and supported" (comments by Lao PDR on the MRC's Technical Review Report of the Proposed Xayaburi Dam Project (retrieved on 4 August 2018 from http://www.mrcmekong.org/topics/pnpca-prior-consultation/xayaburi-hydropower-project-prior-consultation-process/). Hydropower is seen by government officials as being easily aligned with "a clear definition of sustainability", with the challenge mostly being to get the policies and technical guidelines right (https://www.ifc.org/wps/wcm/connect/news_ext_content/ifc_external_corporate_site/news+and+events/news/new+hydro+policy+puts+focus+on+sustainability+in+lao+pdr (retrieved on 4 August 2018)).

While it is not surprising that a government would uncritically support its own policies and projects, it is noteworthy that these narratives prevail in the face of so much evidence to the contrary. Explaining their dominance requires unpacking how the engineering and technological knowledge that is necessary to promote and build large-scale hydro projects is evolving in conjunction with representations of the river, discourses about national identity, institutions for "sustainable development", and narratives of mitigation and limited harm. The policies that emerge from this dynamic create space for certain types of knowledge, and that knowledge, in turn, supports and justifies policy, providing a clear example of co-production.

In the Mekong Basin, an influential array of actors—including construction and electricity companies, international consultants, engineering firms, government ministries, and multilateral organizations—come together to create a world of experts and expertise that generates and promotes knowledge aligned with technological and engineering views of natural resources. This world is circumscribed by epistemological boundaries that are not easily penetrated by outsiders. The river is portrayed as an ecosystem that can sustain large dams, particularly when, according to the project developer for Xayaburi, they are "appropriately designed using the best available technologies". This narrative extends to mitigation, which depends on engineering-based interventions, such as "fish-friendly turbines" (http://www.xayaburi.com/Environment_SG_eng.aspx (retrieved on 5 August 2018)). Uncertainty about the long-term, cumulative impacts and failure to investigate impacts beyond political borders further support technological and engineering-based knowledges, while institutional arrangements privilege state-based knowledge and suppress other understandings.

In the case of the Xayaburi dam, expert technological and engineering knowledge has been invoked throughout the contentious design and construction process. In 2010, the Lao government informed the Mekong River Commission of its intention to move forward with the Xayaburi dam, as required by the Procedures for Notification, Prior Consultation, and Agreement (http://www.mrcmekong.org/home/SearchForm?Search=xayaburi&action_results= %C2%A0%C2%A0%C2%A0 (retrieved on 5 August 2018)). Since Xayaburi is the first dam being built on the mainstream in the lower basin, its construction marks the first time that a member country has had to seek approval by the region's government (which the MRC calls "prior consultation"), although there is no enforcement mechanism to actually stop a country from proceeding with a project [53,60]. Non-state actors have no official say, but they can provide comments. The MRC found that the design was inadequate, violating the MRC's design guidelines for mainstream dams [61]. The Lao government subsequently hired a Finnish engineering company, Pöyry Energy, to review the Xayaburi Dam's compliance with MRC requirements for mainstream dams. Pöyry, which styles itself as "the global thought leader in engineering balanced sustainability for a complex world" [62], recommended that dam construction continue, with additional environmental studies, the installation of additional fish ladders, and other technical design modifications to improve fish passage and sediment and nutrient issues.

In addition to concerns about conflict of interest (Pöyry is a business partner with the Thai company building the dam, and it was hired to do engineering work on Xayaburi after completing the report), the company's recommendations raise many questions. For example, Pöyry acknowledged a lack of baseline data, but claimed that any baseline studies can be conducted and mitigation measures can be designed after construction on the dam is already underway [63]. The report stated that Xayaburi's "fish-friendly turbines," fish ladders, and locks can protect at least 28 species of fish, but research shows there are at least 139 fish species that would be blocked from swimming past the dam [64]. Moreover, scientific evidence from fisheries experts around the world suggests that no technology currently exists that would mitigate the impact of mainstream dams on Mekong fisheries [63]. None of these concerns are reflected in the discussion of environmental safeguards on the official Xayaburi website, where the company, Xayaburi Power Company Limited (XPCL), reassuringly wrote about the "thorough environmental and social impact assessment conducted by leading experts with extensive experience in environmental engineering" (http://www.xayaburi.com/ Environment_SG_eng.aspx (retrieved on 5 August 2018)). Revealingly, the discussion of safeguards continued with the acknowledgement of consultation and feedback from the MRC, but this stated only that Lao PDR met its legal requirements regarding social and environmental safeguards.

Shortly after the Lao government began the prior consultation process, a major report on the environmental and social impacts of mainstream Mekong dams was released. The Strategic Environment Assessment (SEA) report, which was commissioned by the MRC, recommended that decision-making on mainstream dams (including Xayaburi), be deferred for 10 years due to the massive risks and impacts associated with the projects. Findings included significant negative impacts on fisheries and agriculture, growing inequality, increased poverty among fishers and riparian communities, the degradation of longitudinal connectivity, and limited opportunities to mitigate these damages [16]. While both Cambodia and Vietnam argued for further studies, particularly of the transboundary environmental and social impact, Laos rejected the recommendation, demonstrating how political borders can reinforce epistemological barriers in a knowledge controversy. Notably,

Pöyry also concluded that it was not the responsibility of the project developer to assess transboundary impacts in its response to MRC concerns [63].

A similar process of co-production involving political borders, expert engineering knowledge, mitigation strategies, and multinational capital is playing out in the construction of the Don Sahong dam in southern Laos. In this stretch of the river, which is characterized by complex channels, small islands, and enormous biodiversity, there is widespread agreement among ecologists and communities that the dam will have devastating consequences for fisheries [55,65]. The dam is being built on the Hou Sahong Channel, which is the largest and most accessible channel in the Siphandone area, and allows for year-round fish migration for more than 100 species. The dam also threatens the survival of the endangered Irrawaddy Dolphins, whose habitat is located immediately downstream. The dam, being built by Mega First Corporation Berhad of Malaysia and Sino-hydro (a State-owned Chinese company), is called a run-of-river scheme by the developers, but it will create a head-pond inundating about 308 acres of land between the islands of Don Sahong and Don Sadam. (Research on the ecohydrological impacts of so-called run-of-river dams, which are presumed by hydropower advocates to exert much less influence on river functioning than large storage dams, is still nascent. Recent work argues that run-of-river dams are likely to have significant and comparable impacts on river systems compared to those of large storage dams [66]. Most of the 260 MW that is generated by the project will go to Thailand and Cambodia. Even though Don Sahong is just a few kilometers from the Cambodian border, there have been no studies on its potential transboundary impacts. Don Sahong Power Company noted that "the project has clearly demonstrated in the Environmental Impact Assessment (EIA) and in several of the engineering studies there will not be any downstream transboundary impacts on regional sediment transport, water flow, or water quality" (http://dshpp.com/faq/ (retrieved on 4 August 2018)). Moreover, the Lao government claimed that Don Sahong is a tributary dam, and therefore not one that requires agreement by all of the MRC partners [67].

The premise that engineering interventions can mitigate ecological harm is even more apparent in the case of Don Sahong than Xayaburi. The developers propose to mitigate the loss of the Hou Sahong Channel for fish migration by re-engineering nearby Hou Xang Phuek and Hou Sadam channels. The plan is to widen and deepen those channels to enable fish passage. There is no evidence that this will work [55]. According to fisheries biologist Eric Baran, "Mitigation, if any, of the effects on fish migrations in the most intensive freshwater fishery in the world lies in the hands of consultants working for (dam management company Don Sahong Power Company (DSPC)), without any control or oversight by any institution nor scientific organization" [67]. Despite these concerns, the DSPC remains optimistic about success, stating that they have "every confidence that natural channels can be improved to allow fish passage. We have initially targeted two specific channels, but there are more than 30 channels which could be modified" (http://dshpp.com/faq/ (retrieved on 4 August 2018)).

The situation with the Pak Beng hydropower project, which is currently undergoing prior consultation, is similar to what has happened in the cases of Xayaburi and Don Sahong. Pak Beng is proposed on the Mekong mainstream in northern Laos, which is upstream from the Xayaburi dam site. It is a run-of-river project with capacity of 912 MW. It will likely to lead to declines in the abundance and diversity of fish species, as well as "seriously and negatively" impact the Giant Mekong Catfish, including "eventually or possibly leading to extinction" [56]. Upstream and downstream fish passage mitigation measures are "untested and inadequate", and baseline data are seriously lacking [57].

These dam projects reveal, in contrast to a more precautionary principle advocated by scientific studies such as the Strategic Environmental Assessment on Mekong mainstream dams, how engineering and technical knowledge focused on the dam itself is privileged when it comes to future uncertainty. Epistemological boundaries are drawn around the science and practice of mitigation as a type of expert knowledge, with few effective opportunities to contest the assertions put forward by consultants, power companies, and governments. Mitigation science is increasingly used to de-politicize and normalize controversial infrastructure decisions. This results in the legitimization of certain types of knowledge, such as that which goes into "designing river flows" through hydropower projects [68], leaving less space for knowledge about a free-flowing river and the services that it provides.

4. Encountering Boundaries—Contested Epistemologies of Large Dams as Development

More than 157 new plant and animal species were discovered in the Mekong River Basin in 2017 which is both a testament to the region's incredible biological diversity and further evidence of the uncertainty surrounding the impact of large dams on the river ecosystem [69]. The richness of the basin's heterogeneous and overlapping ecological systems sustains the livelihoods and cultures of tens of millions of people. Based on their lived experiences of changes to the ecosystem from recent hydropower projects, communities along the river have repeatedly expressed concerns that align with the assessments of scientists such as Dr. Lilliana Corredor, who testified that hydropower development is "actually increasing poverty and despair, instead of 'improving the standard of living and decreasing poverty' as advertised; is displacing tens of thousands of people in poor communities from their homes, lands, and cultural sites, while offering a dismal compensation, which does not support the people to cultivate food or to fish" (http://www.mrcmekong.org/assets/Other-Documents/ stakeholder-submissions/Open-Letter-to-the-Mekong-River-Commission.pdf (retrieved on 4 August 2018)). A 2013 study by the Inland Fisheries Research and Development Institute (IFReDI) found that changes in the availability of fish and aquatic resources in Cambodia are likely to have major negative impacts in terms of nutrition, income, and social equity. Replacement measures (e.g., livestock and aquaculture) would only partially compensate for the loss of wild fisheries, and would likely be more expensive, and less accessible to the poor (http://ifredi-cambodia.org/research-projects/ (retrieved on 4 August 2018)).

Similarly, a recent statement signed by community members along the river noted: "We have witnessed and experienced the destruction caused by the dams. For us, who live by the river and experience every change in the water systems, there is no question that such dams result in serious negative impacts for present and future generations, and should not be built" (https://globaljusticeecology.org/support-to-the-statement-by-local-people-on-dams-in-the- mekong-region/). Already, upstream dams in China have caused riverbank erosion, the loss of riverbank gardens, and unpredictable flows. Communities have directly experienced the loss of fish species with tributary dams such as the Pak Mun Dam in Thailand (www.livingriversiam. org/index-eng.html (retrieved on 4 August 2018)). At the site of the Xayaburi dam, which was a formerly "self-sufficient community that generated revenues via gold panning and cultivated their own riverbank gardens to produce rice, fruits, and vegetables, villagers are now finding themselves without jobs, very little money, and not enough food" (http://www.ipsnews.net/2013/06/dams- threaten-mekong-basin-food-supply/ (retrieved on 5 August 2018)). In Cambodia, upstream of a dam on the Sesan River, "fishermen waited in vain for the yearly migration in May and June. No more fish to catch. The villagers have moved elsewhere, escaping the rising water and increasing poverty. The only reminder of a once lively Kbal Romeas is the roof of a pagoda that seems to float on the empty water" (http://www.ipsnews.net/2017/11/mekong-dammed-die/ (retrieved on 5 August 2018)). Yet, the scientific and local knowledge that underpins these experiences and perspectives has had very little influence to date on the decision-making processes around the Xayaburi, Don Sahong, and Pak Beng dams.

As scholarship in STS and political ecology has long argued, the privileging of certain kinds of knowledge over others does not automatically inhere to the way that knowledge is produced. Rather, the situation results from a process whereby technological and engineering knowledge, evolving in conjunction with narratives of development, inequitable power relations, and institutional arrangements, creates epistemological barriers that devalue or de-legitimize local and ecological knowledge. The ability to envision and presumably create a better future (i.e., "sustainable development"), offers a powerful rationale for privileging the knowledge and expertise that support such a path, while simultaneously dismissing other visions of sustainable livelihoods as outside the bounds of "expert" knowledge. This, coupled with extremely limited institutional opportunities to contest hydro projects through formal channels, marginalizes certain types of knowledge.

All three dams demonstrate the high value that is placed on a model of development that begins with energy production. The Lao Ministry of Energy and Mines is the primary governmental representative to the MRC, and it promotes the narrative of hydropower as the only feasible means to achieve development. The Ministry explains that the "development of electric power facilities a highly appropriate method of achieving sustainable social and economic development through an electrification program as well as a source of income" (http://www.poweringprogress.org/new/2-uncategorised/3-hydropower-in-lao-pdr (retrieved on 5 August 2018)). The Ministry describes hydropower as a low-cost and optimal use of the country's resources that is driven by the logic that "we are going to build infrastructure because we are poor" [70]. It is often said that hydropower has the potential to turn Laos into the 'Battery of Southeast Asia", and the government hopes to have 12,000 MW in operation by 2020, with two-thirds for export (https://uk.reuters.com/article/laos-energy-minister/interview-laos-hydropower-generation-capacity-to-jump-almost-four-fold-by-2020-idUKL4N0SL0EG20141028 (retrieved on 5 August 2018)). Profits from exports are intended to fight poverty, and there is a clear sense that there is no real alternative. As Hun Sen, Prime Minister of Cambodia noted in response to criticism of the impacts of large dams: "How else can we develop?" (http://www.ipsnews.net/2017/11/mekong-dammed-die/ (retrieved on 5 August 2018)). These sorts of narratives create epistemological boundaries that exclude alternative scenarios.

Yet there are other visions of development for the region, which do not depend on "gambling with our future" (https://savethemekong.org (retrieved on 5 August 2018)). Instead, they prioritize livelihoods and the sustainability of ecosystems, as opposed to moving forward with "ecological time bombs" that could threaten the food security of 60 million people [71]. These scenarios are not based on romanticized views of subsistence livelihoods. Rather, they are models of development that would prioritize protecting Lower Mekong fisheries, which, according to recent estimates, are valued at $17 billion a year, contributing 3% to the combined gross domestic product (GDP) of Vietnam, Cambodia, Laos, and Thailand" [72]. The productivity of fisheries is deeply linked to local knowledge and communities' connection to and interdependence with the river, and there is a need to take seriously the claims of locals who assert, "What we want is our village, our river. The river and forest are not for sale, and especially not our identity and dignity" [70]. In a document titled "Mekong Governments: Listen to the People!", a host of local community organizations and individuals made a similar claim, writing that dams "have resulted in severe changes in the Mekong's ecosystems, endangering life, livelihoods, and the economy of the entire region. Indigenous peoples, women, and children are most affected by these changes. The dams have also worsened the impacts of climate change that we are already facing" (https://globaljusticeecology.org/support-to-the-statement-by-local-people-on-dams-in-the-mekong-region/ (retrieved on 5 August 2018)).

These knowledge claims about the relationship between fisheries, development, and hydropower are increasingly supported by academic assessments, which make clear the tenuous link between large dams and development. For example, drawing on statistical and comparative evidence, and "reference class forecasting", researchers found that the economic, social, and environmental costs of large-scale hydropower megaprojects consistently outweigh the benefits [73]. Their findings suggest that not only is there no evidence that large dams create the sort of development that Laos and other governments are hoping for, but they often actually have negative economic impacts. In light of such findings, the authors suggested that countries should pursue more "agile energy alternatives" [73]. These conclusions have been supported by other research that shows that the overall economic impact of Mekong dams would be negative, due to the devastation of wild-capture fisheries [74]. Similar findings of economic, social, and environmental costs outstripping benefits have emerged from assessments of large dams such as Belo Monte in Brazil [75].

Furthermore, while there is clearly a need for energy, there are many options that would not devastate fisheries and livelihoods. Plans exist for alternative power sources, such as Thailand's Alternative Energy Development Plan, which laid out a path for the government to work at small-scale renewable energy efforts such as biomass and solar, which would have local investment and work with local populations (http://www.unescap.org/sites/default/files/MoE%20_%20AE%20policies. pdf (retrieved on 4 August 2018)). Recent research from the US-based Stimson Center details the promising potential for solar power in the Mekong region. Researchers predict that solar will be available in the next few years at six cents a kilowatt hour, making some of the most-damaging dams unnecessary, since many planned Mekong dams would generate power costing eight or nine cents per unit. This would take place in conjunction with improvements to the current grid transmissions and a rethinking of regional power trade, allowing for overall "less-damaging power infrastructure" [76,77]. Currently, the grid is in such bad shape in Laos that, according to the World Bank, in some places over 20% of the energy supply is lost during distribution (http://www.worldbank.org/en/news/press-release/2015/06/23/lao-pdr-to-improve-electricity-network-with-world-bank-support (retrieved on 4 August 2018)). This makes the devastation of fisheries and livelihoods even more questionable, as they are traded off for large losses of energy. People in the Mekong region clearly want to escape poverty and improve their lives. They have "aspirations for development and modernity" [78], but these aspirations and linkages to national economic development visions vary widely across the region given its ethnic, cultural, linguistic, socioeconomic, and political diversity. Moreover, there is little evidence that the individuals and communities that have been confronted with dams that affect their livelihoods in a myriad of ways accept coercion as a method of promoting a state-determined version of that development, particularly one that places disproportionately high costs on resource dependent communities.

5. Conclusions—New Spaces for Knowledge Production and Transfer?

In December 2017, China announced that it would not proceed with plans to blast islets in the Mekong River near Chiang Rai, Thailand because the project to alter the channel for navigation would harm nearby Thai communities. While this project does not match the scale of a mainstream dam, the blasting plans were vigorously opposed by NGOs, communities, civil society, and scientists [79]. It is notable not only that China is halting the detonations, but that it did so in response to the activities of a coalition of activists (albeit, supported by the Thai state). Furthermore, the opposition was based on local knowledge emerging from experiences with upstream dams in China. As dams were being built in China over the past 20 years, communities began to notice unusual fluctuations in water levels. One consequence was the destruction of a freshwater seaweed that was normally harvested during the dry season [80]. On a similar note, International Rivers commissioned an independent expert review of the social and environmental impacts of the Pak Beng Dam in May 2017, attempting to get information disseminated early in the prior consultation process. Then, in June 2017, the Thai Network of Eight Mekong Provinces filed a lawsuit against the relevant Thai government agencies for their involvement in the Pak Beng Dam on the Mekong River, and the expected transboundary impacts on communities in Thailand. In February 2018, possibly as a consequence of so much opposition, Thailand's electricity authority delayed signing an agreement to purchase power from the Pak Beng dam, which is being built primarily to meet Thailand's energy demand [81].

Despite the Lao and other Mekong governments seeming to be pushing ahead with large dam projects, these examples suggest that the opposition is increasingly engaged in knowledge generation and exchange, with a significant amount of cross-border activity. This often takes the form of organizations such as the Mekong People's Forum, which comprises representatives from local communities, and local, regional, and national NGOs. They come together to present and discuss changes that they have observed to their livelihoods, sharing research about the potential impacts caused by the planned 11 mainstream dams (http://www.meenet.org/mekong-peoples-forum-in-an-giang-vietnam/ (retrieved on 4 August 2018)). Similarly, the Rivers Coalition in Cambodia (RCC), along with community representatives, have initiated a campaign against Don Sahong, conducting a national workshop about the dam, doing radio talk shows, and holding press conferences for national and international media to share their concerns and urge the reconsideration of the development of mainstream dams (http://www.mrcmekong.org/assets/Other-Documents/stakeholder-submissions/Final-010414-Eng-Open-letter-to-the-4-govt-on-DSH.pdf (Retrieved on 5 August 2018)). What's more, in 2014, the Vietnam Rivers Network held 15 consultation workshops in the Mekong Delta with the participation of thousands of people, including farmers, women, and representatives of local communities from 12 provinces in the Delta. It was the first time that community members in the region had heard about Don Sahong and other proposed mainstream dams. People shared concerns about potential impacts to their livelihoods and living conditions, discussing changes in water flow, sediment, and fish migration. There was general agreement that the Mekong River is a shared water resource and common asset among Mekong countries (https://wrm.org.uy/all-campaigns/mekong-governments-listen-to-the-people-statement-to-sign-on/ (retrieved on 5 August 2018)).

Although it is largely undocumented, it is our sense that there are numerous practices of information exchange among rural residents operating throughout the basin. These residents, including farmers and fisherfolk, share knowledge regarding the shifting environmental conditions of the basin through the use of social media as well as knowledge of development projects and their immediate impacts (Ian Baird, personal communication, 27 March 2017). NGOs are also generating knowledge, with organizations such as the NGO Forum for Cambodia conducting community surveys about the process of dam construction. With regard to the Sesan 2 dam, they found that "developers fell far short of Cambodia's legal requirements for obtaining free, prior, and informed consent from affected communities, or even conducting genuine consultations. Many people, the survey found, had not even been given official information about the project" (https://news.mongabay.com/2017/10/if-its-going-to-kill-us-ok-well-die-villagers-stand-firm-as-cambodian-dam-begins-to-fill/ (retrieved on 4 August 2018)).

Still, the politics of knowledge production and transfer in the Mekong basin remain inordinately influenced by the developmental narratives of state actors whose tropes are bolstered by the technological and engineering-based discourses that have been mentioned throughout this article. And of course, these discourses are undergirded by powerful political-economic interests manifesting across a range of state and industry actors who benefit materially from hydropower development. It remains to be seen to what extent knowledge claims challenging the efficacy of the Lao government's three mainstream hydroelectric dams, whatever the source, will actually influence the current dams' designs and operations, or result in the cancellation of future projects. Both the political and epistemological boundaries that we have explored here are at present conducive to enhancing state decision-making power over development initiatives. This insight makes it hard not to conclude that the types of knowledge, however contested, that are currently privileged in public knowledge controversies over hydropower development in the Mekong basin are the product of brute political–economic interests. The states of the region, as exemplified by the Lao government's actions described here, have seen hydropower development as conducive to both economic development and, to the extent that development can be portrayed as "successful", political legitimation. The technical and engineering knowledge that is driving the construction of the Xayaburi, Don Sahong, and Pak Beng dams is a critical component of the entire enterprise.

This raises a particularly daunting challenge for the coalitions of basin residents, academic researchers, NGOs, and others who question the dominant hydropower and developmental narratives of state actors. While the obstacles are formidable, there can be no question that we are seeing a diversification of knowledge bases—including competing scientific claims and the claims generated by local communities, NGOs, and their allies—around which decisions on hydropower dams are made and challenged. At the least, this diversification in turn has created arenas for a broader range of political actors to confront state discourses of hydropower development and, however marginally, influence or mitigate the consequences of dam construction. Communities throughout the basin are reaching across political borders to share experiences and information, and they are simultaneously challenging the epistemological barriers that have historically marginalized their local knowledges. There are clearly practical consequences associated with challenges to these borders and barriers: the livelihoods of millions of people and the health of critical ecosystems are at stake.

At a conceptual level, a focus on knowledge flows around large dams raises critical questions about the base of information on which decisions with wide-ranging socioecological impacts (e.g., the construction of hydroelectric facilities) are made and how the actors making those decisions—typically state officials—sift through sometimes conflicting or incomplete knowledge sets. Yet flows of knowledge concerning the biophysical dynamics of transnational river systems such as the Mekong cannot be studied in isolation from other crucial processes influencing water governance. As we have argued, borders at the geopolitical level and boundaries between different knowledge domains are deeply implicated in the public knowledge controversies emerging from the construction and operation of large dams and other development interventions. These connections are certainly apparent in the case of the Mekong dams that we have highlighted here, but are also influential in transnational river basins throughout the world, and for other kinds of mega-projects. It is our hope that critical engagements with political borders and epistemological barriers in other cases will result in innovative theoretical and political practices that challenge their presumed rigidity in ways that foment the production and flow of knowledge that is more socioecologically just and inclusive.

Author Contributions: C.A.F. and C.S.S. contributed equally to the conceptualization, research, analysis, writing and editing of this paper.

Funding: This research received no external funding.

Acknowledgments: We would like to thank Jonathan Chipman for making the map and Michaela Caplan for research assistance. Thank you also for the very insightful feedback from two anonymous reviewers.

Conflicts of Interest: The authors declare no conflict of interest.

References

1. Kondolf, G.M.; Rubin, Z.K.; Minear, J.T. Dams on the Mekong: Cumulative sediment starvation. *Water Resour. Res.* **2014**, *50*, 5158–5169. [CrossRef]
2. Lu, X.; Kummu, M.; Oeurng, C. Reappraisal of sediment dynamics in the Lower Mekong River, Cambodia. *Earth Surf. Process. Landf.* **2014**, *39*, 1855–1865. [CrossRef]
3. Räsänen, T.A.; Koponen, J.; Lauri, H.; Kummu, M. Downstream hydrological impacts of hydropower development in the Upper Mekong Basin. *Water Resour. Manag.* **2012**, *26*, 3495–3513. [CrossRef]
4. Walling, D.E. The changing sediment load of the Mekong River. *AMBIO J. Hum. Environ.* **2008**, *37*, 150–157. [CrossRef]
5. Baran, E.; Myschowoda, C. Dams and fisheries in the Mekong Basin. *Aquat. Ecosyst. Health Manag.* **2009**, *12*, 227–234. [CrossRef]
6. Dugan, P.J.; Barlow, C.; Agostinho, A.A.; Baran, E.; Cada, G.F.; Chen, D.; Marmulla, G. Fish migration, dams, and loss of ecosystem services in the Mekong basin. *AMBIO J. Hum. Environ.* **2010**, *39*, 344–348. [CrossRef]
7. Orr, S.; Pittock, J.; Chapagain, A.; Dumaresq, D. Dams on the Mekong River: Lost fish protein and the implications for land and water resources. *Glob. Environ. Chang.* **2012**, *22*, 925–932. [CrossRef]
8. Ou, C.; Winemiller, K.O. Seasonal hydrology shifts production sources supporting fishes in rivers of the Lower Mekong Basin. *Can. J. Fish. Aquat. Sci.* **2016**, *73*, 1342–1362. [CrossRef]
9. Thomas, G. *Sambor Hydropower Dam Alternatives Assessment*; Natural Heritage Institute: San Francisco, CA, USA, 2017.
10. Thomas, K.A. The river-border complex: A border-integrated approach to transboundary river governance illustrated by the Ganges River and Indo-Bangladeshi border. *Water Int.* **2017**, *42*, 34–53. [CrossRef]

11. Ziv, G.; Baran, E.; Nam, S.; Rodríguez-Iturbe, I.; Levin, S.A. Trading-off fish biodiversity, food security, and hydropower in the Mekong River Basin. *Proc. Natl. Acad. Sci. USA* **2012**, *109*, 5609–5614. [CrossRef] [PubMed]

12. Piman, T.; Lennaerts, T.; Southalack, P. Assessment of hydrological changes in the lower Mekong Basin from Basin-Wide development scenarios. *Hydrol. Process.* **2013**, *27*, 2115–2125. [CrossRef]

13. Kingston, D.G.; Thompson, J.R.; Kite, G. Uncertainty in climate change projections of discharge for the Mekong River Basin. *Hydrol. Earth Syst. Sci.* **2011**, *15*, 1459. [CrossRef]

14. Hirsch, P. The shifting regional geopolitics of Mekong dams. *Political Geogr.* **2016**, *51*, 63–74. [CrossRef]

15. Denton, J. Despite the Risks, Laos Goes Ahead with Pak Beng Dam. *Asia Times*. 11 February 2017. Available online: http://www.atimes.com/article/despite-risks-laos-goes-ahead-pak-beng-dam/ (accessed on 1 July 2018).

16. ICEM. *MRC Strategic Environmental Assessment (SEA) of Hydropower on the Mekong Mainstream: Summary of the Final Report*; International Centre for Environmental Management: Hanoi, Viet Nam, 2010.

17. Hirsch, P. China and the cascading geopolitics of Lower Mekong dams. *Asia-Pac. J.* **2011**, *9*, 2.

18. Narin, S. 'No Impact' on Fish from Major Hydropower Projects. *PM Claims*. 2017. Available online: https://www.voacambodia.com/a/no-impact-on-fish-from-major-hydropower-projects-pm-claims/4045015.html (accessed on 5 July 2018).

19. Barry, A. *Material Politics: Disputes along the Pipeline*; Wiley Blackwell: Oxford, UK, 2013.

20. Boelens, R.; Shah, E.; Bruins, B. Contested knowledges: Large dams and mega-hydraulic development. *Water* **2019**, *11*, 416. [CrossRef]

21. Waitt, G. Doing Foucauldain discousrse analysis—Revealing social realities. In *Qualitative Research Methods in Geography*, 4th ed.; Hay, I., Ed.; Oxford University Press: Oxford, UK, 2016; pp. 288–312.

22. Yanow, D. *Conducting Interpretive Policy Analysis*; Sage Publications: Thousand Oaks, CA, USA, 2000.

23. International Centre for Environmental Management. Available online: http://icem.com.au (accessed on 7 July 2018).

24. Sneddon, C.; Fox, C. Rethinking transboundary waters: A critical hydropolitics of the Mekong basin. *Political Geogr.* **2006**, *25*, 181–202. [CrossRef]

25. Hirsch, P. Laos mutes opposition to controversial Mekong Dam. 2013. Available online: https://www.chinadialogue.net/article/show/single/en/6509-Laos-mutes-opposition-to-controversial-Mekong-dam (accessed on 10 July 2018).

26. Sassoon, A. Siem Reap Declaration Charts New Future for MRC. *The Phnom Penh Post*. 2018. Available online: https://www.phnompenhpost.com/national/siem-reap-declaration-charts-new-future-mrc (accessed on 9 July 2018).

27. Agnew, J. Still trapped in territory? *Geopolitics* **2010**, *15*, 779–784. [CrossRef]

28. Grundy-Warr, C. B/ordering nature and biophysical geopolitics: A response to Hirsch. *Political Geogr.* **2017**, *58*, 131–135. [CrossRef]

29. Lidskog, R.; Sundqvist, G. When Does Science Matter? International Relations Meets Science and Technology Studies. *Glob. Environ. Political* **2015**, *15*, 1. [CrossRef]

30. Jasanoff, S. *States of Knowledge: The Co-Production of Science and Social Order*; Routledge: London, UK, 2004.

31. Jasanoff, S. Genealogies of STS. *Soc. Stud. Sci.* **2012**, *42*, 435–441. [CrossRef]

32. Lamb, V. "Where is the border?" Villagers, environmental consultants and the 'work' of the Thai-Burma border. *Polical. Geogr.* **2014**, *40*, 1–12. [CrossRef]

33. Chaitrong, W. Hydro Plants Power Laos into Middle-Income Status. *Sunday Nation*. 2017. Available online: http://www.nationmultimedia.com/detail/national/30325617 (accessed on 1 July 2018).

34. Li, T.M. *The Will to Improve: Governmentality, Development, and the Practice of Politics*; Duke University Press: Durham, NC, USA, 2007.

35. Castro, J.E. Water governance in the twentieth-first century. *Ambiente Sociedade* **2007**, *10*, 97–118. [CrossRef]

36. Conca, K. *Governing Water: Contentious Transnational Politics and Global Institution Building*; Massachusetts Institute of Technology: Cambridge, MA, USA; London, UK, 2006.

37. Furlong, K. Hidden theories, troubled waters: International relations, the 'territorial trap', and the Southern African Development Community's transboundary waters. *Political Geogr.* **2006**, *25*, 438–458. [CrossRef]

38. Callon, M. The role of lay people in the production and dissemination of scientific knowledge. *Sci. Technol. Soc.* **1999**, *4*, 81–94. [CrossRef]

39. Latour, B. A textbook case revisited—Knowledge as a mode of existence. In *The Handbook of Science and Technology Studies*, 3rd ed.; Hackett, E.J., Amsterdamska, O., Michael, E.L., Wajeman, J., Eds.; MIT Press: Cambridge, MA. USA, 2007; pp. 83–112.

40. Wynne, B. May the sheep safely graze? A reflexive view of the expert-lay knowledge divide. In *Risk, Environment and Modernity: Towards a New Ecology*; Lash, S., Szerszynski, B., Wynne, B., Eds.; Sage: London, UK, 1996; pp. 44–83.

41. Yearley, S. Making systematic sense of public discontents with expert knowledge: Two analytical approaches and a case study. *Public Underst. Sci.* **2000**, *9*, 105–122. [CrossRef]

42. Forsyth, T. *Critical Political Ecology*; Routledge: New York, NY, USA, 2004.

43. Zimmerer, K.S.; Bassett, T.J. (Eds.) *Political Ecology: An Integrative Approach to Geography and Environment-Development Studies*; Guilford Press: New York, NY, USA, 2003.

44. Baird, I.G.; Flaherty, M.S. Mekong River fish conservation zones in southern Laos: Assessing effectiveness using local ecological knowledge. *Environ. Manag.* **2005**, *36*, 439–454. [CrossRef] [PubMed]

45. Valbo-Jørgensen, J.; Poulsen, A.F. Using local knowledge as a research tool in the study of river fish biology: Experiences from the Mekong. *Environ. Dev. Sustain.* **2000**, *2*, 253–376. [CrossRef]

46. Purcell, M.; Brown, J.C. Against the local trap: Scale and the study of environment and development. *Prog. Dev. Stud.* **2005**, *5*, 279–297. [CrossRef]

47. Gururani, S.; Vandergeest, P. Introduction: New frontiers of ecological knowledge: Coproducing knowledge and governance in Asia. *Conserv. Soc.* **2014**, *12*, 343. [CrossRef]

48. St. Martin, K.; McCay, B.J.; Murray, G.D.; Johnson, T.R.; Oles, B. Communities, knowledge and fisheries of the future. *Int. J. Glob. Environ. Issues* **2007**, *7*, 221–239. [CrossRef]

49. Barry, A. Political situations: Knowledge controversies in transnational governance. *Crit. Policy Stud.* **2012**, *6*, 324–336. [CrossRef]

50. Whatmore, S.J. Mapping knowledge controversies: Science, democracy and the redistribution of expertise. *Prog. Hum. Geogr.* **2009**, *33*, 587–598. [CrossRef]

51. Kuenzer, C.; Campbell, I.; Roch, M.; Leinenkugel, P.; Tuan, V.Q.; Dech, S. Understanding the impact of hydropower developments in the context of upstream–downstream relations in the Mekong river basin. *Sustain. Sci.* **2013**, *8*, 565–584. [CrossRef]

52. Wells-Dang, A.; Soe, K.N.; Inthakoun, L.; Tola, P.; Socheat, P.; Van, T.T.; Youttananukorn, W. A political economy of environmental impact assessment in the Mekong Region. *Water Altern.* **2016**, *9*, 33–55.

53. International Rivers. The Xayaburi Dam. 2011. Available online: https://www.internationalrivers.org/campaigns/xayaburi-dam (accessed on 9 July 2018).

54. Zaffos, J. Life on Mekong Faces Threats as Major Dams begin to Rise. 2014. Available online: http://e360.yale.edu/features/life_on_mekong_faces_threats_as_major_dams_begin_to_rise (accessed on 1 August 2018).

55. International Rivers. The Don Sahong Dam. 2015. Available online: https://www.internationalrivers.org/campaigns/don-sahong-dam (accessed on 10 July 2018).

56. Mekong River Commission. Facts about the Pak Beng Hydropower Project. 2017. Available online: www.mrcmekong.org/assets/Publications/Fact-sheet-of-Pak-Beng-26-Jan-2017.pdf (accessed on 3 July 2018).

57. International Rivers. *Independent Expert Review of the Pak. Beng Dam Environmental Impact Assessment and Supporting Documents*. 2017. Available online: https://www.internationalrivers.org/resources/independent-expert-review-of-the-pak-beng-dam-eia-16488 (accessed on 1 July 2018).

58. Scott, J.C. *Seeing Like a State: How Certain Schemes to Improve the Human Condition Have Failed*; Yale University Press: New Haven, CT, USA, 1998.

59. Sneddon, C.; Fox, C. Water, Geopolitics, and Economic Development in the Conceptualization of a Region. *Eurasian Geogr. Econ.* **2012**, *53*, 143–160. [CrossRef]

60. Hensengerth, O. Where is the power? Transnational networks, authority and the dispute over the Xayaburi Dam on the Lower Mekong Mainstream. *Water Int.* **2015**, *49*, 911–928. [CrossRef]

61. Mekong River Commission. *Procedures for Notification, Prior Consultation and Agreement (PNPCA), Proposed Xayaburi Dam Project–Mekong River, Prior Consultation Project Review Report*; MRC: Vientiane, Laos, 2011.

62. Pohjonen, M. Xayaburi Run-of-River HPP: Poyry Review of the Project. 2012. Available online: https://www.oecdwatch.org/cases/Case_259/1173/at_download/file (accessed on 3 July 2018).

63. Herbertson, K. Sidestepping science: Review of the Poyry Report on the Xayaburi Dam. 2011. Available online: https://www.internationalrivers.org/sites/default/files/attached-files/poyryreview_areasofnon-compliance.pdf (accessed on 9 July 2018).
64. Fawthrop, T. Killing the Mekong Dam by Dam. 2016. Available online: https://thediplomat.com/2016/11/killing-the-mekong-dam-by-dam/ (accessed on 8 July 2018).
65. Baran, E. *Strategic Environmental Assessment of Hydropower on the Mekong Mainstream*; International Centre for Environmental Management (ICEM): Brisbane, Australia, 2010.
66. Csiki, S.; Rhoads, B.L. Hydraulic and geomorphological effects of run-of-river dams. *Prog. Phys. Geogr.* **2010**, *34*, 755–780. [CrossRef]
67. Kemp, M. The Dammed Don: Lao Hydropower Project Pushes Ahead Despite Alarm from Scientists. Mongabay. 2017. Available online: https://news.mongabay.com/2017/01/the-dammed-don-lao-hydropower-project-pushes-ahead-despite-alarm-from-scientists/ (accessed on 10 July 2018).
68. Sabo, J.L.; Ruhi, A.; Holtgrieve, G.W.; Elliott, V.; Arias, M.E.; Ngor, P.B.; Räsänen, T.A.; Nam, S. Designing river flows to improve food security futures in the Lower Mekong Basin. *Science* **2017**, *358*, 1270. [CrossRef] [PubMed]
69. WWF. WWF announces discovery of 157 new species in Southeast Asia. 2018. Available online: https://whnt.com/2018/12/14/wwf-announces-discovery-of-157-new-species-in-southeast-asia/ (accessed on 16 January 2019).
70. Greenstein, G.; Meyer, A. Fork in the River. *Slate*. 2017. Available online: http://www.slate.com/articles/news_and_politics/roads/2017/04/new_dams_will_transform_the_lives_of_tens_of_millions_of_people_in_southeast.html (accessed on 15 July 2018).
71. Laureyn, P. The Mekong, Dammed to Die. 2017. Available online: http://www.ipsnews.net/2017/11/mekong-dammed-die/ (accessed on 1 July 2018).
72. Hunt, L. What is the Value of the Mekong River? *The Diplomat*. 2016. Available online: https://thediplomat.com/2016/01/what-is-the-value-of-the-mekong-river/ (accessed on 13 July 2018).
73. Ansar, A.; Flyvbjerg, B.; Budzier, A.; Lunn, D. Should we build more large dams? The actual costs of hydropower megaproject development. *Energy Policy* **2014**, *69*, 43–56. [CrossRef]
74. Intralawan, A.; Wood, D.; Frankel, R. Research: Economic Evaluation of hydropower projects in the Lower Mekong Basin. 2017. Available online: https://www.mekongeye.com/2017/05/12/research-economic-evaluation-of-hydropower-projects-in-the-lower-mekong-basin/ (accessed on 1 July 2018).
75. De Sousa Júnior, W.C.; Reid, J. Uncertainties in Amazon Hydropower Development: Risk Scenarios and Environmental Issues around the Belo Monte Dam. *Water Alternatives* **2010**, *3*, 249–268.
76. Boyle, D. Solar Surge Threatens Hydro Future on Mekong. *VOA*. 11 April 2018. Available online: https://www.voanews.com/a/solar-surge-threatens-hydro-future-mekong/4341660.html (accessed on 9 July 2018).
77. Eyler, B.; Weathersby, C. Generating Sustainable Energy Solutions in the Mekong River Basin. 2017. Available online: https://www.stimson.org/content/letters-mekong-mekong-power-shift-emerging-trends-gms-power-sector (accessed on 8 July 2018).
78. High, H.; Baird, I.G.; Barney, K.; Vandergeest, P.; Shoemaker, B. Internal resettlement in Laos. *Crit. Asian Stud.* **2009**, *41*, 605–620. [CrossRef]
79. China Steps Back from Islet Blasting Plan Says Thais Living by Mekong Would Suffer. *Bangkok Post*. 2018. Available online: https://www.bangkokpost.com/news/general/1382091/china-steps-back-from-islet-blasting-plan (accessed on 1 July 2018).
80. Bernstein, R. China's Mekong Plans Threaten Disaster for Countries Downstream. *Foreign Policy*. 2017. Available online: http://foreignpolicy.com/2017/09/27/chinas-mekong-plans-threaten-disaster-for-countries-downstream/ (accessed on 19 July 2018).
81. Seiff, A. In the Mekong, Questions Arise over Impact of Favoring Hydropower. *Devex*. 2018. Available online: https://www.devex.com/news/in-the-mekong-questions-arise-over-impact-of-favoring-hydropower-92384 (accessed on 10 July 2018).

water

MDPI

Article

Dams and Damages. Conflicting Epistemological Frameworks and Interests Concerning "Compensation" for the Misicuni Project's Socio-Environmental Impacts in Cochabamba, Bolivia

Paul Hoogendam [1],* and Rutgerd Boelens [1,2,3,4,*]

[1] Department of Environmental Sciences, Wageningen University, P.O. Box 47, 6700 AA Wageningen, The Netherlands

[2] CEDLA Centre for Latin American Research and Documentation, University of Amsterdam, Roetersstraat 33, 1018 WB Amsterdam, The Netherlands

[3] Faculty of Agricultural Sciences, Universidad Central del Ecuador, Ciudadela Universitaria, Quito 170129, Ecuador

[4] Department of Social Sciences, Pontificia Universidad Católica del Perú, Av. Universitaria 1801, San Miguel 15088, Lima, Peru

* Correspondence: hoo@ces-bolivia.com (P.H.); rutgerd.boelens@wur.nl (R.B.); Tel.: +591-714-20820 (P.H.); +31-317-484190 or +31-20-525-3498 (R.B.)

Received: 2 July 2018; Accepted: 3 January 2019; Published: 26 February 2019

check for updates

Abstract: The Misicuni multipurpose hydraulic project was designed to transfer water from a neighboring watershed to the Cochabamba Valley in the center of Bolivia for domestic, hydropower, and agricultural use. The project involved the construction of a 120 m high large dam and a 19 km transfer tunnel, which negatively affected the rural indigenous host communities that were deprived of productive lands, houses, and livelihoods. This article critically analyzes the process to compensate for harmful effects, demonstrating the existence of divergent knowledge systems, interpretations, and valuing of what was affected and how the impacts had to be compensated. The analysis shows that the compensation was fundamentally a process of negotiation about the meaning and the contested commensuration that was implemented in a context of unequal power relations between state institutions and the indigenous population. This led to unfavorable arrangements for the affected communities. The article details the discussions about impacts, knowledge, and values of key elements of the compensation process and highlights how "compensation" was embedded in the wider struggle over territorial control and natural resource governance. The unreliability of the state institutions worsened the negative impacts for the rural communities because the negotiated outcomes were not always materialized.

Keywords: large dams; socioenvironmental impacts; compensation measures; knowledge systems; commensuration; negotiation; territorial control; Bolivia

1. Introduction

The Misicuni Project (the project's official name is Proyecto Multiple Misicuni, translated as Misicuni Multipurpose Project. For reasons of simplicity, in the text we use the colloquial name Misicuni Project) is a long-awaited water transfer project to deliver water to the central Bolivian city of Cochabamba and surrounding municipalities. Construction started in 1997 and operation began in the 2017/2018 period. The multipurpose project consists of a large dam and a transfer tunnel to conduct water from the Misicuni watershed to the Cochabamba valley (Figure 1), where it is used to generate hydropower, after which it is distributed for domestic use and irrigation.

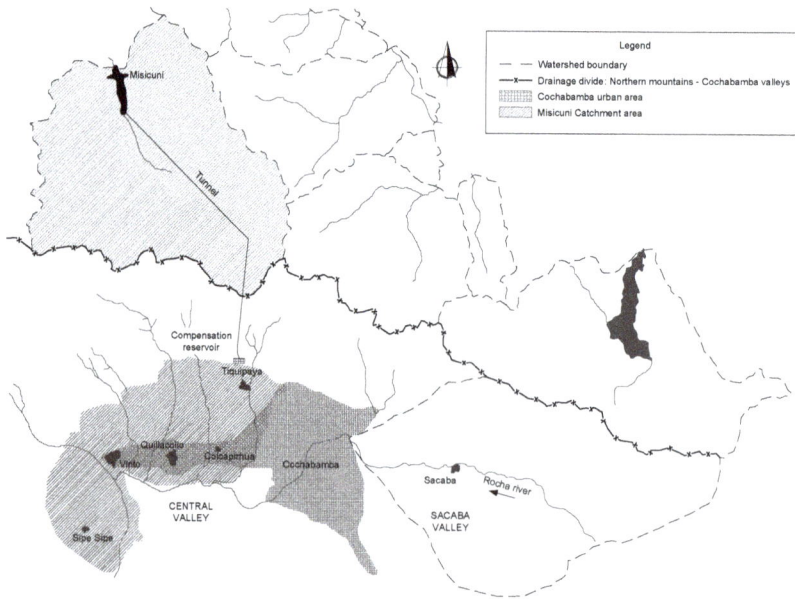

Figure 1. Location of the Misicuni project in relation to the Cochabamba Valley, Bolivia. Elaboration: Paul Hoogendam and Ronald Brañez.

Water delivery from the Misicuni dam is a relief for the inhabitants of the Cochabamba valley. From the 1940s onward, they suffered periods of serious water scarcity. Since then, the Misicuni project became visualized as the main solution to resolving the water shortage and inducing regional development [1,2]. The Misicuni additional water volume is quite substantial compared to actual water availability; the Misicuni water more than doubles the drinking water volume for the over 1,100,000 residents of the Cochabamba metropolitan area and almost doubles irrigation water as well [3,4]. After decades of frustration and doubts about the project's implementation, optimism reigns among city dwellers and irrigators about its outcomes. The Misicuni Company office shares this optimism, pleased with the conclusion of the construction works and the accomplished water storage.

In stark contrast to this, the highland indigenous peasant communities—where the dam was built—persist in protesting against the project. They demand fulfillment of unresolved commitments and claim their share in future benefits [5]. The community leaders express disillusion and deception about the impact on their livelihoods, territory, and communities, and about the way the compensation process has evolved. Their opinions strongly influence neighboring communities who object to additional water transfer projects because of the negative experiences with the Misicuni project.

In this article, we analyze the reasons why the highland communities feel unsatisfied with and deceived by the compensation process and results. We describe the different compensation measures taken and analyze the compensation process, showing how this, even in a context of well-intentioned state organizations, obliged the communities to constantly struggle for their rights and defend their knowledge frameworks related to water and dignified territorial livelihoods (see also the conceptual backgrounding in this special issues' introductory article [6]).

As part of our analytical exercise, we compare the Misicuni events with the framework developed by the World Commission on Dams (WCD) on compensation measures. After their worldwide evaluation of the impacts of large dams, the WCD formulated a series of strategic priorities and policy principles for future dam development, incorporating the changing ideas on human rights, the right to development, and the imperative nature of sustainability [7], as well as including considerations to ensure that negotiation and decision-making processes are fair to the host communities affected by the construction of mayor hydraulic works (see also [8–10]).

This article is based on the first author's responsibility for the water balance study for the Misicuni irrigation component [11] and his over 20 years of working experience as a water professional and action-researcher in technical, social, and political discussions before, around, and since the occurrences of the Cochabamba Water War (2000). Both authors have worked and lived in Andean countries for several decades, interacting with state, peasant, and indigenous organizations in academic and action research. We studied the historic development of the Misicuni project through the revision of technical reports and local and international scientific studies. Review of the Misicuni internal reports, in-depth interviews with the Misicuni technical staff responsible for compensation measures, and group interviews with former and actual leaders of the affected communities provided us with detailed information of the Misicuni compensation policy, measures, and outcomes. We triangulated their opinions with a detailed review of newspaper articles about the conflicts that emerged during the construction works, followed by renewed interviews with responsible professional staff and community leaders on their current interpretation of historical events. Personal conversations with peasant leaders in the main surrounding areas (where additional transfer projects are proposed) provided us with the interpretations of neighboring communities about the compensation measures taken by the Misicuni project. All interviews took place in 2017 and 2018.

The structure of the article is as follows. In the second section, we describe the Misicuni project's history, its main hydraulic works, and the biophysical and socioeconomic setting of the indigenous communities where these works are located. We also show how preparatory studies presented an almost virgin territory, minimizing the very existence of these communities and the project's socio-environmental impacts. This led to a rather limited compensation plan, which during implementation was fiercely contested by the communities. In the third section, we develop our theoretical framework to analyze the Misicuni compensation process. We start with the comprehensive framework of the WCD for compensation measures and extend the analysis to matters of authority, power, and knowledge in negotiation processes regarding territorial change. As we argue here, the issue of "commensuration" is crucial. In the fourth section, we apply these analytical viewpoints to scrutinize three critical compensation issues in the Misicuni project: expropriation of land, compensation for construction aggregates, and mitigation of dried wetlands. We show how in each of these issues, the compensation process was a field of contestation and negotiation about knowledge, meaning, and value. Finally, in our conclusions, we argue that a crucial part of compensation processes is the unequal struggle over decisions that must be made on what is lost, what is to be compensated, the meaning and values of displaced items, beings and relationships, and the measurement units for compensation. These decisions are made within a context of unequal economic, political, institutional, and discursive power relationships and are marked by particular worldviews and epistemologies within the wider struggle about territorial control and natural resource governance.

2. History and Imaginaries of the Misicuni Project

2.1. The Misicuni Multipurpose Project

In the first half of the 20th century, the population of Cochabamba city and its neighboring municipalities started to suffer from water shortage, which over the last decades became more severe due to accelerated population growth. One of the main proposals to resolve water scarcity was to transfer water from the nearby Misicuni watershed located in the mountain range north of the city, a natural drain to the Amazon River. By the end of the 1950s, a first serious substantive outline of the Misicuni multipurpose water transfer project was made. It encompassed three dams located 1600 m higher than the city, a transfer tunnel from the Misicuni watershed to the Cochabamba Valley, and a hydropower plant at the valley floor. The transferred water would generate hydropower and then be distributed for domestic use and irrigation. Power supply and irrigation were considered to be elementary conditions for regional modernization through agricultural intensification and industrial development [1,2].

In the early 1970s, a prefeasibility study was done to determine the project's costs. This study confirmed the need to construct the dams, transfer tunnel, hydropower plant, drinking water distribution network, and a full-fledged 5000-hectare irrigation system. The high costs of the project and the then political preference for investment in agriculture and power generation in the oriental departments of Bolivia jeopardized the construction of the Misicuni project for decades [2]. A series of popular mobilizations from the 1970s to the 1990s kept the proposal for the Misicuni project alive, making it synonymous with departmental development and regional identity [12–14]. Finally, popular and political pressure made state and international development banks fund its construction [15].

For its implementation, the Misicuni Company was founded—an autonomous public enterprise responsible for contracting preparatory studies (hydrology, geology, engineering, environmental impact, etc.), construction works, and supervision. As a public enterprise under control of central government, the Misicuni Company had a great scope to request cooperation of other state institutions in order to solve project related problems.

Misicuni construction works took place in subsequent phases. Between 1997 and 2005, the transfer tunnel was excavated, and perforation activities took place from both ends, which led to the first period of intensive contact with the indigenous communities at the upstream end of the tunnel. After tunnel delivery, administrative and financial problems slowed further implementation, but eventually, in 2009, the state contracted a consortium to build the dam and complementary hydraulic works in the Misicuni valley. Eventually, by the end of 2016, the dam was ready to collect water, and by the end of 2017, it entered into operation.

Overall, the following works were constructed (see Figure 1): A 120 m high dam with an overall water volume of 180 million cubic meters and an inundation area of 470 hectares; a 19.4 km tunnel from the Misicuni reservoir to the Cochabamba valley; A 120 MW hydropower plant aside a 350,000 m^3 compensation reservoir to temporarily store the processed water for later distribution among domestic use and irrigation.

From a social and political perspective, the Misicuni project suffered a history of ups and downs that are extensively documented in [15]. She characterizes the project as "vernacular modernism" since it was promoted and defended by coalitions of city dwellers and Cochabamba Valley peasant irrigators as opposed to "high modernism" projects that used to be activated by a modernizing state elite [16,17]. However, from our examination, we concluded that the popular interest in and defense of the project did not result in a more protective stand towards affected host communities. Similar to the position and actions of dominant mega-hydraulic proponents in other cases around the world (e.g., [18–26]), the popular supporters (city inhabitants and valley peasants) ignored the fate of the affected highland indigenous communities.

The urge to construct the Misicuni project increased during the 2000 Cochabamba Water War due to the pressure of a coalition of city dwellers and surrounding peasant irrigators. On the one hand, urban population and their political representatives saw the Misicuni project as a final and conflict-free solution to the drinking water scarcity when compared to the alternative of drilling extra wells in the nearby valley, which was highly contested by the surrounding municipalities. On the other hand, peasant irrigators saw the Misicuni project as a solution to their irrigation water problems and a way to avoid city dwellers' claims over existing water sources over which peasant families held customary rights (e.g., [27–29]). For them, Misicuni resolved both practical and normative issues that were threatened by the urban-steered capitalization process that led to the Water War [30,31]. Neither of the two groups included in their analysis the fortune of the highland communities [13].

It is striking to see that in the extensively documented Cochabamba Water War portrayed with images of social justice for the rural and urban poor (e.g., [28,32–34]), the far poorer indigenous families living in the inundation area of the Misicuni dam were mainly ignored (with few exceptions like [12,15]). Many critical studies that emphasized the negative effects of water privatization and modernization embraced the Misicuni project because of it being a solution supported by the "water warriors" from Cochabamba city and neighboring irrigators, ignoring issues of social injustice for the indigenous highland communities. This uncritical support was due to scholars/activists' unfamiliarity with the Misicuni valley and its inhabitants. In the popular and academic imaginary, the Misicuni dam was to be constructed in no-one's land [35,36]. As Nixon [37] (p. 75) stated, " ... through the invention of emptiness (...) 'underdeveloped' people on 'underdeveloped' land can be rendered spectral uninhabitants whose territory may be cleared for the staging of the national theatrics of mega dams".

2.2. The Socio-Environmental Context of the Dam and Reservoir Site

In reality, however, almost 200 families inhabit the 470 hectares inundation area, making up a total population of around 1200 people. These families are part of eight rural indigenous communities: Patapampa, Misicuni, Uyuni, Sivingani, Cochamayu, Aguadas, San Isidro, and Putucuni (Figure 2). Their communal territories consist of adjacent land strips that lead up from the valley bottom to the upper parts of the surrounding mountains. Their spatial organization is a localized version of the vertical dominion of agro-climatological zones common to most Andean regions. Access to different altitudinal zones enables communities to combine distinct climatic conditions and natural resource qualities.

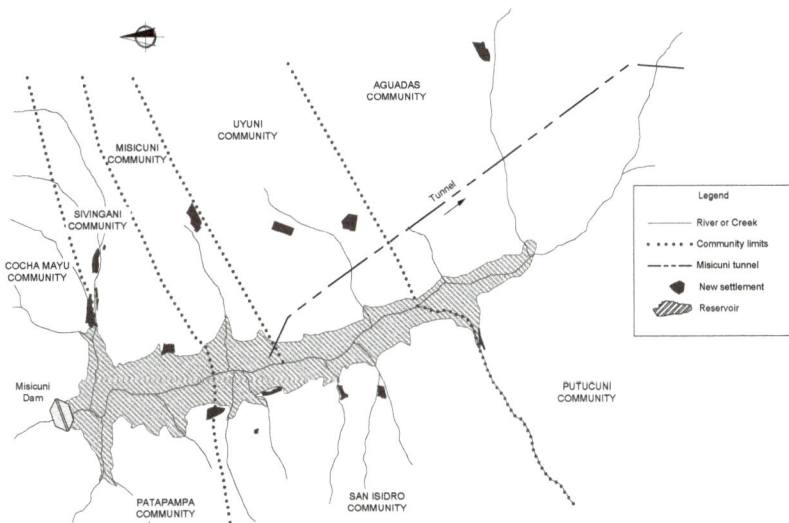

Figure 2. Communities in the Misicuni inundation area. Based on [38]; elaboration: Paul Hoogendam and Ronald Brañez.

The spatial organizations of these communities are alike—the valley bottom near the river, relatively protected from harsh climate conditions, is used for houses, agricultural plots, and wetlands. The lower mountainsides are covered by scattered plots on steep slopes, whereas the upper mountains are used for sheep, alpaca, and llama grazing. Agricultural production changes with altitude—the valley floor plots are used to grow potatoes and oca (Oxalis tuberosa), whereas the steeper mountainside plots are used for cold resistant crops like oat, barley, and bitter potatoes (Solanum curtilobum, used to produce freeze-dried potatoes that can be conserved for years). Soil qualities differ strongly by altitude—the flat plots near the river have more clayey, relatively deep soils with moderate to good fertility because of decades of incorporating organic material and manure, whereas the slope plots have stony, shallow, and chemically poor topsoils.

The highland families are organized in indigenous peasant communities with leadership rotating among all members. The main language spoken is Quechua, although most people, especially the young, also speak Spanish. Collective action covers many communal needs, like building the main road to the eight communities. Before the Misicuni project, the communities hardly had access to basic services, but water was available from many natural springs and the Misicuni River. There were no tap or sanitation systems, no electricity, and no telephone coverage. Two communities had schools up to 8th grade (12 years old) and one community had a very precarious health center but without regular attendance. Until the arrival of the Misicuni project, the region lacked municipal support for community improvement.

The living conditions in the Misicuni communities are extremely harsh and all families are poor to very poor. Although families engage in agricultural production and husbandry, the nearly 3800 m altitude, the low temperatures, and the hail and frost risks make production unreliable. Most of the produce is used for household consumption; only a minor part is sold at regional markets. Though seasonal labor migration is part of most families' economic strategy, the main resource base of their livelihood is their land.

2.3. The Compensation Process

The studies and discussions about the Misicuni project hardly mentioned the communities in the Misicuni river basin. The first environmental assessment study stated that the project would flood the living area of approximately 20 families and might cause some erosion around the construction site [39]. The official 1996 environmental impact study mentioned that construction would entail "clearance of approximately 140 hectares of cultivable land" and "resettlement of approximately 400 people currently living in the flood zone". Keeping up with a "philosophy of environmental protection", the Misicuni Company promised to "resettle and compensate the population, inform them about the project, and respond immediately to any grievance, complaint, or protest" [40] (p. 47). Clearly, the aim of the environmental studies was to present the project's negative impacts as almost negligible and repairable via compensation, resettlement, and information (see also [23]).

Soon after starting tunnel excavation, the Misicuni Company began a campaign to measure land and landed improvements (houses, corrals, fences, trees). The families initially cooperated but, due to the uncertainty about compensation, shortly after protested and demanded basic agreements about future actions. After a period of negotiation, in 1998, the Misicuni Company signed a general agreement with the communities' leaders that determined six conditions to proceed to future land transfer: emission of land titles, evaluation and compensation, a social base line study, evaluation of houses and improvements, relocation of the cemetery, and technical assistance for agricultural and husbandry production.

The emission of land titles was necessary for the Misicuni Company to obtain official property rights. Therefore, families' land titles were to be formally registered and then transferred to the Misicuni Company (for the politics and complexities of such formalization, see [41]). A university consultancy unit contracted by the Misicuni Company realized the valuation of the land, identifying several categories of land quality and their respective values [38]. The mean value paid for was 870 USD per hectare, making up a total of 409,000 USD [42]. No value was assigned to land without productive use, house parcels, wastelands, and riverbeds.

Compensation for the houses was done likewise. A survey measured and valuated the houses of 175 families. For 110 families, replacement houses were constructed on uphill sites appointed by the communities, while 65 families preferred to receive compensation in cash. Most of the latter later regretted their preference, since the new houses cost 10 times more than the compensation money granted. The project also paid in cash for corrals and other landed improvements. Furthermore, the project transferred three churches, a health center, and a cemetery to higher lands. Between 2009 and 2013, the Misicuni Company helped in moving all family belongings and paid for Catholic services for reburying human rests [42].

After signing the 1998 general agreement, at first, the local population no longer opposed tunnel construction. Their quietness changed when reservoirs and springs started to dry out due to underground water flows into the recently excavated tunnel. To avoid protests, the Misicuni Company constructed community drinking water systems. However, these did not resolve the problems of watering animals and irrigation. Even so, communities' complaints were diluted after finishing tunnel construction in 2005, more so because they remained living in their homes and cultivating their already expropriated lands.

This situation lasted until the beginning of the dam construction in 2009. Before reinitiating construction works, experts contracted by the Misicuni Company updated the Environmental Impact Study [43]. This time, the description of the local communities was more precise, but the assessment of impacts on the indigenous population was, at most, naive. In its socioeconomic paragraph, the study foresaw "in general, positive impacts (for the local population), because of all year water availability and improvement of inhabitants' living conditions" [43] (p. 103), thus making a direct association between new reservoir water availability and increased well-being while ignoring that the local population had always lived in water abundance. Contrary to analyzing the local population as those affected by the project, the document even blamed the "local floating population" (expulsed from their inundated homes) as a possible source of "proliferation of poor houses, of bad appearance, without good sanitation services" and for being a pollution risk to the reservoir. Most strikingly, it stated that the local population would "negatively impact the scenery of the reservoir, for its esthetical quality will be spoiled by the incursion of *elements foreign to the natural environment*" [43] (p. 118, our italics).

Thus, the environmental experts turned the local population from "impacted" into "impactors" and from local inhabitants into foreigners, while at the same time transforming the dam and reservoir into "natural elements" of an Andean scenery that need to be conserved. Consequently, the study did not propose any mitigation measures beyond the already completed compensation process for the future inundation of land and houses, thus stimulating institutional blindness and ignoring the need for resources for the Misicuni Company's future interactions with the local communities.

This double neglect was the basis for the difficult relation between communities and the Misicuni Company. During construction works, the communities had to claim and protest for any demand to be heard. From 2009 onwards, they protested, among other causes, for the repair of fallen bridges, compensation for three extra hectares of land at the dam site that were not transferred in the earlier process, health care, electrification, a fish-culture tourism project, roads to the new hamlets, construction of two new bridges, and assistance for lost crops. They even had to threaten with roadblocks and encampment occupation to enforce their claims.

During this period, the local communities became far better organized in their position vis-a-vis the Misicuni Company and other state institutions involved. This was enhanced by concrete negative changes in their livelihood conditions. Dam-construction works obliged them to leave their homes, whereas their lands were excavated for aggregate exploitation. It was also result of their experience. The community leaders learned that the Misicuni Company reacted slowly to their demands, promised more than they could accomplish, and often argued that claims were beyond the company's responsibility, redirecting their demands to other (unreliable) state institutions.

To obtain stronger commitment from the institutions involved, in 2011, the peasant leaders proposed the conformation of a high-level commission, demanding participation of the departmental governor, the municipal mayor, members of parliament, and peasant union's representatives, aiming to streamline responsibilities and assure compliance of concrete agreements. Although originally the Misicuni Company and the departmental government did not agree to its conformation, they afterwards recognized its functionality since many compensation measures required complex political and institutional coordination.

Together with the conformation of the high-level commission, in 2012, the Misicuni Company set up a socioenvironmental unit to coordinate activities with and for the communities and introduced a mitigation trust fund for resettlement and environmental measures. This fund was used to quickly respond to all kinds of claims (avoiding long administrative procedures), thus tempering communities' protests. According to a Misicuni Company technician, "without the trust fund, the construction would never have been completed" (interview Misicuni technician, 2 April 2018).

3. Dams and Damages, Rights and Risks, Negotiation and Power

3.1. Dams and Damages

Large dams are often seen as fundamental for the provision of water and energy to an ever-growing part of the global population. Whereas from the 1950s until 1970s, such major hydraulic works were greatly embraced as positive developments, per se, awareness grew after the 1980s about the adverse consequences of dam construction and hydraulic control (see the overview in the introductory paper to this special issue, [6]). Human relocation, loss of sustainable livelihood, and damage to ecosystems prevail among the main negative outcomes mentioned. In some cases, the number of people affected is stunningly high, running into tens or hundreds of thousands for a single dam, with extreme cases as the Three Gorges Dam in China, where, according to official figures, 1.13 million people (but probably many more) have been displaced.

It is overly common that affected communities are not rightly compensated, leading to what is shamefully referred to as "resettlement poverty", as, for example, at Sudan's Merowe dam, where 50,000 people were violently displaced by the government and then experienced a 10–65% increase in poverty over two years because of poorer soil fertility and water access [16]. Such statements give credit to the general conclusion that large dam projects induce development and marginalization and give benefits and burdens in differential ways for different groups of people [7,44–48]. The burdens are often seen as inevitable for the sake of development and its need for water and hydropower. As Johnson stated in her description of the dramatically unjust history of Guatemala's Chixoy Dam, the loss of livelihoods of the affected communities "are casualties in the climate change opportunism accompanying efforts to build global 'clean, green energy' systems" [49] (p. 180) (see also [25]).

3.2. The WCD Framework on Compensation for Dam Damages

One of the important issues for the perception about benefits and burdens is how communities and people affected by dam construction are compensated for their losses and inconveniences. An important reference in this respect is the WCD's recommendations about reparation, restitution, and restoration of livelihoods and land compensation for relocated host communities [7]. These recommendations build on existing agreements and policies at national and international levels (among others: 1986 UN Declaration on the Right to Development, 1992 Rio Declaration on Environment and Development).

The comprehensive summary of the WCD Fifth Policy principle about compensation for affected populations reads:

"Rather than benefiting from them, many of those affected by dams are aware only of their negative impacts. To redress the balance, a process of joint negotiation with such groups is required, based on recognition of rights and assessment of risks. The aim of these negotiations is to agree on legally enforceable mitigation and development provisions, which recognize entitlements that improve livelihoods and quality of life. States and developers are responsible for resettling and compensating all affected people and satisfying them so that their livelihoods will be improved by moving from their current situation. Legal means, such as contracts and accessible recourse at national and international levels, should be used to ensure that responsible parties fulfill their commitments to agreed mitigation, resettlement and development provisions". [50] (p. 14)

This policy principle attends political, substantial, and procedural elements. At the political level, it underlines the need to recognize the rights of affected host communities and obliges that the overall outcome for the affected population should be an improvement of their living conditions and livelihood.

At the substantial level, it draws attention to the correct understanding of what is lost and should be compensated for—not just resources but livelihood, not just land but territory. The WCD acknowledges that displacement refers to both "physical displacement" and "livelihood" displacement (or deprivation): "the inundation of land (. . .) also affects the resources available for (. . .) productive activities. In the case of communities dependent on land and the natural resources base this often results in the loss of access to traditional means of livelihood, including agricultural production, fishing, livestock grazing, fuelwood gathering and collection of forest products" [7] (p. 103). It also implies that the compensation policy should include all mitigation measures needed to protect affected communities from involuntary risks related with their new environment.

On the procedural level, the policy principle underscores the need for a decision-making process based on the pursuit of negotiated outcomes conducted in an open and transparent manner and inclusive for all legitimate actors involved, agreement on the way to implement the measures decided upon, and the possibility to demand full compliance of negotiated agreements and commitments. The WCD recommends setting up a multi-stakeholders platform that discusses the possible impacts on the host communities, establishes a Mitigation, Resettlement, and Development Action Plan, defines mechanism for dispute resolution, supervises the work of the mitigation and development office, and sets up an independent field monitoring team for continuous monitoring of implementation [7].

3.3. Compensation as a Political but De-Politicized and "Equalizing" Construction

The proposal of the WCD, while recognizing the contested nature of defining "impact" and deciding on "compensation", is at the same time a pragmatic way to socially engineer towards solutions and institutional arrangements. It highlights the importance of access to information, agreeing on institutional frameworks, debating the contents and images of (and contradictions between) expert and "lay" knowledge, and the access to concrete resources and decision-making power.

Indeed, in mega-hydraulic project development and dam compensation negotiations, a crucial part of the struggle is over decisions that must be made on what is lost, what is to be compensated, the meaning and values of displaced items, beings and relationships, and the measurement units for compensation. All of these decisions are marked by particular worldviews and epistemologies, often divergent normative and moral frameworks regarding the relationships and issues at stake, and the possibility of devising and recognizing shared rules and norms.

A problem in projects such as Misicuni is that the arena of divergent knowledge systems in which the definition of "rightful compensations" takes place is characterized by highly unequal economic, political, and discursive power relationships (as elaborated more generally in the introductory paper of this special issue). As in many places of the world, water expertise and the corresponding policy and project decision-making privileges in Bolivia are largely reserved for those who are political-economically selected to hold water knowledge, speak water truths, and exercise water authority (see also [48,51,52]). "Rightful" water authority, rather than following from actually knowing local water cultures and territorial realities, importantly originates in economic structures, cultural politics, and gender divisions. In turn, legitimate hydraulic expertise, territorial planning, and the labeling of "efficient and rational water development" banks on their formal accreditation by officialdom and powerful economic interest groups [53–55]. Dam engineers and the mega-hydraulic projects they work on commonly symbolize the denial of connections between power and knowledge, while their hidden moralism of "good water governance", "water efficiency", and overall modernist progress is pervasive. This, in conjunction with the status of being a representative of scientific reason, makes the large dam development expert into a powerful political actor in territorial transformation and compensation processes. Behind the mask of neutrality, the social conventions and political choices that are basic to building large dam schemes—as in Misicuni—are depoliticized, justifying far-reaching interventions and territorial transformations (see also [56]). Implicitly, the affected indigenous communities that accept the rules of mega-hydraulic water management and territorial modernization get the label of the "compatible poor"—they are worthy of receiving charitable compensation guided by expert-based decisions. By contrast, indigenous and peasant communities that do not accept the rules and regulations of modern hydro-territorial re-patterning are on the wrong track not just socially, institutionally, and productively (by sticking to "bad practices"), but also ethically; these "incompatibles" are the cause for their own poverty and backwardness. Progress will unfortunately but rightly undermine or take away their water and territorial rights.

No matter the progressive or conservative background of the ruling elites and water administrators, such modernist-moralist background importantly colors the compensation negotiation process that is set up in dam projects such as Misicuni. The question of who may or may not express their interest, which interests, and how to frame these, is part of the discussion. In repeated instances, peasants and construction workers are denied to do their say for the sake of general interest. Time and again, project officials renew their belief in an imaginary, universal, expert-planned model of "modern water management" and "rational territorial ordering" to control irregularities, correct incapacities, and subdue Andean nature and peoples' stubbornness to efficiently deliver water and energy to the urban majorities and industrial areas (e.g., [13,23,56–61]). As the Misicuni case manifests, the urge to morally decide what is right and what is wrong comes from a desire to establish the universal substance, values, and norms of large-scale water and territorial planning expertise and from a need to legitimize the expert community's own epistemic position as neutral and apolitical, thus legitimizing decision-making and shaping water policy agendas [51,62,63].

A fundamental, mostly unconscious challenge and effort of mega-hydraulic compensation programs is the issue of commensuration, which is "the expression or measurement of characteristics normally represented by different units according to a common metric (. . .) Commensuration transforms qualities into quantities, difference into magnitude" [64] (p. 315,316). Most professional studies and academic investigations, even when dedicating attention to the economic facets of this equalization process (in terms of commodities and prices), neglect enormous social and cultural importance. As these authors argue, "commensuration can render some aspects of life invisible or irrelevant (. . .) Commensuration changes the terms of what can be talked about, how we value, and how we treat what we value. It is symbolic, inherently interpretive, deeply political . . . " [64] (p. 314,315).

Similar to all other compensation programs (see also the introduction paper [6]), the Misicuni project can be understood as a huge cultural and epistemological "purification endeavor". Particular information is given particular meaning in accordance with formal and (universalistic) expert notions, while inconvenient facts and knowledge, or peoples' territory and livelihood understandings, may be actively sidelined or overlooked [52,65–67].

A crucial aspect of commensuration processes is the pressure to present all issues at stake as commensurable, even when the affected population claim their special nature and incommensurability (see [68]). Whenever they decide to engage in conversations about compensation, they will have to express their interests in a common metric. "Negotiation requires commensurating with the enemy: it requires comparing the cherished with the reprehensible in ways that make the former less distinctive, less incomparably valuable than it once was. Not surprisingly, movements that stake their identities on incommensurables—radical democracy, heavenly truths, and native lands, for examples—face a dilemma even coming to the bargaining table" [64] (p. 337).

The fundamental challenge for affected communities and grassroots alliances who claim for repair and compensation, therefore, is to negotiate not just the issues and amounts to be compensated but also the very terms of "comparing" and "equalizing" the meaning and values of things, beings, relations, processes, and contexts. As the introduction paper states, "beyond the conflict over the material means of production and the socio-political/hydro-technological re-patterning of humans and non-humans in dam-affected territories, there is the struggle over the control of the means of knowledge production, as well as the battle over who controls societal power to determine what counts as 'normal', legitimate and valid knowledge" [6].

As the next sections' evidence of the Misicuni case manifest, though strongly influenced by unfair (but changing) power relations, the "knowledge of socionatures", their commensuration in the negotiation process, and the compensations that sprout thereof are importantly determined by interacting epistemologies and require affected communities' learning in the struggle. This transdisciplinary co-creation of knowledge involves both confrontation and mutuality among the water user, policy, and scientific communities (see also [6,48,69,70]).

4. Compensation Issues and the Struggle about Meaning and Value

4.1. "Expropriation of Land" Versus "Lost Livelihoods"

The main impact of the Misicuni project was the loss of agricultural, pasture, and housing land at both sides of the river, from near the valley floor up to 130 m above it. In total, the Misicuni Company had to liberate 470 hectares, nearly 75% of which was in active use for farming, grassland, and housing. The river flow, riverbeds, and wastelands occupied the other 25%.

Since the inundated area was in use by local communities, it had to be transferred to the state (represented by the Misicuni Company) through a process of expropriation. According to the 1884 Expropriation Law, the expropriated party is entitled to receive in exchange for the expropriated object a compensation equivalent to its economic value. This value is to be established by two experts, one appointed by either party, with the eventual help of a settling third expert. The Misicuni Company contracted the Cochabamba University for valuation. Their work consisted in marking the plot boundaries, defining land qualities, and identifying constructions and landed improvements (walls, corrals, fences). The expert team identified five land quality categories and established differentiated values. In the absence of a referential land market, they defined extremely low prices (from 400 to 2000 USD per hectare). In total, the project paid only 409,000 USD for land compensation, averaging around 2000 USD per family. The highest amount a family perceived was less than 8000 USD (interview with Misicuni technician, 2 April 2018).

These figures illustrate how the forced commodification of agricultural plots was highly disadvantageous for the highland indigenous families. Experts set extremely low prices to the land based on its low agricultural productivity and the lack of local land markets. They did not consider the high value of the land for the Misicuni project, nor the fact that it was the crucial resource for the communities' livelihoods. To compare, at the Cochabamba valley floor during the same time period, the cheapest agricultural land was valued 50,000 USD per hectare. Consequently, with compensation money, affected families could not purchase land elsewhere to produce for self-subsistence.

Some families protested against the low prices but eventually accepted and signed the transfer agreements, albeit not completely at free will. Community leaders mentioned that the Misicuni Company created division among the communities to weaken their bargaining position ("First they talked in our organization, but then they started individual talks, telling us that the others had already accepted", interview with community leader, 8 March 2018) and threatened them to accept the offer ("If you do not take this money, you will simply lose it. It will return to the state (. . .). The dam will be constructed anyway. The police may come, or they will militarize this area", interview with community leader, 6 March 2018).

In retrospect, community leaders concluded that they were not prepared for the negotiations with the Misicuni Company and did not know their rights (collective interview, 5 March 2018). Almost all decisions were induced or imposed by the Misicuni Company—claiming legal arguments, officials decided what could or could not be compensated, what was to be measured, who defined its value, and what could or could not be discussed. In their retrospection, leaders repeatedly mentioned their lack of control over the compensation process and expressed their ignorance and distrust as to the outcome of decisions made. As one leader expressed, "We made a mistake measuring our land (agricultural) plot by plot. We should have measured our land as a whole (including wasteland, tracks, etc.). For in the end, we lost everything". Another added, "We wanted to do another study (on the value of our lands), but we did not have the money for it". A third one explained, "The Misicuni Company threatened our lawyer not to interfere with the compensation process".

The indigenous families hardly had any experience with government intervention and did not know their legal rights. Their disadvantageous position made them accept the compensation money. However, communities quickly learned from their earlier disappointing experience. In 2010, when the dam height was changed from 85 to 120 m, the Misicuni Company needed to expropriate three extra hectares. When constructors invaded the not yet compensated land, local communities blocked the road and negotiated a far higher price for the extra unproductive steep land, considering its crucial importance for the project 8.080 USD per hectare, 10 times higher than their productive (interview with Misicuni technician, 6 May 2018).

Their leaders' main critique on the expropriation process is, however, that it only considered the (presumed) commodity value of land and not its use value for families' livelihoods. The expropriated plots were by far their most productive agricultural land, improved through decades of human investment in soil fertility and structure, and the main resource to sustaining their agriculture-based livelihood. In contrast, higher up on the valley's steep flanks, plots do not join similar conditions in terms of slope, water, fertility, and climatic conditions, and are not suited for all crops that form part of the household diet, and topographic conditions make access far more difficult. In the years after relocation, crops on the hillside plots suffered from drought, hail, frost, and deceases, affecting especially the potato production, the main staple and most important cash crop. After relocation, family income from agricultural production reduced considerably [71–73]. In 2016, resettled families demanded the transfer of top soil material from the valley bottom to the slope plots to improve productive conditions to sustain their livelihoods, but this demand was not approved.

Similar to cases described by Dye [16] and Hidalgo, Boelens and Isch [58], the Misicuni compensation process was an example of top-down thinking, valuing supposed experts' knowledge over engagement with the local community within the logic of contracting experts for rapid assessments. While the Misicuni Company technicians and university consultants focused on physical size and monetary value, indigenous families were concerned with their ability to continue their livelihoods (see also [24,74]). Community leaders framed the destruction of livelihoods as the main unfairness in the compensation process. This opinion was even stronger under younger leaders, many of whom blamed their parents for having sold their family patrimony for a miserable sum of money, depriving their children of a sufficient resource base for subsistence.

4.2. Extraction of Construction Aggregates: Deceit, Deception, and Loss of Faith

Another critical issue in the compensation process is related to the aggregates for constructing the dam body and auxiliary works, which were extracted from the subsoil of the recently expropriated family plots. It was only after they sold their lands that indigenous families became aware of the resource wealth they had handed over, whereas project officials knew from the start that the area that was to be inundated was also the site for extracting construction materials. In fact, the dam design had changed from a concrete to a rockfill body in view of the abundant presence of nearby available rockfill.

Both the recent constitution [75] and the specific law on aggregate administration and regulation (Ley 3425, 2006 and its bylaw [76]) determine the right of indigenous communities to participate in the benefits of non-renewable resources within their territory. In the case of aggregates, community projects (such as riverbanks conservation, irrigation projects, productive improvements, etc.) deliver the benefits. In view of these recently created rights, community leaders interpreted the silence about the plan to collect aggregates from the expropriated fields as a deliberate deceit by the Misicuni Company. Even worse, they see it as a cynical detail that the best aggregates were situated precisely in those terrains that were expropriated at zero cost since they were categorized as "nonagricultural use". They blame themselves and the Misicuni Company for not having received a better deal. A community leader framed his disappointment as follows: "About the aggregates, we simply did not know. We were sleeping. So much material they have used from our land. They already knew but did not tell us. As far as we knew, the aggregates would come from Cochabamba, but in the end, it was all taken from our land" (community leader, 2 March 2018).

From the part of the officials, silencing the issue of the aggregates dormant under the community soils was a conscious decision. Since land valuing was based on topsoil evaluation only, and the available budget was established considering these values, they had no budget to pay higher prices because of aggregate presence. Nonetheless, one of the technicians acknowledged that the deliberate deceit undermined the local population's confidence in the Misicuni officials; it generated a jealous anger about the lack of compensation for what proved to be an extremely valuable resource for dam construction. He also argued, however, that over time, the Misicuni Company more than compensated for the aggregate use by financing projects to improve living conditions (conservation measures, roads, etc.). Still, peasants' deception persists—they argue that taking away their resources for free has importantly reduced the total project costs to the benefit of the government, but they had paid the costs.

4.3. Drying Springs and Wetlands

Before dam construction, the Misicuni watershed counted with numerous water sources. Water sprang from the mountainsides through hundreds of small wells, creating a myriad of ponds and flows. Especially in the communities Aguadas ("Watery"), Uyuni, and Putucuni, water was abundant. On moderate slopes, this water abundance gave way to the creation of wetlands (bofedales) with a variety of hydrophytic plants, which formed an essential part of the fodder for the llama and alpaca herds. Almost 6% of the total Misicuni watershed consisted of wetlands [43].

During the excavation of the tunnel, several local springs diminished their flow and even completely dried out, which made communities protest and claim for restoration. At first, project planners and contractors' engineers strongly denied their claims, arguing that tunnel excavation did affect uphill springs. Even though the geological study determined that the tunnel trajectory passed through two geological faults with rock fractures and possibly high permeability rates, no evaluation studies were done regarding the future impacts of the tunnel on the mountainous water bulb, uphill reservoirs, and springs. Ironically, the experts' denial of the local communities' worry and claims coexisted with their own observation that "there was an extreme lot of water in the tunnel" (interview geotechnical expert, 3 May 2018)—so much that, at the upper end of the tunnel, water had to be pumped out continuously to permit perforation activities.

From the local peasants' perspective, it was obvious that all this water was drained from the veins that fed existing springs; they had never dried out before and suddenly all depleted. Experts did not have any data to deny this local historic knowledge, although their expert opinion on the non-interference between tunnel digging and wetlands drying was expressed as "factual knowledge", but in fact they had no long-time data to sustain it. Their image of prestigious and objective knowledge bearers was deployed to defend the Misicuni Company's interests and deny peasants' claims. To diminish protests, in 2003, the Misicuni Company constructed drinking water systems for the new hamlets as part of the resettlement agreement and as compensation for lost waters.

After years of unfruitful, unrealistic expert argumentation and in view of abundant evidence, experts started to acknowledge the influence of the tunnel on local water sources and flows. The empty tunnel functions as an open mountain vein, drawing water from nearby permeable layers. The 2009 Environmental Impact Study mentioned the drying up of more than 200 water sources in the socioeconomic description chapter, but this was not officially recognized at the level of engineering studies—an example of how institutional knowledge is not equally shared and acknowledged by its units. Eventually, the Misicuni Company decided to implement a measurement campaign, combining field measurements, observations, and testimonies, leading to the conclusion that nearly 230 springs dried out because of the company's activities. After the communities included the mitigation of lost water sources in the 2012 demands lists, the Misicuni Company constructed two small dams and distribution networks in Putucuni and Uyuni to extenuate the most serious manmade water problems (interview with Misicuni technician, 6 May 2018).

4.4. Closure Experiences

Towards the end of construction works, in July 2016, the communities proposed a list of pending and new demands to the departmental authorities: a network of roads and bridges to reach the new hamlets aside the reservoir; an irrigation system drawing on reservoir water; transport of rich soil from the valley bottom to their new plots; greenhouses for seed production; solar panels; river and reservoir access for livestock; fish farms; scholarships and additional teachers for the valley's new public boarding high school; 600 additional homes for the communities' youth; and a well-stocked, first-rate hospital. This list is a clear indication of persisting needs, as well as an opportunity-driven proposal to perceive whatever they could while there were still construction activities going on.

Since water accumulation in the reservoir started in 2017, construction activities were reduced to a minimum, which made local communities worry about the fulfillment of earlier promises and rethink their demands strategy. The next citations show their preoccupation:

"In August 2016 we organized a road block to demand for roads and bridges. At that moment we figured out that at the beginning of the dam construction we should have included everything, all our needs, compensations, roads, bridges, houses for everyone, so that at the end of the dam construction all agreements would have been finished as well."

"The dam is ready (. . .) now the works for us are not sure. When the reservoir is filled with water, there will be landslides that may affect our roads and animal paths. We want a signed agreement through which they guarantee fulfillment of repairs, but they say 'let's first see what happens, then we will talk'. It is not like that . . . "

"However, we cannot complain strongly, for instance going to the press, because they say that if we cause problems, there will be no new houses for 2018. You know, we are negotiating for 2018, that is also a thing . . . "

These comments show their distrust towards the Misicuni Company and other state institutions responsible for fulfilling the latest agreements, thus concluding that they should have formulated their demands earlier and claimed contractual obligation to comply with them before the end of the construction phase (which is one of the crucial recommendations in the WCD policy document [7]). The reaction of the Misicuni Company was to avoid upcoming obligations in light of their own uncertain institutional future, using delaying tactics and pressure to reduce further demands.

In 2017, another important shift took place in the communities' demand strategy—from then on, they started claiming a share in project revenues. As a community leader stated, "From last year onward, we are thinking about participation in the benefits of the Misicuni project. Something must come back to our communities. The Misicuni staff say that there are no royalties, you are talking in vain about royalties . . . "—according to Bolivian law, royalties can be stipulated for nonrenewable natural resources only—" . . . But we cannot permit such a huge construction with so many perjuries. We agree that compensation may come in works," (interview with community leader, 8 March 2018).

To operationalize this proposal, the affected communities plea for a new general agreement to grant additional compensation. In their view, the original agreement was related to the dam construction period only, providing compensation for inundated lands and relocation of their homes. They demand a new general agreement to regulate a sustainable livelihood perspective, either in money or in kind. Unfortunately, the first experiences with in kind support are not yet very optimistic; the departmental government implemented a fishery cum ecotourism project but almost without local participation—only a few families took part, mainly because there was no clarity in how the benefits of the new common resource would be distributed.

5. Discussion and Conclusions

The history of the Misicuni project gives clear examples of the costs and burdens that large hydraulic modernization projects pose on the host communities that live in the sites where the main construction works and reservoirs are planned. Although the Misicuni project is a case of relatively "good government intentions" constructed mainly by a peasant-favorable regime, it resulted in indigenous communities (although partly compensated) being deceived and perjured by the projects' interventions.

This outcome contrasts with the expectations about the progressive nature of the Misicuni project, the pro-rural-community nature of Bolivia's popular government in place since 2006, and the fact that the Misicuni project was an example of "vernacular modernism" [1]. As Laurie, Andolina and Radcliffe [12] already warned, the progressive city-rural beneficiary coalition would not necessarily solidarize with highland indigenous communities that had to be displaced. In that sense, the history of the Misicuni project is as traditional as that of many other hydraulic mega-works. Independent of whom may be the promoters, large hydraulic works put burdens on host communities; final outcomes for them depend on the process of contestation, compensation, institutional responsibilities put in place, and the understanding of what must be compensated to deliver better living conditions for the locally affected.

Initially, the government adopted principles of decision making in almost exclusively top-down ways and valued expert understanding only; local knowledge and experiences were left out of decision making. This denial ensured uncritical thinking about the negative livelihood impacts that were likely to leave the poorest even worse off. Local knowledge and understanding and related local demands were considered only after repeated and often violent protests.

The Misicuni approach to compensation fell strongly behind the WCD recommendations and did not resemble the approach suggested in the WCD Mitigation, Resettlement, and Development Action Plan [7]. In fact, the Misicuni Company had no systematic approach to dealing with negative impacts on the local population and it studied adverse impacts from an environmental angle only. Framing the impact on local communities as part of the environmental impact had a two-fold effect: first, urban environmental experts neglected the impacts on the local highland population and envisioned from an urban perspective only the positive impacts for the city (and "nature"), and second, they accused the local population of being "future contaminating outsiders" [43] (p. 118). The combination of invisibility and accusation led to the absence of measures to protect the local communities from upcoming harms and to a lack of initiatives that could have favored the marginal position of these highland communities. Similar to what Lynch concluded, it also "created an environment where the kind of careful social and environmental research that needed to shed light on potential impacts was not conducted" [77] (p. 11), maintaining the systematic knowledge gaps between project officials and indigenous communities regarding real impacts.

The lack of a systematic plan made the compensation process a tedious story of recurrent opposition of the indigenous population towards the Misicuni Company. Opposition was firstly against direct negative impacts, but along the process became more and more related to territorial authority, compliance mechanisms, and a procedural relation between the host communities, Misicuni Company, and other state institutions. The mutual learning experience led to the successful conformation of a high-level commission, which through monthly meetings improved compliance, although contractual obligations that could have assured clear beneficial results during the construction phase were never established [7]. During the whole process, the affected population was obliged to prove negative consequences, bargain on compensation issues, and control compliance, contrary to what might be expected.

This disadvantageous and vulnerable position of the indigenous communities was manifested in their invisibility during the compensation process. Their claims were largely disregarded and their existence "unimagined". The Misicuni Company, supported by other state institutions and authorities, imposed its vision on the need for modernization, especially after proclaiming hydropower as a mayor state policy to accomplish Bolivia's role as an energetic center of Latin America. The public framing of the project as a need for regional and national development enabled the avoidance of the generation of solidarity with (or empathy for) the highland communities' demands. Members of parliament and even the national president openly criticized highland communities for delaying the project by claiming their rights, just as they criticized construction workers for striking when they did not receive their pay [78]. Local communities have not been able to create sympathy for their position nor broaden their social alliances further than the surrounding mountain communities.

As stated, originally the Misicuni Company considered only compensation for land and housing parcels. In this process, different interpretations arose regarding what was to be compensated and how issues had to be valued. Experts reduced the compensation to measurable and commodifiable units, assessing land value on virtual market prices, whereas the local population valuated it as part of their livelihood resources. In the end, the legal definition of compensation was imposed, resulting in small monetary compensations per family and depriving them of their resource base. Misicuni illustrates the WCD conclusion that "cash compensation (. . .) even when paid on time, has usually failed to replace lost livelihoods (. . .) They have often been forced to resettle in resource depleted and environmentally degraded areas around the reservoir (. . .) Absence of livelihood opportunities forces affected people to abandon resettlement sites and migrate" [7] (p. 107). The later expropriation of three extra hectares is a clear example of a change in price setting, consciousness, and claim-making capacity; local leaders achieved that it be based on "project necessity" more than on "market" value. Their claim was highly informed by their learning about the aggregates issue. The Misicuni Company got access to the huge aggregate banks by just compensating "worthless" land and withholding the indigenous population from information on the value underneath their plots.

In later issues on impacts and compensations, similar contradictions in knowledge, perception, and valuing occurred. For instance, lost multipurpose springs were replaced with tap systems for human consumption. Only after repeated complaints did communities manage to acquire access to small reservoirs and irrigation systems. To this respect, a common confusion comes to the fore regarding what is to be considered as compensation measures. Whereas project officials upheld that living conditions have improved importantly because of the project, it seems opportunistic to consider the provision of drinking water and sanitation networks, roads, health services, education, electricity, and communication as part of compensation measures—these are basic services to be provided by local or departmental governments. The Misicuni Company helped to accelerate access to these provisions, which should be regarded as a historical debt pay-off rather than a compensation for dam construction.

During the construction of the Misicuni dam, important learning processes took place, combining the issues of contested "meaning" and "values" in the dam-development epistemological battlefield, the depth of the knowledge on the issues at stake, the procedures to deal with them, and how to interpret the opponents' behaviors. For the indigenous leaders, it has meant an intense learning process on state interests, functioning, and reliability. A major problem of this process was that it followed events and almost never permitted them to anticipate later actions. They learned that they had to struggle for justice, even though mitigation measures were at stake. This strengthened the jealousy towards valley inhabitants who received multimillion advantages in water and energy supply, whereas the poorer indigenous communities were deprived from house and land and were involved in a decades' long battle against injustice. This insight, at last, led them to demand for project revenue related co-benefits in the form of yearly royalties or a construction/productive-projects fund based on the recognition of their territorial authority over natural resources. This is the pending issue to be resolved in the operation phase of the Misicuni Multipurpose Project.

Finally, the Misicuni case shows that "compensation" is a politically contested and fiercely fought social construction and not a shared objective decision making process, as suggested by the WCD principles. The modernization project defended by state institutions and Cochabamba's urban and rural groups evidently contradicts with the host communities' interests. The outcome of the struggle to compensate for their losses is, in the end, determined by the power balance between the affected and the beneficiaries, and thus depends on either's base and instruments of power. While state institutions make use of formal, legal, and institutional norms and rules and impositions, the indigenous population principally builds on their territorial presence, dynamically rooted and collectively enforced norms, and the physical threat to affect water provision to the city.

Author Contributions: Conceptualization, P.H. and R.B.; Methodology, P.H.; Formal Analysis, P.H. and R.B.; Investigation, P.H.; Writing—Original Draft Preparation, P.H. and R.B.; Writing—Review & Editing, P.H. and R.B.

Funding: This research received no external funding.

Acknowledgments: We thank the Misicuni community leaders for the information shared with us and David Alconcé for his careful translation of all Quechua interviews and meetings. We also thank the Misicuni officials for sharing their views and insights with us on the issues analyzed in this article.

Conflicts of Interest: The authors declare no conflict of interest.

References

1. Hines, S. The power and ethics of vernacular modernism: The Misicuni Dam Project in Cochabamba, Bolivia, 1944–2017. *Hisp. Am. Hist. Rev.* **2018**, *98*, 223–256. [CrossRef]
2. Vera Varela, R. *Misicuni: ¿la frustración de un pueblo? (Misicuni: frustration of a city?)*; Aliaga: Cochabamba, Bolivia, 1995.
3. Hoogendam, P. *Diseño Conceptual del Componente Riego de Proyecto Misicuni (Conceptual design of the Irrigation Component of the Misicuni Project)*; Dirección de Gestión de Agua del Gobierno Departamental: Cochabamba, Bolivia, 2017.
4. Ministerio de Medio Ambiente y Agua (MMAyA). *Plan Maestro Metropolitano de Agua y Saneamiento de Cochabamba, Bolivia (Metropolitan Master Plan for Water and Sanitation)*; Ministerio de Medio Ambiente y Agua: Cochabamba, Bolivia, 2014.
5. Editorial Staff. Amenazan con bloqueo en Misicuni por falta de obras (Roadblock threat in Misicuni for lack of compensation works). *Los Tiempos*, 25 January 2018.
6. Boelens, R.; Shah, E.; Bruins, B. Contested knowledges: Large dams and mega-hydraulic development. *Water* **2019**, *11*, 416. [CrossRef]
7. World Commission on Dams. *Dams and Development: A New Framework for Decision-Making*; Earthscan: London, UK, 2000.
8. Instituto Nacional de Estadística. *Población de la Región Metropolitana Kanata (Population of the Kanata Metropolitan Region)*; INE: La Paz, Bolivia, 2017.
9. Moore, D.; Dore, J.; Gnawable, D. The world commission on dams + 10: Revisiting the large dam controversy. *Water Altern.* **2010**, *3*, 3–13.
10. Sneddon, C.; Fox, C. Struggles over dams as struggles for justice: The World Commission on Dams (WCD) and anti-dam campaigns in Thailand and Mozambique. *Soc. Nat. Resour.* **2008**, *21*, 625–640. [CrossRef]
11. Proyecto de Enseñanza e Investigación en Riego Andino y de los Valles. *Balance Hídrico de la Producción Agrícola en el Valle Central de Cochabamba (Water Balance of the Agricultural Production in the Cochabamba Central Valley)*; Universidad Mayor de San Simón: Cochabamba, Bolivia, 1999.
12. Laurie, N.; Andolina, R.; Radcliffe, S. The excluded "Indigenous"? The implications of multi-ethnic policies for water reform in Bolivia. In *Multiculturalism in Latin America*; Sieder, R., Ed.; Palgrave: London, UK, 2002.
13. Laurie, N.; Marvin, S. Globalisation, neo-liberalism and negotiated development in the Andes: Water privatisation and regional identity in Cochabamba. *Environ. Plan.* **1999**, *31*, 1401–1415. [CrossRef]
14. Oporto, H. Cuando la política hace Aguas -de la "guerra del agua" al futuro sombrío de los servicios públicos. In *Agua y Poder*; Oporto, H., Salinas Gamarral, L.F., Eds.; Fundación Milenio: La Paz, Bolivia, 2007.
15. Hines, S. *Dividing the Waters: How Power, Property and Protest Transformed the Waterscape of Cochabamba, Bolivia, 1879–2000*; University of California: Berkeley, CA, USA, 2015.
16. Dye, B. The return of 'high modernism'? Exploring the changing development paradigm through a Rwandan case study of dam construction. *J. Eastern Afr. Stud.* **2016**, *10*, 303–324. [CrossRef]
17. Scott, J. *Seeing Like a State: How Certain Schemes to Improve the Human Condition Have Failed*; Yale University Press: New Haven, CT, USA, 1998.

18. Bakker, K.; Hendriks, R. Contested Knowledges in Hydroelectric Project Assessment: The Case of Canada's Site C Project. *Water* **2019**, *11*, 406. [CrossRef]

19. Del Bene, D.; Scheidel, A.; Temper, L. More dams, more violence? A global analysis on resistances and repression around conflictive dams through co-produced knowledge. *Sustain. Sci.* **2018**, *13*, 617–633. [CrossRef]

20. Duarte-Abadía, B.; Boelens, R.; du Pre, L. Mobilizing water actors and bodies of knowledge. The multi-scalar movement against the Río Grande Dam in Málaga, Spain. *Water* **2019**, *11*, 410. [CrossRef]

21. Dukpa, R.D.; Joshi, D.; Boelens, R. Hydropower development and the meaning of place. Multi-ethnic hydropower struggles in Sikkim, India. *Geoforum* **2018**, *89*, 60–72. [CrossRef]

22. Fox, C.; Sneddon, C. Political Borders, Epistemological Boundaries, and Contested Knowledges: Constructing Dams and Narratives in the Mekong River Basin. *Water* **2019**, *11*, 413. [CrossRef]

23. Hidalgo-Bastidas, J.P.; Boelens, R. Hydraulic order and the politics of the governed: The Baba Dam in coastal Ecuador. *Water* **2019**, *11*, 409. [CrossRef]

24. Huber, A. Hydropower in the Himalayan Hazardscape: Strategic Ignorance and the Production of Unequal Risk. *Water* **2019**, *11*, 414. [CrossRef]

25. Lynch, B. What Hirschman's hiding hand hid in San Lorenzo and Chixoy. *Water* **2019**, *11*, 415. [CrossRef]

26. Teräväinen, T. Negotiating water and technology—Competing expectations and confronting knowledges in the case of the Coca Codo Sinclair in Ecuador. *Water* **2019**, *11*, 411. [CrossRef]

27. Beccar, L.; Boelens, R.; Hoogendam, P. Water rights and collective action in community irrigation. In *Water Rights and Empowerment*; Boelens, R., Hoogendam, P., Eds.; Van Gorcum: Assen, The Netherlands, 2002; pp. 1–21.

28. Crespo, C.; Fernández, O. *Los Campesinos Regantes de Cochabamba en la Guerra del Agua: Una Experiencia de Presión Social y Negociación (Cochabamba's Peasant irrigators in the Water War: An Experience of Social Pressure and Negotiation)*; Universidad Mayor de San Simón: Cochabamba, Bolivia, 2001.

29. Gerbrandij, G.; Hoogendam, P. *Aguas y Acequias; los Derechos al agua y la Gestión Campesina de Riego en los Andes Bolivianos (Water and Canal; Water Rights and Farmer Managed Irrigation in the Bolivian Andes)*; Plural Editores: La Paz, Bolivia, 1998.

30. García Orellana, A.; García Yapur, F.; Quitón Herbas, L. *La «Guerra del Agua» Abril de 2000, la Crisis de la Política en Bolivia (The Water War, April 2000, Political Crisis in Bolivia)*; FUNDACIÓN PIEB: La Paz, Bolivia, 2003.

31. Perreault, T. Custom and contradiction: Rural water governance and the politics of usos y costumbres in Bolivia's irrigators' movement. *Ann. Assoc. Am. Geogr.* **2008**, *98*, 834–854. [CrossRef]

32. Assies, W. David versus Goliath in Cochabamba: Water rights, neoliberalism, and the revival of social protest in Bolivia. *Latin Am. Perspect.* **2003**, *30*, 14–36. [CrossRef]

33. Crespo Flores, C. El movimiento nacional del agua boliviano. De la resistencia a la cooptación (2000–2007) (The Bolivian national water movement. From resistance to co-optation). In *Modelos de Gestión del agua en los Andes, Actes y Mémoires de l'Institut Francais d'Études Andines*; Poupeau, F., González, C., Eds.; Instituto Francés de Estudios Andinos: Lima, Perú, 2010; Volume 6, pp. 111–132.

34. Kruse, T. La "Guerra del Agua" en Cochabamba, Bolivia: Terrenos complejos, convergencias nuevas (The "Water War" in Cochabamba, Bolivia: Complex issues, new convergences). In *Sindicatos y Nuevos Movimientos Sociales en América Latina*; Garza Toledo, E., Ed.; Consejo Latinoamericano de Ciencias Sociales (CLACSO): Buenos Aires, Argentina, 2005.

35. Nixon, R. *Slow Violence and the Environmentalism of the Poor*; Harvard University Press: Cambridge, MA, USA; London, UK, 2011.

36. Watts, M.J. Antinomies of community: Some thoughts on geography, resources and empire. *Trans. Inst. Br. Geogr.* **2004**, *29*, 195–216. [CrossRef]

37. Nixon, R. Unimagined communities: Developmental refugees, megadams and monumental modernity. *New Form.* **2009**, *69*, 62–80. [CrossRef]

38. Centro de Levantamiento Aeroespacial para Sistemas de Información Geográfica (CLAS). *Inventario de Tierras de Misicuni para Compensación (Misicuni Land Evaluation for Compensation)*; CLAS/UMSS: Cochabamba, Bolivia, 1999.

39. Empresa Misicuni. *Estudio de Evaluación de Impacto Ambiental Proyecto Misicuni (Misicuni Project Environmental Impact Study)*; Empresa Misicuni: Cochabamba, Bolivia, 1992.

40. Empresa Misicuni. *Estudio de Evaluación de Impacto Ambiental Proyecto Misicuni; Versión Final (Misicuni Project Environmental Impact Study; Final Version)*; Empresa Misicuni: Cochabamba, Bolivia, 1996.
41. Sosa, M.; Boelens, R.; Zwarteveen, M. The influence of large mining: Restructuring water rights among rural communities in Apurimac, Peru. *Hum. Organ.* **2017**, *76*, 215–226. [CrossRef]
42. Empresa Misicuni. *Resúmenes ejecutivos de los Reasentamientos Misicuni (Misicuni Reallocations Summary)*; Empresa Misicuni: Cochabamba, Bolivia, 2009.
43. Empresa Misicuni. *Actualización del Estudio de Evaluación de Impacto Ambiental Proyecto Misicuni Fase I (Update of the Misicuni Project Phase I Environmental Impact Study)*; Empresa Misicuni: Cochabamba, Bolivia, 2009.
44. Duarte-Abadía, B.; Boelens, R.; Roa-Avendaño, T. Hydropower, encroachment and the re-patterning of hydrosocial territory: The case of Hidrosogamoso in Colombia. *Hum. Organ.* **2015**, *74*, 243–254. [CrossRef]
45. Johnston, B.R. Chixoy dam legacies: The struggle to secure reparation and the right to remedy in Guatemala. *Water Altern.* **2010**, *3*, 341–361.
46. Menga, F.; Swyngedouw, S. *Water, Technology and the Nation-State*; Routledge: London, UK, 2018.
47. Molle, F.; Mollinga, P.; Wester, P. Hydraulic bureaucracies and the hydraulic mission: Flows of water, flows of power. *Water Altern.* **2009**, *2*, 328–349.
48. Zwarteveen, M.Z.; Boelens, R. Defining, researching and struggling for water justice: Some conceptual building blocks for research and action. *Water Int.* **2014**, *39*, 143–158. [CrossRef]
49. Johnston, B.R. Large-scale dam development and counter movements: Water justice struggles around Guatemala's Chixoy Dam. In *Water Justice*; Boelens, R., Perreault, T., Vos, J., Eds.; Cambridge University Press: Cambridge, UK, 2018; pp. 169–186.
50. International Institute for Environment and Development. *Dams and Development: A New Framework for Decision-Making Overview of the Report by the World Commission on Dams, Issue Paper 108*; Drylands Programme, IIED: London, UK, 2001.
51. Hommes, L.; Boelens, R.; Maat, H. Contested hydro-social territories and disputed water governance: Struggles and competing claims over the Ilisu Dam development in southeastern Turkey. *Geoforum* **2016**, *71*, 9–20. [CrossRef]
52. Long, N. Actors, interfaces and development intervention: Meanings, purposes and powers. In *Development Intervention. Actor and Activity Perspectives*; Kontinen, T., Ed.; University of Helsinki: Helsinki, Finland, 2004; pp. 14–36.
53. Boelens, R. Cultural politics and the hydrosocial cycle: Water, power and identity in the Andean highlands. *Geoforum* **2014**, *57*, 234–247. [CrossRef]
54. Boelens, R. *Water, Power and Identity. The Cultural Politics of Water in the Andes*; Routledge: London, UK, 2015.
55. Vos, J.; Boelens, R. Sustainability standards and the water question. *Dev. Chang.* **2014**, *45*, 205–230. [CrossRef]
56. Boelens, R.; Hoogesteger, J.; Swyngedouw, E.; Vos, J.; Wester, P. Hydrosocial territories: A political ecology perspective. *Water Int.* **2016**, *41*, 1–14. [CrossRef]
57. Boelens, R.; Hoogesteger, J.; Baud, M. Water reform governmentality in Ecuador: Neoliberalism, centralization and the restraining of polycentric authority and community rule-making. *Geoforum* **2015**, *64*, 281–291. [CrossRef]
58. Hidalgo, J.P.; Boelens, R.; Isch, E. Hydro-territorial configuration and confrontation. The Daule-Peripa multipurpose hydraulic scheme in Coastal Ecuador. *Latin Am. Res. Rev.* **2018**, *53*, 1–18. [CrossRef]
59. Hommes, L.; Boelens, R. Urbanizing rural waters: Rural-urban water transfers and the reconfiguration of hydrosocial territories in Lima. *Polit. Geogr.* **2017**, *57*, 71–80. [CrossRef]
60. Hommes, L.; Boelens, R. From natural flow to 'working river': Hydropower development, modernity and socio-territorial transformations in Lima's Rímac watershed. *J. Hist. Geogr.* **2018**, *62*, 85–95. [CrossRef]
61. Kaika, M. Dams as symbols of modernization: The urbanization of nature between geographical imagination and materiality. *Ann. Assoc. Am. Geogr.* **2006**, *96*, 276–301. [CrossRef]
62. Hoogesteger, J.; Verzijl, A. Grassroots scalar politics: Insights from peasant water struggles in the Ecuadorian and Peruvian Andes. *Geoforum* **2015**, *62*, 13–23. [CrossRef]
63. Hoogesteger, J.; Boelens, R.; Baud, M. Territorial pluralism: Water users' multi-scalar struggles against state ordering in Ecuador's highlands. *Water Int.* **2016**, *41*, 91–106. [CrossRef]
64. Espeland, W.; Stevens, M.L. Commensuration as a social process. *Annu. Rev. Sociol.* **1998**, *24*, 313–343. [CrossRef]

65. Dupuits, E.; Baud, M.; Boelens, R.; de Castro, F.; Hogenboom, B. Transnational grassroots movements defending water commons in Latin America: Professionalisation, expert knowledge and resistance. *Jnl Ecol. Econ.* **2019**. submitted for publication.
66. Long, N.; Long, A. *Battlefields of Knowledge. The Interlocking of Theory and Practice in Social Research and Development*; Routledge: London, UK; New York, NY, USA, 1992.
67. Valladares, C.; Boelens, R. Extractivism and the rights of nature: Governmentality, 'convenient communities', and epistemic pacts in Ecuador. *Environ. Politics* **2017**, *26*, 1015–1034. [CrossRef]
68. Duarte-Abadía, B.; Boelens, R. Disputes over territorial boundaries and diverging valuation languages: The Santurban hydrosocial highlands territory in Colombia. *Water Int.* **2016**, *41*, 15–36. [CrossRef]
69. Bebbington, A.; Humphreys-Bebbington, D.; Bury, J. Federating and defending: Water, territory and extraction in the Andes. In *Out of the Mainstream. Water Rights, Politics and Identity*; Boelens, R., Getches, D., Guevara-Gil, A., Eds.; Earthscan: London, UK; Washington, DC, USA, 2010; pp. 307–327.
70. Schlosberg, D. Reconceiving environmental justice: Global movements and political theories. *Environ. Politics* **2004**, *13*, 517–540. [CrossRef]
71. Editorial Staff. Piden ayuda por muerte de camélidos (Plea for assistance because of camelidae starving). *Los Tiempos*, 6 December 2014.
72. Editorial Staff. Se reduce producción de papa en Misicuni (Misicuni Potato production reduced). *Los Tiempos*, 23 December 2014.
73. Editorial Staff. Comunarios se quejan de pérdidas por presa de Misicuni (Peasant complain about crop los due to Misicuni dam). *Los Tiempos*, 18 January 2016.
74. Dukpa, R.; Joshi, D.; Boelens, R. Contesting Hydropower dams in the Eastern Himalaya: The Cultural Politics of Identity, Territory and Self-Governance Institutions in Sikkim, India. *Water* **2019**, *11*, 412. [CrossRef]
75. Plurinational State of Bolivia. *Constitución política del Estado (Nations' Political Constitution)*; Plurinational State of Bolivia: La Paz, Bolivia, 2009.
76. Plurinational State of Bolivia. *Law 3425 on Administration and Regulation of Aggregates*; Plurinational State of Bolivia: La Paz, Bolivia, 2006.
77. Lynch, B. *The Chixoy Dam and the Achi Maya: Violence, Ignorance, and the Politics of Blame*; Mario Einaudi Center for International Studies Working Paper No. 10-06; Cornell University: Ithaca, NY, USA, 2006.
78. Editorial Staff. Presidente critica huelga en Misicuni (President criticizes Misicuni strike). *Los Tiempos*, 4 October 2013.

water

MDPI

Article

Contesting Hydropower Dams in the Eastern Himalaya: The Cultural Politics of Identity, Territory and Self-Governance Institutions in Sikkim, India

Rinchu Doma Dukpa [1,*], Deepa Joshi [2] and Rutgerd Boelens [1,3]

1   Department of Environment Sciences, Water Resources Management Group, Wageningen University and Research, P.O. Box 47, 6700 AA Wageningen, The Netherlands; rutgerd.boelens@wur.nl
2   Water Governance and Feminist Political Ecology, Center for Water, Agroecology and Resilience, Coventry University, Priory St, Coventry CV1 5FB, UK; deepa.joshi@coventry.ac.uk
3   CEDLA Center for Latin American Research and Documentation; and Department of Geography, Planning and International Development Studies, University of Amsterdam, Roetersstraat 33, 1018 WB Amsterdam, The Netherlands
*   Correspondence: rinchu.dukpa@wur.nl; Tel.: +31-317-484-190

Received: 9 July 2018; Accepted: 23 November 2018; Published: 26 February 2019

check for updates

Abstract: In India's Eastern Himalayan State of Sikkim, the indigenous Bhutia communities, Lachungpas and Lachenpas, successfully contested all proposed hydropower projects and have managed to sustain an anti-dam opposition in their home regions, Lachung and Lachen. In this paper, we discuss this remarkable, un-researched, effective collective action against hydropower development, examining how identity and territory influence collective action through production, creation and application of vernacular knowledge systems. The role of the Dzumsa, a prevailing traditional system of self-governance among the Lachungpas and Lachenpas, has been central in their collective resistance against large dams in Lachung and Lachen. Our findings show that contrary to popular imageries, the Dzumsa is neither an egalitarian nor a democratic institution—rather, it is an exercise of an "agonistic unity". The Dzumsas operate as complex collectives, which serve to politicize identity, decision-making and place-based territoriality in their struggle against internal and external threats. Principles of a "vernacular statecraft" helped bringing the local communities together in imperfect unions to oppose modernist designs of hydropower development. However, while such vernacular institutions were able to construct a powerful local adversary to neoliberal agendas, they also pose high social, political and emotional risks to the few within the community, who chose not to align with the normative principles of the collective.

Keywords: hydropower development; politicized collective identity; territory; collective action; agonistic unity; vernacular statecraft; Dzumsa; North Sikkim

---

1. Introduction

Since 2003, over 168 large dams for hydropower development have been proposed in the Eastern Himalayan Region of India [1,2]. The push for hydropower development in the north-eastern region of India (see Figure 1) by both Central and State Governments, have made these developments highly conflict prone [1,3,4]. Several major contentious projects (such as the 520 MW Teesta Stage IV, 500 MW Teesta Stage VI and 300 MW Panam in Sikkim; the 2000 MW Subansari Lower HEP in Assam; the 1500 Tipaimukh Dam in Tripura; the 2880 MW Dibang Multipurpose Project and Tawang I & II in Arunachal Pradesh, etcetera) have been stalled, delayed or are waiting for clearance across Northeast India [5–8], often characterized by prolonged struggles between dam opponents and proponents. Yet, the business of hydropower development continues as usual, with many official attempts to fast-track, facilitate and revive old and new hydropower projects across the north-eastern region [3–5].

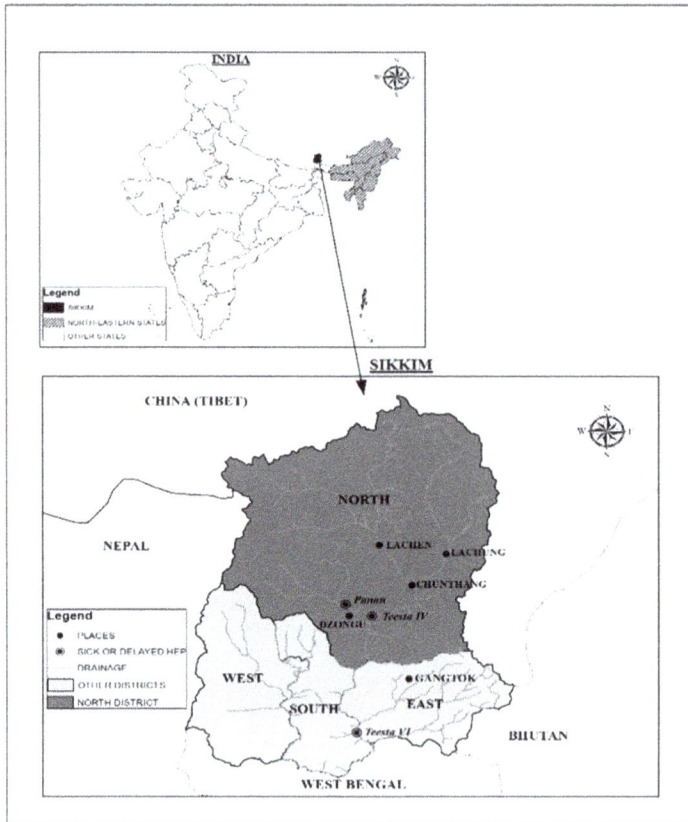

Figure 1. Delayed hydropower projects, fast tracked in Sikkim. Source: Own elaboration, adapted from GoS websites. Map not to scale.

India's most well-known anti-dam movement, the Narmada Bachao Andolan (NBA) (or Save Narmada Movement) began contesting the controversial 1450 MW Sardar Sarovar Dam on the Narmada River in 1989. Nonetheless, after three decades of resistance, which captured global attention—the Sardar Sarovar Dam was completed in 2006: a stark reminder of the powerful nexus among Government (Central, State, Local), power companies and other pro-dam advocates, who are able to pursue dam development against all odds. However, the success of the NBA movement is less about the outcome and more about the process—creating "space for India's faceless and nameless displaced" to voice and influence attention to "equitable development alongside economic growth" [9] (p. 382). It has brought to the public domain "the hitherto closed and protected discourse on mega development projects . . . opening new vistas for environmental movements" [10] (p. 25).

The north-eastern region of India, where hydropower projects are being rolled out currently [7,11–13], is predominantly inhabited by diverse tribal communities. Tribal autonomy, traditional political institutions, cultures, socio-economic practices and landscapes are constitutionally protected under special provisions guaranteed by Article 371 of the Indian Constitution. Ironically, large dam development, promoted officially as instrument for "development", often happens against the wishes of many local tribal communities. This explains why unsuccessful contestations against large dams in the tribal north-eastern region are occasionally violent, resulting in some cases in the death of anti-dam protesters [14,15]. It is in this alarmingly pessimistic scenario that we draw attention to the intriguing case of how two small tribal communities, the Lachungpas of Lachung and the Lachenpas of Lachen (in North Sikkim) powerfully contested and managed within a short time frame of a few years, to cancel all the five hydropower projects proposed in their area. Regardless of the external advocacy for large dams, these two closely associated tribal communities successfully mobilized; and to this day maintain a unanimous anti-dam position. Ironically, it is the neighboring region of Dzongu, inhabited by tribal Lepchas, that literature and media consider as the epicenter of anti-dam movement in North Sikkim. Even though a few dams have been scrapped in Dzongu this is remarkable since—different from Lachung and Lachen—two mega dam initiatives have been implemented there with little resistance.

As we will discuss below, a place-based identity precedes all other identities in the case of the Lachungpas and Lachenpas. In addition, the small minority of Lachungpas (1478 in Lachung) and Lachenpas (1314 in Lachen) are amongst the few tribal groups in India, and the only ones in Sikkim who have a traditional, territorial system of self-governance known as the Dzumsa (or Dzomsa). Outsiders to these areas assume that the Lachungpas and Lachenpas are a "collective entity" united by a common Dzumsa system. It is believed that this is what enabled the community to "kick-out" hydropower companies from their respective areas. In analyzing the nature of collective action among the Lachungpas and Lachenpas and the assumed collective resistance against hydropower development in these regions, our paper unpacks complex ground realities, pointing evidence to how territory, identity and traditional governance come together to forge "agonistic unity" and "vernacular statecraft" [16].

Academic studies in various disciplines discuss how identity triggers collective action or vice versa [17–22]. Many scholars argue that within a maze of identity(ies) experienced by individuals and/or groups, there is a "specific" identity which is key to enabling collective action and/or that place (or territory) and identity [23,24] are closely intertwined in protecting and strengthening cultural values, norms, shared interests and traditional territories (e.g., [25–27], see also [8,28,29]). In other words, territories or places are key markers of identity [30–32]. We have engaged here with the theoretical framework proposed by Klandermans and colleagues [17–19,33] (p. 5)—how politicized collective identity is "the engine of collective action". They outline three processes through which collective identity politicizes, triggers or mobilizes collective action, which we discuss in Section 3. Here we point out that theoretical analyses of collective action rarely pay attention to how collective actions are sustained over time and/or how consensus is maintained in any society, which is anyway divided by many fractures—class, age, gender, ethnicity, religion—to name a few. Colloredo–Mansfeld's [16] work on "vernacular statecraft" and the creation of "agonistic unity" is particularly useful in understanding how and why the Lachungpas and Lachenpas collectively and successfully protested against hydropower development in their respective areas.

Our findings reveal that territorially exclusive and ethnically cohesive collectives like the Dzumsa do not automatically or easily coalesce as a response to outsider-imposed agendas and interventions. Rather, collective action is mobilized by some individuals who politicize the notions of territorial collective identity inside and/or outside existing institutional systems, in this case, the Dzumsas. When communities are fractured into polarized groups, these vernacular institutions also become highly politicized, as they are often the means to coerce divided communities into a collective front or unity, which is nonetheless "agonistic". Here principles of "vernacular statecraft" can become highly contentious. We discuss how traditional systems and practices of shamanism (Chya) coercively bring back dissenters to "agonistic unity". As we explain, the local imposition of collective territoriality and identity notions (deploying, amongst others, fear-driven practices as the Chya) make these highland tribal communities in North Sikkim successful in maintaining their unanimous anti-dam position. Such virtues of cohesion, collective identity and action are not without contradictions. Moreover, these practices are also fundamentally at odds with liberal, modern notions of individual civil liberties.

We have concluded that identities are not always rooted to land, territory, culture or even indigeneity, rather they are strategic, fluid, political actions that serve to defend a particular group from "outsiders/others" and(or) to protect specific claims and interests [34,35]. The united anti-dam stand by the Lachungpas and Lachenpas is much more than just the voicing of socio-economic and environmental concerns relating to large dams. Their resistance is really about (re)claiming territory, (re)asserting collective identity, reiterating collective action, and valuing as well as using non-official, non-centralized knowledge and modes of knowing (see [24]).

In the subsequent sections, we detail the political history of identity construction in Sikkim, to explain how the exclusive Lachungpa/Lachenpa identity came into being in the first place and sustained via the Dzumsas. A short review of key concepts in Section 3 is followed by the study area and methodology in Section 4. Our findings are described in Section 5. Section 6 gives an overview of discussions and Section 7 presents our conclusions.

2. At the Background: Identity Dynamics in Sikkim

To understand the anti-dam resistance in Lachung and Lachen, it is necessary to comprehend the historical, political, cultural and economic context that determines individual or collective routes to protest. Schendel's work on "Zomia" [36] or Shneiderman's on the "Himalayan Massif" [37] both describe the Himalayan region as an invisible, transnational area, " ... marked by a sparse population, historical isolation ... and linguistic and religious diversity" [38] (p. 187). Before notions of nation-state crafted definite geo-political borders in the so-called Himalayan Zomia or Massif (encompassing Nepal, Bhutan, Indian States of Jammu & Kashmir, Himachal Pradesh, Uttaranchal, Sikkim, and Arunachal Pradesh, and China including the Tibetan Autonomous Region [37]), these regions, more than being "boundary, border", were like "a zip-per" stitching together various "densely textured cultural fabrics" [39] (p. 2). The Himalayan State of Sikkim, landlocked by Bhutan in the west, Tibet in the north, Nepal in its east and India in the South (before the 1975 merger) (see Figure 1) exhibits typical "Zomian" characteristics. This explains why "society here is a constellation of multiple identities" [40] (p. 1), resulting from diverse as well as entangled "geographical, linguistic, racial, national, cultural and religious mixtures, commonalities, fluidity with neighboring" regions [37] (p. 290).

2.1. "Sikkimese"—A Newly Created Identity?

The oral history of Sikkim, based on myths, legends and folklore [41], goes back to the 13th century, when a blood-brotherhood-treaty was signed between the Tibetan prince Khye-Bumsa and the Lepcha Chief Thekong-thek [40,41] in North Sikkim. The treaty sealed friendship between the Tibetans (who referred to themselves as the Lhopos) with the Lepchas of Sikkim (who referred to themselves as the Rongs) [40–42]. Nonetheless, modern documented history of Sikkim begins with the consecration of the Chogyal (righteous King), a Lhopo descendent in 1642 AD, leading to the establishment of the Namgyal Dynasty with a Lhopo ancestry in Sikkim (1642–1975). Sikkim's ties with Tibet thus go a long way into history and were "sustained through matrimonial, religious and trade activities including administrative support from Tibet" [41] (p. 72). Both the Lhopos and Rongs comprised of numerous clans or tribal groups, who identified themselves on the basis of their affiliation to specific territory of origin or places of habitation. In fact, the term Lhopo refers to people of South Tibet, while Rongs meant "mother's (nature's) loved one" [42] (p. 77).

It was the Nepalese who initiated the use of singular terms generalizing the diverse clans of Lhopos as "Bhotiya" meaning from "Bhot" (Tibet) and Rongs as "Lapcho" referring to people living in a heap of stone or the stone house [40] in Sikkim and across other Himalayan regions. Although Sikkim has no similar historical ties with Nepal, Nepali presence in Sikkim predates the arrival of British in Sikkim in the late 1880s as noted in the first population census of Sikkim recorded in 1891 [42]. These generic terms gained legitimacy in time. After Sikkim officially became a protectorate of the British colony of India (1889–1947) with the appointment of the first British Political Officer—John

Claude White [41], there was a systematic in-migration of Nepali laborers into Sikkim, facilitated by the British. The terms, Bhotiya and Lapcho (or "Lapcha" in Parbatiya dialect of Nepal, where Lap meant speech and Cha meant unintelligible i.e. unintelligible speaker who could not adopt the Parbatiya language [40]) transitioned into Bhutias and Lepchas and this is how diverse groups belonging to these two generalized categories are known officially and colloquially in Sikkim. The immigrant Nepalese is also a generic category that subsumes diverse Nepali ethnic groups (such as Limbo, Khambu-rai, Yakhas, Sunuwars, Mangars, Gurungs, Tamangs, Bhujels, Thamis, Bahuns, Chettris, Kamis, Damais, Sarkis, Thakuris, Jogis, Sanyasis, Majhis and Newars in Sikkim) [42]. As we discuss below, this framing of identity by ethnicity is certainly not nuanced and does little to help explain deep-rooted and complex identities.

Following the merger of the Kingdom of Sikkim with the Republic of India in 1975, the Bhutias and Lepchas were pronounced as Scheduled Tribes under the Constitution (Sikkim) Schedule Tribes Order—derived from clause (1), Article 342 of the Indian Constitution in 1978 [41,42]. This recognition entitles these communities, privileges and protections accorded to (all) recognized indigenous tribal groups by the Indian Constitution. This GoI accreditation is also extended to all Bhutia and Lepcha communities living outside of Sikkim in the neighboring state of West Bengal, as well as Tibetan communities across the other Indian Himalayan regions of Laul-Spiti, Kumaon, Garwal referred to as the Bhotiya, Bhot or Bhoti [43,44]. In Sikkim, the prefix "Sikkimese" term was thus added to distinguish local inhabitants from ethnically similar outsider others (see [13,43,44]). This happened also because Sikkim's merger with India led to a massive in-migration of "outsiders" from all over India [41,42]. The influx of a skilled and educated outsider population evoked a conscious construction and imbibing of the Sikkimese identity, constructing what has become a sharp wedge between the Sikkimese and the non-Sikkimese. As the Sikkimese prefix came to be adopted by the later migrated Nepalis, who became the majority population in Sikkim, the minority Bhutias and Lepchas furthered their innate Sikkimese-ness, constructing more nuanced (and exclusionary) identities such as "local" and "public" implying different identities and privileges politically (see [13]). Such contentious identity-territoriality frictions define politico-ethnic fragmentations in this small Himalayan State. It is in this context, that place affiliated "Lachungpa" and "Lachenpa" terminologies are relevant, reasserted and reiterated.

2.2. The Lachungpas and Lachenpas of North Sikkim and Their Dzumsas

A general understanding is that Sikkimese–Bhutia groups inhabiting the valley regions of Lachung and Lachen in North Sikkim, located at an altitude of over 2500 m masl along the Indo-China border, are referred to as Lachungpas and Lachenpas respectively [41,45,46]—"pas" meaning "people of" in Tibetan. However, in practice, not all-Bhutia groups of Lachung and Lachen are considered as Lachungpas and Lachenpas despite decades of residence in the region. This includes Tibetans who settled in the Lachen and Lachung regions before the Chinese occupation of Tibet in the early 1950s, Tibetan refugees who settled here post the closure of Sino-Indo border after 1962, long-term resident nomadic herders—the Dokpas and some Sherpas—all with Tibetan ancestries. It makes it difficult to know how the Lachungpas and Lachenpas distinguish themselves from other Tibetan settlers and refugees, Sherpas, Dokpas of Bhutanese origin. There are many similarities between these different highland communities living in the Lachung and Lachen region: a centuries old transhumance practice i.e., migrating seasonally from one ecological zone to another (into higher Himalaya) for agricultural and pastoral activities; trade ties with Tibet [45], socio-economic and cultural commonalities that come with geographic proximity. However, an exclusive hallmark that differentiates the Lachungpas and Lachenpas from others in the region as well as across Sikkim is their traditional administrative system of local self-governance called the Dzumsa and membership in it.

Ironically, the Dzumsas have a feudal origin. The institution was set up by the Chogyal monarchy to establish authority and ensure "structural cohesiveness" for collecting land tax in the distant, far-flung regions of Lachung and Lachen [45–47]. In time, the Dzumsas also took responsibility for settling local disputes, overseeing fulfillment of cultural and religious obligations, etcetera. When monarchy was abolished in Sikkim in 1975, following Sikkim's merger with India, the Dzumsas of Lachung and Lachen were retained and later, conferred recognition within Sikkim (via Sikkim Panchayat Amendment Acts, 1982/1993/1995) [41]. This brought the Dzumsas at par with the Gram Panchayat—the third tier of local village self-governance system under the Panchayati Raj Institution as imposed in the rest of Sikkim [41]. Further Amendments (2001) protected the Dzumsa's customary laws, uncodified in nature, making the two Dzumsas uniquely official as well as traditional [41,42]. These unwritten customary laws bestow enormous power on the Dzumsas—making the Dzumsa rigid and flexible in executing its functioning—in contrast to Gram Panchayats that are strictly based on GoI and GoS guidelines. One of the key features of the Dzumsa is its social structure: all male heads of Lachungpa/Lachenpa households are members of their respective Dzumsa committees and thus influence the dynamics of decision making as well as the execution of the responsibilities and functions of the Dzumsa. This is hailed by many researchers as one the most traditional models of democracy [41,45,46] and participation. In addition, unlike Gram Panchayats that have affiliations to political parties and where decisions are influenced by party-ideologies or agendas, the Dzumsas are deliberately politically neutral. Therefore, while individual affiliation to political parties are allowed, public displays of such affiliations are banned in Lachung and Lachen.

Elders in Lachung and Lachen explain that in earlier times, membership of the Dzumsa was open to all households resident in these regions. However, post-merger with India, the geopolitically sensitive border regions of Lachung and Lachen were the site of significant defense and infrastructural development by the GoI. This resulted in a huge influx of outsiders, including Indian Army and Border Relief Organization personnel and various categories of construction workers employed on military projects (see [13]). This made the Lachungpas and Lachenpas increasingly conscious about protecting and preserving their territory-affiliated identity and their institutions. The nomination of the first ever Minister from Lachung in the Government of Sikkim in the early 1980s and the candidate's use of the Lachungpa suffix (and not Bhutia) was a conscious re-affirmation of the place-affiliated identity. Thus, while the generalized terminology Bhutia is used by (especially younger) Lachungpas and Lachenpas in official documentation (the term brings constitutionally assigned Scheduled Tribes protections, entitlements and privileges), the older generation mostly do not use the Bhutia title. They (and the younger generation too) attach the exclusive Lachungpa or Lachenpa as a suffix after the term Bhutia to reassert their "real" identity. Today, apart from the Bhutia–Lachungpas and Bhutia–Lachenpas, other resident communities are not Dzumsa members, nor are they considered to be Lachungpas or Lachenpas in Lachung and Lachen respectively. This benefits those who were granted Dzumsa membership decades ago by virtue of their residency in the region or through marriage to Lachungpa/Lachenpa. Dzumsa membership is not a privilege for all inhabitants and expresses unequal rights. As we will discuss in Sections 5.1 and 5.2, currently, the Dzumsa is an exclusive, exclusionary institution, but before we explain this, we present a brief literature review on some selected concepts and theoretical frameworks to ground our paper.

3. Conceptual Notions—The Plurality of Identity

The notion of identity is complex and ambiguous, understood in myriad ways. Identity is multi-faceted [31], a social construction [48], a social process [49], a social product [50], a collective phenomenon [31], a fundamental condition of social being [51], etcetera. It is hard to pin down one's identity, being a composite of behaviors and factors, a collection of beliefs about oneself. Weinreich defines identity as a relational construct joining a person's past, present and future self-images, "… the totality of one's self-construal, in which how one construes oneself in the present expresses the continuity between how one construes oneself as one was in the past and how one construes oneself as one aspires to be in the future" [52] (p. 1). Escobar [23] (p. 203) notes that identity is an "articulation of difference(s)" that are both "dialogic and relational", which is why identity is not fixed, continuing to evolve throughout the lifespan and multiple experiences of any one individual [53]. Or as Massey [54] (p.5) says, identities "are not rooted or static, but mutable ongoing (re)productions". At the same time,

identities are also not entirely fluid. Weinreich and Saunderson [55] note that identities constitute "
... a structural representation of the individual's existential experience, in which the relationships
between self and other agents are organized in relatively stable structures over time ... with the
emphasis on the socio-cultural milieu in which self relates to other agents and institutions" [52] (p. 1).

Identity exists not only at an individual, but also at relational and collective levels [56] giving rise
to a plurality of individual and collective identities, which can be political, social and cultural. Whether
identity is socially constructed [48], discovered, ascribed by others or dominant institutions [51],
or acquired by oneself [35], according to Castells [48] (p. 7), identity derives meaning and relevance
when "social actors ... internalize or acknowledge" these constructs. Thus, at any point in time,
"individuals have multiple identities, which may not always work in the same direction" (i.e., may be
conflicting) and collectively, any society is often fractured [17] (p. 2). This explains the complexity of
cooperation, solidarities, conflicts and exclusions. While some argue that identity plays an important
role in collective action, identity is not the only factor that influences collective action [57]. Other
factors—such as perceived threats [58], perceived injustices [20], grievances [17,20], efficacy [20],
economic interests and motives [17], norms [20], social embeddedness [19], emotions [18,20,22],
appeals [17], moral and(or) inner obligations [20,59], leadership structures [22], etcetera, also influence
collective action.

3.1. Collective 'Politicized' Identity and Collective Action

While individual identities are entirely diverse, it is shared interests and beliefs that converge to
enable collective motivational interests. Here, a sense of "sameness ... manifest(s as) ... solidarity,
shared disposition or consciousness, or in collective action" [60] (p. 7). Collective identity is thus better
described as the identity of an individual as a group member [61]—serving "psychological functions"
that relate to basic needs of the group such as belongingness, distinctiveness, respect, meanings and
agency [62]. However, not all collective identities are salient at the same time; depending on contextual
circumstances, collective identities can acquire or lose their relevance, position and status [61].

Klandermans [17] and many others [20–22], argue that collective identity (or identity in general)
"become(s) the engine of collective action" only when politicized [18] (p. 5). Klandermans has
outlined three processes for the politicization of collective identity: 1. awareness of shared grievances;
2. identification of an external adversary (against which/whom claims and grievances can be levied);
and 3. obtaining the support of a legitimate, authoritative third party [17–19,33]. A politicized collective
identity often instigates a strong internal, moral obligation to concerned individuals to participate
in collective action [59]. As we discuss below, Klandermans analysis makes a close fit in helping
unpack the construct of a Lachungpa/Lachenpa identity and the relative politicization of it against the
hydropower agenda.

Melucci points out that for any grievance to be explosive, there must be a breaking point or critical
threshold where conflictual reaction is triggered [22]. As he notes, "when norms or shared values are
threatened by some form of imbalance or crisis, the response through which an attempt is made to
re(establish) social order is centered around a common belief which, while often fictitious, mobilizes
collective energies" [22] (p. 14). Further, as Boelens and Claudin [63] argue, in "adverse economic
conditions, competing political influences, and the hegemonic powers that surround and penetrate
... , it is a challenge to maintain and reproduce a 'community' ... " to ensure, "the collective defense
of a community's material–economic foundations ... creating and reaffirming shared norms, values,
rights, and symbols" [63] (p. 1071). They state that, while collective institutions are (mostly) rational,
the "rules, relations, and behaviors" that mobilize collective action are not necessarily established
rationally [63] (pp. 1070–1071). Strategies driving politicized collective action are often not about
rational calculations. Rather they are driven strongly by emotions, feelings and perceptions [22,64].
Boelens and Claudin detail how these strategies may be the "outgrowths of historical and contemporary
events, of context-specific trial-and-error, of opportunities and limitations on power, and of neighboring
and supralocal institutions that are incorporated" [63] (p. 1071). Critically analyzed, these processes

debunk the often, "dogmatic myths of romanticized, rationalistic, or economist" narratives of collective action [63] (p. 1071). This explains why collective action depends deeply on trust, emotion, connect and cooperation among participants, spurred by shared understandings, experiences and identities [65].

Going beyond altruistic views, Boelens and Claudin continue to explain how collective action, not just outwardly but also internally, rests on harsh struggle to shape collective rules and orientations—inwardly, these institutions constitute "both an arena of power struggles and conflict negotiation, and a collective entity" [63] (p.1071). In the same vein, Colloredo-Mansfeld (in "Fighting like a community . . . ") [16] and Boelens and Zwarteveen (writing on water justice collectives) [66], follow Chantal Mouffe's notions of "agonistic spaces and relationships" [67]. Colloredo-Mansfeld describes how collective action rests on "agonistic unity" [16], in other words, a unity that exists despite of diverse differences (see also [68]). An agonistic unity is often mobilized via techniques or "organizational measures or strategies developed by leaders . . . to administer, persuade and at times coerce residents to move towards a collective purpose" [16] (p. 7). Colloredo-Mansfeld termed this mode of arriving at consensus against the odds as vernacular statecraft. Indeed, as our research findings show, a politicized, sustained collective action against hydropower projects in Lachung and Lachen provides evidence of an agonistic unity and vernacular statecraft.

To establish and sustain effective collective action in situations of competing interests or in high-risk context where participants might face repercussion for their actions, social embeddedness of the conflict in supportive institutions is hugely strategic [19,21,69]. These institutions not only provide relevant resources but also make the "benefits of participation and the cost of non-participation as high as possible" [61] (p. 588). Certainly, institutions do not always politicize collective identity and mobilization [21,70]—there are multiple ways by which "power works within communities" [71] (p. 258). Nonetheless, as we discuss below, the Dzumsas provided resources such as information and funding, and forced the community by making the benefits of participation and, especially, the costs of non-participation utterly high.

As observed by Boelens and Claudin [63] (p. 1071), maintaining and reproducing "community", its material-economic foundations and norms, values, rights and symbols, is closely interlinked with notions of territory and territoriality, which we briefly describe below.

3.2. Territory and Territoriality

In the research regions, both identity (as Lachungpas and Lachenpas) and its social embeddedness in the traditional institution (Dzumsas) had a strategic connect to place/territory. This weave between identity, institution and place finds resonance in the views that territories are not just formal nation state, province or other legal-administratively demarcated regions [72]. Rather, territories are geographically demarcated and cultural-politically bound spaces, constructed around and by socio-spatial authority. In a broad sense, territories link social, physical and symbolic entities: they entwine ecological systems, legal-administrative arrangements, technical-physical infrastructures, political discourses, and socio-economic livelihoods. Or as Swyngedouw and Boelens [73] (p. 117) say, "territory is the socio-materially constituted and geographically delineated organization and expression of and for the exercise of political power".

Similarly, Antonsich [74] (p. 425) argues that territory is "the socio-spatial context where the living together is produced, organized and negotiated". Territories are dynamic, historically shaped, contested and permanently negotiated. As Hommes, Boelens and Maat state: "They evolve out of social encounters and are the effect of social relations' material inscriptions that define what spaces look like and how, in turn, connected social relations are organized . . . The making of territory is an interactive and continuous process that emerges from imaginaries about what a territory in its judicial, political, economic, social, cultural, affective and physical aspects, should look like" [75] (p. 3), (see also [76–78]). Importantly, therefore, this broad concept of territory includes blatant and subtle everyday struggles, disputes about discourses, and battles around the use and recognition of divergent knowledge systems. Consequently, battles over local territorial constructs and territorial governance forms deeply constitute and interact with identity and knowledge formation and re-creation. Territory thus has profoundly divergent meanings [26]. Territories come with "limits" and "otherness" and these demarcations are often determined by identity [79]. Territories thus are markers of identity (and vice-versa) and more often than not, enablers or disablers of processes of social exclusion perpetuating "lack or denial or resources, rights, good and services, and the ability to participate in normal relationships and activities" to some over others [80] (p. 25). Territories are therefore not static, rather, they are continually contested and actively negotiated [51].

Agnew and Oslender refer to territory as the popular acceptance of classification of space (e.g., ours versus yours), as a way of communication regarding a sense of place, and as a concept to express enforcing control over space (such as by barrier construction, interception, surveillance, policing and judicial review) [81]. As such, territoriality is usually put into practice in a number of different but complementary ways. Often, protecting physical demarcations of territory through "territoriality" serves to protect, preserve and strengthen identity and associated cultural values [26] by "affect(ing), influence(ing) or control(ing) people, phenomena, and relationships, by delimiting and asserting control over a geographic area" [82] (p. 19). In sum, identity, territory and territoriality are deeply entangled and often inseparable [31].

4. Study Area and Methodology

4.1. The Study Area: Cancelled Hydropower Projects in Lachung and Lachen

Lachung (altitude 2600 masl) means "small-mountain" and Lachen (altitude 2700 masl) "big-mountain" in Tibetan [45]. The two regions are approximately 60 km apart from each other and located in the North District of Sikkim. Based on the information displayed on official display boards in the local health offices in the two areas, Lachung has a population of 1478 Lachungpas (in 420 households) and 370 non-Lachungpas (in 72 households) while Lachen has 1314 Lachenpas (in 216 households) and 126 non-Lachenpas. (The total number of non-Lachenpa households in Lachen was not mentioned on the community notice board). Lachung and Lachen are administratively categorized by the GoI and GoS as "restricted" areas and remain under heavy military surveillance because both these valleys regions have mountain passes that connect Sikkim with Tibet, [45]. Travel permits including No-Objection Certificates for research activities are required to enter these areas. However, lately tourism has emerged as a booming local industry in both Lachung and Lachen.

In Lachen, two large hydropower projects, the 320 MW Teesta Stage I and the 330 MW Teesta Stage II, part of the "cascade" dams (i.e., the series of six hydropower dams—Teesta Stage I, II, III, IV, V and VI that were conceived as early as the 1970s) were cancelled after public protest (see Figure 2). Additionally, the 210 MW Lachen HEP and the 75 MW Talem Chu planned by multiple Independent Power Producers (IPPs) after the 2003 Hydel-Initiative Announcement by GoI were also cancelled. In Lachung, a 99 MW Lachung HEP, originally planned two dams in different sites was cancelled following local contestations. It is important to note that these valley regions are fully electrified by micro-hydel projects (3 MW Lachung Small HEP and the 3 MW Chatten HEP in Lachen) developed in the late 1980s. Another 3 MW Rabom HEP implemented in Lachen was damaged and declared non-functioning by the 2011 earthquake. Nonetheless, energy is a vital need in these high-altitude cold regions that faces frequent power cuts. It is therefore surprising that the large-scale hydropower development planned here with the promise of free electricity and other developmental gains, was fiercely opposed.

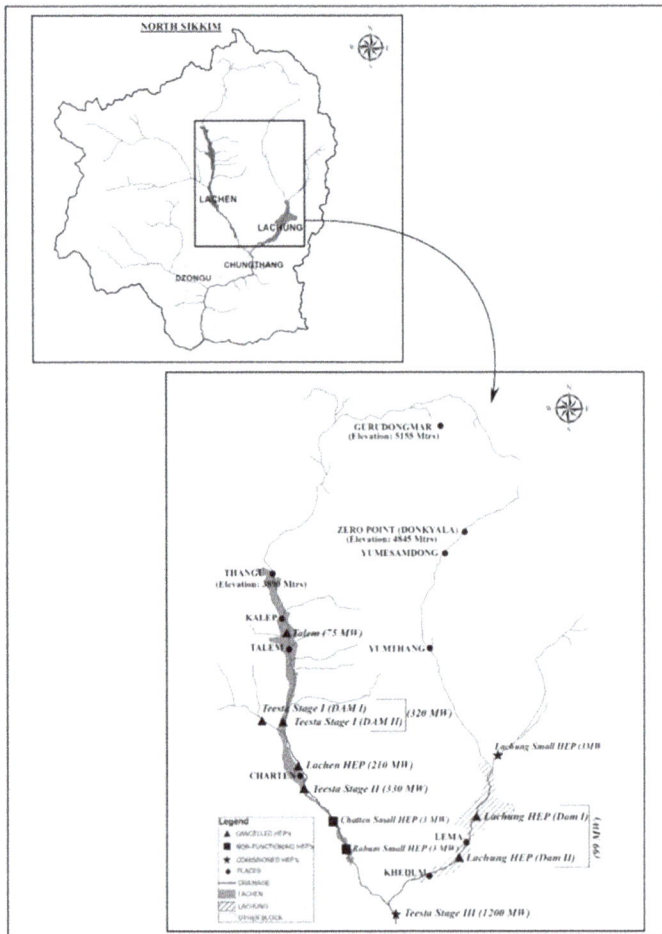

Figure 2. Hydropower Dams in Lachung and Lachen. Source: Own elaboration, adapted from GoS websites. Map not to scale.

4.2. Methodology

This paper draws from ethnographic research (see [83–86]) with diverse data collection methods, such as observations, semi-structured-interviews, focus group and individual discussions etcetera. The first author-researcher had been in the study area since mid-2015, first in the neighboring region of Chungthang (mid 2015–early 2016). The fieldwork in Lachung and Lachen was a continuation of the research in North Sikkim. This set the ground for meeting the Lachungpas and Lachenpas through mutual contacts (from Chungthang). Fieldwork for the current paper was conducted from late September 2016 up to February 2017 in Lachung, and resumed from end May 2017 to early October 2017 in Lachen. In the months spent in each area, familiarization with the place and its people was done through living in Lachung and Lachen and by "deep hanging out" [84,85]. The first author-researcher, being a woman and a non-local in the study area, conducting research, speaking to large numbers of male strangers, initially aroused suspicion and distrust in the study area. However, being a Bhutia herself, having family in Sikkim, with stays for long periods of time with local host families—gave the researcher some degree of familial connection and allowed her to be seen as an *afnai-manchey* (one of us). Attending socio-cultural and religious festivals, taking transects walks alone or with host family members and newly made friends, visiting touristic places enabled to be in the public gaze long enough to be considered a "regular". After some degree of trust and familiarity was gained, data collection was initiated with different groups of the local families residing in Lachung and Lachen: farmers, private business entrepreneurs, government employees, the unemployed, etc. Meetings with Lachungpa and Lachenpa residents, and with Tibetan, Dokpa, Sherpa, Lepcha and Nepali families, took place on an everyday basis, to strengthen confidence. The male head of Lachungpa and Lachenpa households were also members of their respective Dzumsas, who regularly attend Dzumsa meetings. In all 47 individuals were interviewed personally over a period of time (repeated meetings) and multiple discussions were conducted with a much larger number of others. Given the blanket opinion of "no dams in Lachung and Lachen", initial interactions here did not begin with questions about dams and dam resistance. Nonetheless, the purpose and nature of the research was made known to those who opposed dams blatantly as well as latently. Two power company officials were interviewed—not from the power company that the Lachungpas and Lachenpas threw out (see Section 5) but agents working at other operating hydropower projects nearby, in Chungthang and Mangan.

The next sections present our findings. We begin with a brief description of the structure and characteristics of Dzumsa and relate these with our core findings, following which, we discuss how the announcement to develop hydropower projects in the two areas triggered an agonistic unity and how the Dzumsa enabled the reordering of the collective by ensuring high risks of exclusions to those who challenged the vernacular alliance.

5. Hydropower Development and the Politicization of Identity, Territory and Dzumsa

5.1. Dzumsa: Structure and Decision-Making

Colloquially, the term, Dzumsa has three literal meanings, "a gathering place"; "an institution in charge of administrating and organizing activities within a given territory" and "the general council of villagers composed of household heads" [47] (p.95). Lachungpas and Lachenpas have their own (separate) Dzumsas, and this institution is only accessible to male head of households among the Lachungpas and Lachenpas. Unlike Gram Panchayats, administratively both Dzumsas are composed of (and chaired by) Pipons, who are normally the village-headmen. An inner core Dzumsa committee includes Gyapons (elderly males to assist Pipons), Gyembos (male members who function as messengers), Chuitimpas (male monks to assist Pipons), Tsipos/Chipons (male accountants) and Machays (male cooks). All other male heads of Lachungpas and Lachenpas households are Dzumsa members. Lachungpa and Lachenpa women are only occasionally allowed to attend Dzumsa meetings—in exceptional situations, when the male-head of household is absent or unable to attend (with a valid reason). Also, "others" residing in Lachung and Lachen are not a part of the Dzumsa, even though they are governed by Dzumsa norms and conditionalities. They do not participate in the collective decision-making.

In earlier times, Pipons were selected by the Chogyals and this post continued as a hereditary appointment in the Pipon's family. Post-monarchy, individuals who were identified as reliable were nominated by the Dzumsa members and often succession continued to follow along hereditary lines [45–47]. After monarchy was abolished post 1975, both Dzumsas began incorporating various other methods for the nomination—elections, a lottery system or simply hand raising. In Lachung, the Pipons are always elected or selected from two places—Lema and Khedum, and the Pipons from these two places officiate as Pipon 1 and II on a rotational basis. In Lachen individual Lachenpas securing the highest and second highest votes becomes Pipon-I and Pipon-II, respectively. Regardless, the two Pipons (I and II) are bestowed with equal power and functions and are responsible to dispense administrative functions and lead socio-cultural activities. Currently, in Lachung, Pipons are elected for a fixed two-year term and are appointed not by election but through a public lottery system. Pipons in Lachen continue to be elected through voting and are elected only for a year but unlike in Lachung, outgoing Pipons can be re-elected and continue for as many years should they garner votes.

Although extremely rare, Pipons can be ousted from their posts by the Dzumsa members any time if they failed to carry out their duties vis-a-vis the wellbeing of the community and the place. The plans to develop hydropower projects led to (such) an unprecedented removal of the Pipon(s) in Lachung, while in Lachen too, the Pipons were threatened with possible removal from the post. This power of the Dzumsa members to elect or nominate and dispose their representatives at any time and for any issue makes the Dzumsa different from Gram Panchayats, where village representatives must be elected and have a fixed five-year term. Additionally, the two Dzumsas are by choice non-political and do not allow individual party affiliation of the Dzumsa members to influence the functions and powers (as is prevalent in the Panchayat institutions in Sikkim). In fact the public displays of political party affiliation by means of flags and political canvasing were banned in both Lachung and Lachen as such acts were perceived as threats to the collective public unity and peace in the region.

Dzumsa meetings in Lachen and Lachung are called by the Pipons through Gyapons and decision-making is through unanimous consensus. However, the Pipons also have the exclusive power to take unilateral decisions on both urgent critical as well as mundane issues, which speak to the wellbeing of the place and the people. When Pipons fail or hesitate in making critical decisions, the Chuitimpas or Gyembos are consulted to assist arriving at a decision. Once decisions are made, they are relayed to the Dzumsa members, who often go by what their representatives have agreed. It is interesting to note that Dzumsa members can deliberate on and contest the decisions taken by the Dzumsa representatives. However, in turn, if the Pipon considers these arbitrations to be invalid or unreasonable, the persons making these deliberations can be fined. If the Dzumsa representatives cannot make a decision, then all Dzumsa members collectively deliberate until a majority agrees on the decision.

The Dzumsa plays a critical role in these communities—making decisions on a wide array of everyday issues that can be socio-economic, environmental, cultural, religious, law and order, etcetera. All decisions which the community must abide by. This is why, although administratively Dzumsas and its equivalent Gram Panchayats dispense the same functions, the customary and traditional laws of the Dzumsa recognized by the GoS expands the power and function of the Dzumsa beyond that of the Gram Panchayats within their territory. It allows them added power and legitimacy to impose coercive actions on the Lachungpas and Lachenpas but also on the non-Dzumsa members (i.e., the Tibetan, Sherpa, Dokpa or Nepali) like fines, impose new rules and regulations, social exclusions, boycotts, including settling of grave disputes.

Thus, while the Dzumsa is eulogized as egalitarian and democratic by many researchers, it is rather hierarchical, masculine and exclusionary in its structure and operation [46,47] (p. 35). The Dzumsa was not always so closed as it is today. An elderly Lachenpa recalled, "I had heard that in the old times, Dzumsa had very few members. Of-course then our population was also very low, yet, still Dzumsa meetings were not compulsory and any one (outsiders) could join it. The members registration was so low that one had to seek people to join Dzumsa and constantly request people to undertake collective work". Sikkim's merger with India in 1975 and the subsequent marking of territory (land settlement surveys in 1978/1979 under GoI) as well as the nature and extent of translocal developments have contributed to the reassertion of a territorial collective identity among the Lachungpas and Lachenpas.

In the section below, we look at how hydropower development threatened the agonistic union of this traditional self-governing body and how the Dzumsa members resorted to 'vernacular statecraft' to restore their collective identity and institution.

5.2. Hydropower Intervention and Politicization of Collective Identity

Ways of living and governance in Lachung and Lachen conform to an uncodified customary and traditional system, which is deeply exclusionary. The State push for hydropower development took place in a context that has historically been politically suspicious and antagonistic. In the words of a male Lachenpa respondent: "If the 'company' (hydropower project) comes, they will bring with them thousands of outsiders, whose presence will dilute our existence. Our land, culture, tradition, old practices and identity are at stake. We will be outnumbered. We will be forced to relax our existing Dzumsa rules and laws to pave easy way for such developments. This way, our age-old laws, rules and regulations will slowly lose their relevance".

It was reported that the GoI announcement of the 50,000 MW Indian Hydroelectric Initiative in the North District of Sikkim in 2003 [87] led to an urgent Yul-Dru-Sum meeting in Chungthang between the Lepchas of Dzongu, the Lepchas and Bhutias living in Chungthang and the Bhutias living in Lachen and Lachung. (In Tibetan "Yul-Dru-Sum" translates to Yul meaning three, Dru meaning together and Sum meaning places i.e., the people of three places—Lachung and Lachen as one entity, Dzongu and Chungthang). This meeting was also attended by Lepchas from nearby project-affected-areas outside the administrative boundaries of North Sikkim. A momentous unanimous decision was made amongst the two indigenous groups (Lepchas and Bhutias) to not allow any power-companies in the region. A 45-year-old Lachen resident who attended the meeting recalled: "There were around 50 to 60 people that day. We discussed in detail the pros and cons of hydropower development and concluded that if such companies entered, we would be left with just the Sikkim-Subject land documents but with no land. We would have sickles in our hands but no land to farm. We agreed to all say 'no' to the company". However, just a year later in 2004, the Yul-Dru-Sum pact was violated in Chungthang, where 80% of the Lepchas welcomed the 1200 MW Teesta Stage III HEP development (see [13]). Another Lachenpa respondent who had attended the meeting felt that this happened, because, "Unfortunately we did not translate the decision to a written agreement. The Lepchas played a nice game. First, they said no, and then they negotiated for a higher amount of compensation money before saying yes".

Since 2004, the GoS had started to issue Letters-of-Intent (LoI) to power-companies, mostly Independent Power Producers (IPPs), which gave these organizations the right to access protected and reserved areas to initiate detailed surveys and investigations. The LoI also gives power-companies both the right and responsibility for contacting local communities and obtaining local consensus for planned development interventions. Private corporations are particularly skilled in making promises of development and economic gains; this is precisely what had happened in the case of the Teesta Stage III project planned in Chungthang (see [13]). As it turns out, like the Lepchas, some Lachungpas too (even if briefly) had faltered on the Yul-Dru-Sum agreement, although the Lachenpas had honored the decision.

Despite the promise to say no to hydropower development and the skepticism among Lachungpas and Lachenpas regarding such developments, an independent private power company was able to rupture the collective decision in Lachung. Talks for a 99 MW HEP hydropower project by Polyplex India Private Limited—an independent power company—went ahead here in 2005/2006 with the support of a few powerful Lachungpas, who held important government positions and lived in the capital Gangtok, as well as by a (then) Pipon of Kedum in Lachung, who gave his consent to a private company to undertake surveys along the riverbanks. According to the Pipon, he was gifted cash to distribute amongst the people of Lachung for allowing Polyplex company to begin the survey for two dams in Lachung: "I asked my people to accept the money as a gift from God and enjoy it". All Dzumsa members in Lachung had indeed accepted the money initially. "The Pipon was a well-respected man and powerful as well, his brother has been in politics for a long time. We believed him when he said that the survey would be undertaken along the rivers and that land would not be touched. Believing in him, each Lachungpa household head accepted twenty thousand Indian Rupee (equivalent to less than 300 Euro) that he distributed on behalf of the company. But when we saw that they were also assessing our land and mountains, we intervened. People might accuse us of selling out, but trust me, we didn't".

At around this time, young educated Lachungpa youth started to raise concerns about the potential impacts of such development. Initially, these concerns were not considered by the Dzumsa representatives. In fact, the Pipon of Lachung (who had distributed the money) refused to grant the youth an "emergency" meeting with Dzumsa members. According to a youth activist, "We were denied Dzumsa meeting by the Dzumsa representatives. Despite the restrictions placed on us, for the first time in our life, we disobeyed the norm and formal processes associated with the Dzumsa. We announced an emergency meeting publicly on a loudspeaker. Thankfully, people turned up the next day and we could place our concerns in front of everyone".

This emergency meeting led to a direct confrontation between these youth with the (then) Pipon from Khedum. The Pipon and a few of his aides were accused for a lack of transparency and money embezzlement. While such intervening in the Dzumsa's authority was unprecedented, it nonetheless, eventually led to the majority of the Dzumsa members supporting this accusation as the youth were equipped with critical questions and proofs, which led to the ousting of the Pipon from his post, and later on from Dzumsa itself. This process was supported by the other Pipon from Lema but contested by some Lachungpas who still supported the ousted Pipon—bringing much conflict within the community. The Dzumsa members were split between the Pipon of Khedum who had favored dams and the Pipon of Lema who has supported the youth—creating animosity and distrust between the once amicable inhabitants of Lachung. A Lachungpa laments, "We were so polarized initially that when one youngster from Lema attended a public meeting in Khedum in disguise to listen to their discussions about hydropower dams, he was unfortunately caught and brutally beaten."

According to the ousted Pipon, a 45-year-old Lachungpa: "I was the Pipon of Khedum in 2010. The company informed me during my tenure that the sites selected for the project earlier were not correct, and they only wanted to see where the first survey had been done. Just on that premise, Dzumsa members kicked me out of the Pipon post." However, a Lachungpa Dzumsa member added that: "It was forbidden to even talk about the company in Lachung, forget about entertaining their calls or talks. The second ousted Pipon did not consult us, or bring the matter to us, so we kicked him out of the post. We will remove anyone from that position who does *gaddhar* [betrays] to us and our place." Hydropower issues trigered the Lachungpas to mobilize and assert their voice and might; and not always in the most positive ways. Some days after the Pipon had been expelled, a violent confrontation took place when the Polyplex Company began drilling tests in the nearby mountains. "They started to dig through our mountains and take our stones. That was it! It was evening, these people were camping in tents. We burnt their tents, shouted at them and kicked them. Some of them were cooking food. We kicked their pots of rice, hurled them into trucks, drove them outside of Lachung and threatened them to never come back. Eventually, we regretted that we had attacked poor laborers, who were just doing what they were tasked by the company". These violent protests continued in Lachung, where company vehicles were damaged and local residents (Lachungpas and non-Lachungpas) working for the company were threatened to quit working or be ousted from Lachung.

The first expelled Pipon of Kedum, however was unshaken by the stand against him and became a prominent dam supporter. Being powerful and politically well-connected, his expulsion from Dzumsa was re-negotiated and he was allowed to retain his Dzumsa membership, from where he lobbied harder for the dam projects. In his words, "I managed to transport people in 45 to 47 vehicles from Lachung to the District Collector's Office at Mangan, where I confronted the anti-hydropower people from Lachung. I answered every charge levied against me, and finally I asked the District Magistrate to 'welcome' the company back to Lachung and continue their work. That the company didn't go back, is not my fault. It simply shows that it was a weak company".

Although it was in Lachen that the CWC had started the planning for hydropower development, Lachen was the last of the three regions where dams were announced. The time lag between the first dam planning processes in Chungthang and then in Lachung allowed the Lachenpas to observe and understand how coercion in dam development takes place. By the time the power companies went to Lachen to get an agreement on two hydropower projects, the Lachenpas felt they understood the politics of dam development.

The Lachenpas adhered to an absolute "no" right from the very beginning. Unlike in Lachung, where one Pipon and his supporters became local mediators within the Dzumsa for the private company, in Lachen, a powerful collective of Government officials, the power-company, National Hydroelectic Power Corporation (NHPC) as well as Ministers went directly to meet the Dzumsa members of Lachen and seek their approval for multiple hydropower projects in 2005. At this meeting, the Dzumsa members of Lachen remained firm—their verdict was a unanimous "no": "We knew this powerful group of individuals were coming to talk about hydropower development. The Minister accompanying them was a Lachungpa, so we told him directly, take your proposal to your own place, Lachung. The Dzumsa had called a meeting a day before, where we had deliberated and collectively decided to say no. When the group came for the Dzumsa meeting and sat in the Dzumsa hall, we closed all the doors and latched it. This might have intimidated them. The moment our Minister started talking about the company, people shouted . . . We were hostile and managed to scare them off". This development also resulted in significant tensions between the Lachungpas and Lachenpas. The Lachenpas managed to dethrone the Lachungpa Minister. "We were so angry at our Minister. He being a Lachungpa, brought the group here. A devastation masked as development for us here! We considered him our own and had supported him for a long time. However, after that meeting, we told the Chief Minister of Sikkim that all of us at Lachen would no longer support his party and would join the opposition party if the Lachungpa Minister continued in the cabinet. Soon thereafter another person was given the ticket to represent North Sikkim". The power companies (NHPC and a few IPPs such as Hima Giri) tried approaching the Dzumsa many times and eventually managed, like in Lachung, to convince one of the Pipons there to speak on their behalf. However, this did not yield any positive outcome.

A deep distrust, even paranoia, for power companies took root among the Lachungpas and Lachenpas. In fact, company representatives were forcibly asked to leave Lachen when he had visited Lachen as merely a tourist. Another Lachungpa recalled: "Our people have become fearful about the company. I was once urgently called by our elders who told me that some company people had sneaked into Lachung with their instruments and were taking pictures of our rivers. I rushed, to find the intruders surrounded by our village people. As it turned out, these were mere tourists, using a tripod to take pictures near the river". Below we describe the various strategies adopted by the Lachungpas and Lachenpas to counter not just the power companies but the distrust, disunity and animosity that had crept within and in-between their communities—where the power of the Dzumas becomes imperious and unlike Gram Panchayats, becomes "binding" in determining pro- and anti-dam positions among the Lachungpas and Lachenpas.

5.3. Agonistic Unity: Dzumsa, Anti-Hydropower Resistance and Vernacular Statecraft

In Lachung, following the brief tryst with the Polyplex company and the internal fractures, an oath was taken by many Dzumsa members at Thomchi Gumpa in 2010, their main monastery, to *never* allow a hydropower company to enter Lachung. At this oath ceremony, fears of social exclusion and boycott from Dzumsa were announced for any remaining pro-dam supporters. This was followed by a series of new conditions set for Dzumsa members of Lachung: firstly, only those individuals who contested all hydropower projects would be eligible to become Pipons; secondly, all Lachungpas were restricted from talking about or to "company persons", at least within the borders of Lachung. In fact many Lachungpas even collected and submitted their land documents to the anti-dam faction of Dzumsa members to not fall prey into selling their lands to power companies or the pro-dam faction. In Lachen, despite the unanimous decision of no-dams, Dzumsa members came up with similar new rules for its larger residents: no one was allowed to lobby for any company; no hotel owners were allowed to host hydropower "company" people, even if they came in as tourists; no shops were allowed to sell anything, including water to the company people, no one was allowed to talk or negotiate with company. If anyone was found doing these, they would be boycotted from the society and sanctioned out of the Dzumsa.

However, both in Lachung and Lachen—these restrictions were capped by something far more potent. The local cultural practice of "Chya" (referred to as "Chya-Kyapshe" in Lachung and "Ma-Chya" in Lachen) and colloquially known as "Kalo Puja" in Nepali (translating to black ceremony) was announced as a "last weapon". The Chya is a dreaded public ritual usually undertaken in stealth—when the perpetrator is not known or when the intentions of certain individuals go against the wellbeing of the larger community and their place(s). For the Lachungpas and Lachenpas, mountains and glacial lakes surrounding the two valley regions hold great significance as they are considered to be the abode of their Lhasung(s) (guardian-deities), revered and feared in these high mountain knowledge systems. Chya involves invoking these very local-deities to make a collective curse to punish "unknown" or dangerous perpetrators. This ceremony is performed only after the majority of the Dzumsa members are in consensus. Initiated by *paus* (sorcerers), this collective cursing is performed with great faith and belief that the perpetrators are punished through an ultimate death for their acts against place and people. The Ma-Chya of Lachen is believed to also be passed on to the perpetrator's future generations. These rituals are thus deeply feared by all in Lachen and Lachung: "We believe in Chya and take it very seriously. People have died unexpectedly in Lachung and Lachen after Chya was performed. Healthy people suddenly contact grave diseases and die, and the cycles of misfortune continue for future generations. The Chya only works on the guilty, this is the greatest feat of its relevance". It is important to note that when a collective decision is made to perform Chya on known or unknown individuals, there is little room to oppose or not engage in this process. This can have repercussions of social exclusion and boycott.

When the power company gained entry in Lachung, Chya-Khyapse was reported to have been performed against those who "sold-out" and/or embezzled funds. In Lachen, where no company was allowed, Ma-Chya was still performed as a deterrent to all Lachenpas from succumbing to the pressure or lure of money. A Lachenpa elderly explained: "The Pipon from Lachen who had some hand in getting the Ministers and company people here was on his way to Gangtok, when he turned ill and died a year later. He was right here with us when Ma-Chya was performed. One could argue that he was suffering from a disease, however, no one had expected him to suddenly die. His death is said to be the effect of Ma-Chya. We hear now that the first expelled Pipon of Lachung who welcomed the company is so scared of the Chya performed on him, that he has been performing one religious ceremony after another to negate the effects of the Chya". It is important to add here that recently, Chya has been banned in Lachung at the request of the Lachung Rinpoche (learned monk) as the Buddhist practice of compassion does not support destruction and ills wished upon another. However, the practice continues in Lachen.

The knowledge and practice of Chya not only deters all hydropower development, it remains a powerful vernacular statecraft against dissent with the Dzumsa. As a Lachungpa youngster states: "Hydropower companies are powerful and supported by the entire government machinery and bureaucracy. There is constant pressure and lobbying by the advocates of power projects. Money plays an important role in these processes. In the face of these challenges, Chya is our only way of preventing this destruction from development from happening". So great is this fear, that numerous attempts by power companies to enter the region have failed. As a Lachenpa reported: "Even after all these events in Lachen, there are continued attempts to woo our Pipons. One of our Pipons was invited to a five-star hotel (Mayfair) in Gangtok and offered India Rupee 9 to 10 crores (equivalent to a little over 1 million in Euro) to work towards public consensus to the dam building agenda. These tactics have not worked".

In fact, government officials who powerfully lobbied on behalf of the power companies elsewhere in Sikkim and showed no hesitation in coercing local communities to agree to these plans, are cautious in doing the same in Lachung and Lachen. Here they try to resort to logical narratives of gains and benefits from hydropower development and yet, when they fail to convince the Lachungpas and Lachenpas, they say, that hydropower development did not happen here, because most of the areas that will be impacted are forest areas—hence individuals will not receive compensation/s. "It is simply a clever cost/benefit analysis case. They [Lachung-pas/Lachenpas] know very well that most of the land for hydropower development is in forest areas, meaning less compensation. That's why they resist these plans and are successful in doing so". Nonetheless, the power companies and their advocates keep trying to pursue hydropower development.

The consequences of resisting hydropower development have not been easy for the Lachungpas and Lachenpas. This has resulted in official forms of punishment and coercion, mostly done through job transfers of government employees vocal against the hydropower development plans. A young Lachungpa laments, "I have seen first-hand the politically motivated transfers of our people to faraway places, far away from their families". These developments have only served to strengthen resistance. Giving the example of Tibetans, an old Lachungpa states: "The Tibetans are just refugees for the rest of the world. If our land goes, what will happen to our identity? This is why we have to protect this place for us to stay rooted and come back here no matter where we might occasionally go". This process of exercising territoriality continues. In 2015, a traditional "dress code" was made mandatory in both Lachung and Lachen. Married women in Lachen, and women above 15 years in Lachung have to wear their traditional clothes in their respective regions; while men are required to wear these for all social occasions, especially at funerals.

6. Discussion

The findings from this research illustrate how identity constructs and cultural politics play out dynamically in the resistance to hydropower projects by the indigenous Lachungpas and Lachenpas in North-Sikkim. Here, we summarize a couple of key issues for further discussion.

First, resistance to large-scale hydropower development by the Lachung and Lachen communities go far beyond a mere battle against the dreadful material (socio-economic and environmental) impacts that come with mega hydropower development in fragile mountain ecologies. Although these issues were of concern, also because the two communities increasingly rely on a booming tourism industry—where keeping the landscape scenic is vital to people's livelihoods, it was not just concern about possible material losses that led to a collective position of resistance. The Lachungpas and the Lachenpas were deeply concerned about how these new developments would impact upon territoriality—the defense of place (territory), of place-based institutions and the community's unique collective identity as well as their meanings, values and modes of living and knowing. All of these issues are not just central to the ways of being and living for the Lachungpas and the Lachenpas, they are also a powerful means of exercising, asserting and reiterating (collective) identity in the fractured political context of governance in Sikkim. For the Lachungpas and Lachenpas, articulation of their personal and collective identity, of being different, sprouts importantly from their ideas of demarcating territory and constructing (or exercising) territoriality. Territory and territoriality is indeed a key marker of the collective Lachungpa/Lachenpa identity construction. It is their collective identity, interlaced with their historic territorial systems and practices, that has enabled the Lachungpas and Lachenpas to define, project and deploy other kinds of boundaries—moral, cultural, ethnic, economic, political, including their loyalty and solidarities among themselves and against the others. Inwardly and outwardly, territory and identity entwine through the two Dzumsas in their attempt to affect, influence and control people, phenomena, and relationships; thereby demarcating and asserting control over their cultural-political geography. Shared territorial or place-based identity in Lachung and Lachen is the basis around which individual interests are translated into group interests and collective action—which manifests in the projection of a united front against the "others" and "outsiders".

Second, the theoretical analyses discussed in Section 3 explain how an externally imposed agenda of development and the various bearers of these plans and proposals became adversaries which helped both trigger as well as politicize collective identity. The centrality of the Dzumsas in this struggle over territoriality, in making the "benefits of participation and the cost of non-participation as high as possible" [61] (p. 588) cannot be overstated. The role of the Dzumsas in enforcing an agonistic unity is therefore crucial. At the crux of collective identity and associated territorial projections are the Dzumsas, which act as binding authorities and steering institutions to engage in locally-particular consciousness, morality and collective action. Collective measures such as threats of social boycott, community expulsion, no-display of political flags, etc., would not be implemented had it not been for the Dzumsa. These measures helped execute a conscious responsibility to keep the Lachungpas/Lachenpas together. Nothing is as sacrosanct as the maintaining of unity amongst the respective Lachungpa and Lachenpas Dzumsa members. As much as the Dzumsas draw their power from the members to acquire and maintain unity, the members draw their power, agency and voice from the Dzumsas, giving rise to a symbiotic dependence on each other. This symbiotic relationship sustains the effectiveness of Dzumsa officials vis-à-vis its members and also provides particular checks-and-balance to each other. In Sikkim, if not for the traditional and customary Dzumsas, the Lachungpa/Lachenpa identity would have been lumped into the broader "Bhutia" identity losing its hallmark distinction. The Dzumsa gives legitimacy to the exclusive collective identity of the Lachungpas and Lachenpas.

Third, as we note, these virtues of cohesion, collective identity and action are not without contradictions—they pose high social, political and emotional risks to those within this community, who for various reasons, might choose to not align with the normative principles of the collective. In addition, territories confining the two Dzumsas and their members within its demarcated areas are socially and politically accessible only to the Dzumsa male members and office bearers. This is not a pluralistic, inclusive institution—rather it operates by restricting intervention or interference from others—making and marking identity and territory are the core functions of the Dzumsa. As we noted, there have been changes in the Dzumsa's structure, functions and customary laws—but these have all been in tune with the interest and wellbeing of the two Dzumsas and its members—rooted in a "local-first" philosophy. This makes the Dzumsa partially exclusionary even within the Lachungpa/Lachenpa community—excluding women, youth, other long-term local residents from its decision-making membership. Even though excluded, they must conform to the institution's norms and dictates. In addition, as seen in the case of the hydropower project development plans, the Dzumsas served to expel its leaders (the Pipons), announce and enact boycotts and the dreaded Chya ceremony against its own members. This is how, as Colloredo-Mansfield analyzed, vernacular statecraft operates and sustains an agonistic unity, which while imperfect, is hugely effective in countering powerful translocal impositions. In continuing to be the sole recognized local institution deciding every socio-economic, environmental and religious affair of the Lachungpas and Lachenpas and their wellbeing—far more intense than the Panchayat System prevalent in the rest of Sikkim, the Dzumsas continue to emerge and evolve as the center of the everyday life-worlds. The degree of involvement is also what reveals the exclusionary side of this all-inclusive institution on grounds of gender, indigeneity, ethnicity, and rationality. It reveals the authoritarian side of the Dzumsa, since loyalty towards the collective and protecting unity and wellbeing of the locality has priority over any other issue.

The fourth and final point we make here is on the plurality of identity. In addition to the divides by gender, age and ethnicity that we discussed above, the Lachungpas and Lachenpas while being a tightly-knit community are nonetheless two groups with particular forms of self-identification. Shared understandings of place and territory, shared cultural values, beliefs and identities allow for a remarkable solidarity between the two groups and yet, when the situation demands, this collective trust and emotion can also turn into expressions of being different. The same can be said for internal dynamics of the Lachungpas and Lachenpas themselves and how historically, loyalty and solidarity is evoked through threat, fear and coercion—when the stakes are high. These dynamics of identity are multiple and complex—like nested matryoshka dolls. Resistance against large dams in Lachen and Lachung illustrates this complex politics of identity and place-based territoriality, whereby indigenous identity is both a culturally rooted and a politically strategic construct. This complexity of identity is unfortunately missing in many analyses of collective resistance against hydropower development in the region.

7. Conclusions

In this paper we have discussed how identity constructs, territoriality and cultural politics by the indigenous Lachungpas and Lachenpas inform resistance to hydropower in North-Sikkim. Resistance is deeply related to the defense of territory, collective identity, and local meanings, values and modes of living and knowing. This defense is strategically organized around the traditional system of self-governance, the Dzumsas. These execute a fundamental responsibility to keep the Lachungpas/Lachenpas together. Dzumsas draw their power from the members to build unity, and members draw their agency and voice from the Dzumsas. The Dzumsa gives legitimacy to the collective identity of the Lachungpas and Lachenpas. Dzumsas mark the Lachungpas and Lachenpas in terms of their distinct history, culture, traditions, their bounded, protected geographical areas including their exclusive tribe status.

Despite local (and official) discourses pretending to "conserve" local indigenous identity, neither this collective identity, nor its triangular relationship with territory and Dzumsas, are fixed and static. Lachungpas and Lachenpas identities are rooted in history, local culture and permanence in the territory, but equally shaped by confrontation with "the outside" (which obviously comes to form part of local identity, culture and indigeneity). From merely being associated with place/location like it did initially, the Lachungpa/Lachenpa identities have today transitioned to encompass and project their territory, distinct culture, ways of living, traditions, traditional institutions including politics at all levels—individual, relational and collective. Therefore, as we have shown, Lachungpas and Lachenpas identities are both "real and rooted" as well as "real and strategic"; often, they are consciously shaped and reshaped, as political actions that serve to defend against/from "outsiders/others" (e.g., hydropower agents) and to protect specific territorial claims and local interests.

The politically-responsive, territorially-exclusive and ethnically-cohesive Dzumsa institutions allow the Lachungpas and Lachenpas to assert enormous political strength. They reshape identity and redefine (ancient) rules and sanctions and whenever necessary (re)create "convenient past", exclusionary relationships or deploy strategic cultural beliefs—evident in a sustained, unanimous "no-hydropower" defense message in the region. Continuously maintained and updated collective identity enables them to engage in fierce, successful collective actions. In times of modernist commensuration through large-scale hydropower development (imposing a common metric to determine "value", "progress", "development", and "efficient hydro-territorial knowledge" [24,88]), the Dzumsas strategically respond with incommensurate cultural–political notions of animated mountains and sacred territory, such as manifested in the Chya ritual practice. This way, the Lachungpas and Lachenpas effectively engage in the battlefield of culture, knowledge and identity, defending and at the same time reshaping their collective identity and territory.

This cohesion, collective identity and collective action and forced normativity, however, make the Dzumsa both an all-inclusive and disciplinary as well as an exclusionary institution, in terms of gender, indigeneity, ethnicity, and rationality. Loyalty towards the collective and protecting unity and wellbeing of the locality has priority over any other issue. This, combined with deeply cultivated notions of territoriality, also informs the Lachungpas/Lachenpas' strict, unanimous no-hydropower stand. "Over my dead body" assertions today by the Lachungpas and Lachenpas, *in extremis* manifested through the enforcement of Chya, perhaps will keep hydropower development at bay. The irony of the Lachen and Lachung case, one that may be witnessed in many other territories that face modernist encroachment, is that powerful local exclusionary institutions seem to be able to construct forced unity—forms of vernacular statecraft—that effectively counter translocal exclusionary institutions and projects, based on neoliberal agendas of development.

Author Contributions: R.D.D., D.J. and R.B. conceptualized and wrote the paper. R.D.D. framed the overall research project and undertook the field investigation.

Funding: This research was funded by Department of International Development (DFID) through The Netherlands Organisation for Scientific Research (NWO) Grant Number W.O7.68.413.

Acknowledgments: This paper would not be possible had it not been for the people of Lachung and Lachen for allowing us, random researchers to be a part of their struggle in their guarded spaces and trusting us with their stories. This paper is dedicated to the Lachungpas and Lachenpas of North Sikkim. We are sincerely grateful to all the anonymous reviewers for their constructive comments that went into strengthening our paper. We are also thankful to Jaime Hoogesteger, Wageningen University, for his crucial reading suggestions that have helped shape our paper.

Conflicts of Interest: We declare no conflict of interest. The funders had no role in the design of the study; in the collection, analyses, or interpretation of data; in the writing of the manuscript, and in the decision to publish the results.

References

1. Joy, K.J.; Mahanta, C.; Das, P.J. Hydropower Development in Northeast India: Conflicts, Issues and Way Forward. 2013. Available online: https://tinyurl.com/yc5fhf5u (accessed on 23 November 2014).
2. Huber, A. Hydropower in the Himalayan Hazardscape: Strategic Ignorance and the Production of Unequal Risk. *Water* **2019**, *11*, 414, doi:10.3390/w11030414.
3. Vagholikar, N.; Das, P.J. Damming Northeast India: Juggernaut of Hydropower Projects Threatens Social and Environmental Security of Region, Kalpavriksh, Aaranyak and ActionAid India: Pune/Guwahati/New Delhi. 2010. Available online: https://tinyurl.com/y779olgf (accessed on 10 November 2014).
4. International Rivers. Available online: https://tinyurl.com/yarppgxs (accessed on 5 February 2016).
5. Economic Times. Available online: https://tinyurl.com/yd87b6b3 (accessed on 25 November 2016).
6. Assam Times. Available online: https://www.assamtimes.org/node/18252 (accessed on 15 January 2017).
7. International Rivers. Available online: https://tinyurl.com/y72aywkr (accessed on 22 May 2017).
8. Huber, A.; Joshi, D. Hydropower, Anti-Politics, and the Opening of New Political Spaces in the Eastern Himalayas. *World Dev.* **2015**, *76*, 13–25. [CrossRef]
9. Narula, S. The Story of Narmada Bachao Andolan: Human Rights in the Global Economy and the Struggle against the World Bank. New York University Public Law and Legal Theory Working Paper. 2008. Available online: http://lsr.nellco.org/nyu_plltwp/106 (accessed on 6 August 2016).
10. Nepal, P. How Movements Move? Evaluating the role of Ideology and Leadership in Environmental Movement Dynamics in India with Special Reference to the Narmada Bachao Andolan. *Hydro Nepal* **2009**, *4*, 24–29. [CrossRef]
11. Menon, M.; Vagholikar, N. Environmental and Social Impacts of Teesta V Hydro Electric Project, Sikkim: An Investigative Report. 2003. Available online: https://tinyurl.com/ydebeh7na (accessed on 10 November 2014).
12. Huber, A. Contesting Dams and Democracy: State-Society interactions over Hydropower Development in Sikkim, Northeast India. Master's Thesis, Wageningen University, Wageningen, The Netherlands, 2012.
13. Dukpa, R.D.; Joshi, D.; Boelens, R. Hydropower development and the meaning of place. Multiethnic hydropower struggles in Sikkim, India. *Geoforum* **2018**, *89*, 60–72. [CrossRef]
14. Thethirdpole. Available online: https://tinyurl.com/y88hh82v (accessed on 3 May 2016).
15. The Indian Express. Available online: https://tinyurl.com/ybcg7m2l (accessed on 3 May 2016).
16. Colloredo-Mansfeld, R. *Fighting Like a Community: Andean Civil Society in an Era of Indian Uprising*; University of Chicago Press: London, UK, 2009.
17. Klandermans, P.G. Identity Politics and Politicized Identities: Identity Process and the Dynamics of Protest. *Political Psychol.* **2014**, *35*, 1–22. [CrossRef]
18. Van Stekelenburg, J.; Klandermans, B. The social psychology of protest. *Curr. Sociol.* **2013**, *61*, 886–905. [CrossRef]
19. Klandermans, B.; Van der Toorn, J.; Van Stekelenberg, J. Embeddedness and Identity: How Immigrants Turn Grievances into Action. *Am. Sociol. Rev.* **2008**, *73*, 992–1012. [CrossRef]
20. Zomeren, M.V.; Postmes, T.; Spears, R. Towards an integrative Identity Model of Collective Action: A quantitative research synthesis of Three Socio-Psychological Perspective. *Psychol. Bull.* **2008**, *134*, 504–535. [CrossRef]
21. Scholtens, J. The elusive quest for access and collective action: North Sri Lankan fishers' thwarted struggles against a foreign trawler fleet. *Int. J. Commons* **2016**, *10*, 929–952. [CrossRef]

22. Melucci, A. *Challenging Codes: Collective Action in the Information Age*, 1st ed.; Cambridge University Press: New York, NY, USA, 1996.
23. Escobar, A. *Territories of Difference: Place, Movements, Life, Reeds*, 1st ed.; Duke University Press: Durham, UK; London, UK, 2008.
24. Boelens, R.; Shah, E.; Bruins, B. Contested Knowledges: Large Dams and Mega-Hydraulic Development. *Water* **2019**, *11*, 416, doi:10.3390/w11030416.
25. Sawyer, S. *Crude Chronicles: Indigenous Politics, Multinational Oil, and Neoliberalism in Ecuador*; Duke University Press Books: Durham, UK; London, UK, 2004.
26. Delaney, D. *Territory a Short Introduction*, 1st ed.; Blackwell Publishing: Hoboken, NJ, USA, 2005.
27. Anguelovski, I.; Alier, J.M. The 'Environmentalism of the Poor' revisited: Territory and place in disconnected glocal struggles. *Ecol. Econ.* **2014**, *102*, 167–176. [CrossRef]
28. Duarte-Abadía, B.; Boelens, R. Disputes over territorial boundaries and diverging valuation languages: The Santurban hydrosocial highlands territory in Colombia. *Water Int.* **2016**, *41*, 5–36. [CrossRef]
29. Valladares, C.; Boelens, R. Extractivism and the rights of nature: Governmentality, 'convenient communities', and epistemic pacts in Ecuador. *Environ. Politics* **2017**, *26*, 1015–1034. [CrossRef]
30. Holsti, K.J. Territoriaalisuus (territoriality). *Politiikka* **2000**, *42*, 15–29.
31. Storey, D. Land, territory and identity. In *Making Sense of Place: Multidisciplinary Perspectives*; Convery, I., Corsane, G., Davis, P., Eds.; Boydell Press: Woodridge, UK, 2012; pp. 11–22.
32. Hoogesteger, J.; Boelens, R.; Baud, M. Territorial pluralism: Water users' multi-scalar struggles against state ordering in Ecuador's highlands. *Water Int.* **2016**, *41*, 91–106. [CrossRef]
33. Simon, B.; Klandermans, B. Politicized Collective Identity—A social psychological analysis. *Am. Psychol.* **2001**, *56*, 319–331. [CrossRef] [PubMed]
34. Cohen, P.H. Strategy or identity: New theoretical paradigms and contemporary social movements. *Soc. Res.* **1985**, *52*, 663–716.
35. Huddy, L. From Social to Political Identity: A Critical Examination of Social Identity Theory. *Political Psychol.* **2001**, *22*, 127–156. [CrossRef]
36. Schendel, W.V. Geographies of knowing, geographies of ignorance: Jumping scale in Southeast Asia. *Soc. Space* **2002**, *20*, 647–668. [CrossRef]
37. Shneiderman, S. Are the Central Himalayas in Zomia? Some scholarly and political considerations across time and space. *J. Glob. Hist.* **2010**, *5*, 289–312. [CrossRef]
38. Michaud, J. Editorial—Zomia and beyond. *J. Glob. Hist.* **2010**, *5*, 187–214. [CrossRef]
39. Fisher, J. Introduction. In *Himalayan Anthropology: The Indo-Tibetan Interface*; Fisher, J., Ed.; Mouton Publishers: The Hague, The Netherland, 1978; pp. 1–3.
40. Sinha, A.C. Sikkim, Institute of Developing Economies: Chiba. 2005. Available online: https://tinyurl.com/y875gq8t (accessed on 12 July 2015).
41. Government of Sikkim. *Gazetteer of Sikkim, Edition 2013*; Home Department Government of Sikkim: Gangtok, India, 2013.
42. Government of Sikkim. *Human Ecology and Statutory Status of Ethnicity Entities in Sikkim: Report of the Commission for Review of Environmental and Social Sector Policies, Plans and Programmes (CRESP)*, 1st ed.; Department of Information and Public Relations: Gangtok, India, 2008.
43. Paul, L.M. *Sikkimese Ethnologue: Languages of the World*, 16th ed.; SIL International: Dallas, TX, USA, 2009.
44. Gohain, S. Mobilising language, imagining region: Use of Bhoti in West Arunachal Pradesh. *Indian Sociol.* **2012**, *46*, 337–363. [CrossRef]
45. Bhasin, V. *Ecology, Culture and Change: Tribals of Sikkim Himalaya*, 1st ed.; Inter-India Publications: New Delhi, India, 1989.
46. Bhasin, V. Social Organisation, Continuity and Change: The Case of the Bhutia of Lachen and Lachung of North Sikkim. *J. Biodivers.* **2012**, *3*, 1–43. [CrossRef]
47. Bourdet-Sabatier, S. The Dzumsa of Lachen: An example of a Sikkimese political institution. *Bull. Tibetol.* **2004**, 93–104. Available online: http://www.thlib.org/static/reprints/bot/bot_2004_01_04.pdf (accessed on 28 March 2017).
48. Castells, M. *The Power of Identity*, 2nd ed.; Wiley-Blackwell: Oxford, UK, 2010.
49. Paasi, A. Region and place: Regional identity in question. *Prog. Hum. Geogr.* **2003**, *27*, 475–485. [CrossRef]
50. Wise, J.W. Home: Territory and identity. *Cult. Stud.* **2000**, *14*, 295–310. [CrossRef]

51.   Taylor, C. *Sources of the Self: The Making of the Modern Identity*, 1st ed.; Harvard University Press: Cambridge, MA, USA, 1989.

52.   Weinreich, P. The operationalisation of identity theory in racial and ethnic relations. In *Theories of Race and Ethnic Relations*; Rex, J., Mason, D., Eds.; Cambridge University Press: Cambridge, MA, USA, 1986.

53.   Erickson, E.H. The Concept of Identity in Race Relations: Notes and Queries. *Daedalus* **1966**, *95*, 145–171.

54.   Massey, D. Geographies of Responsibility. *Geografiska Annaler* **2004**, *86*, 5–18. [CrossRef]

55.   Weinreich, P.; Saunderson, W. *Analysing Identity: Cross-Cultural, Societal and Clinical Contexts*, 1st ed.; Routledge: London, UK, 2002.

56.   Sedikides, C.; Brewer, M.B. Individual, relational, and collective self: Partners, opponents, or strangers? In *Individual Self, Relational Self, and Collective Self*; Sedikides, C., Brewer, M.B., Eds.; Psychology: Philadelphia, PA, USA, 2001; pp. 1–4.

57.   Opp, K.D. Collective identity, rationality and collective political action. *Ration. Soc.* **2012**, *24*, 73–105. [CrossRef]

58.   Cakal, H.; Hewstone, M.; Guler, M.; Heath, A. Predicting support for collective action in the conflict between Turks and Kurds: Perceived threats as a mediator of intergroup contact and social identity. *Group Process. Intergroup Relat.* **2016**, *19*, 732–752. [CrossRef]

59.   Alberici, A.I.; Milesi, P. Online discussion, politicized identity, and collective action. *Group Process. Intergroup Relat.* **2016**, *19*, 43–59. [CrossRef]

60.   Brubakar, R.; Cooper, F. Beyond "identity". *Theory Soc.* **2000**, *29*, 1–47. [CrossRef]

61.   Klandermans, B. Mobilization and Participation: Social Psychological Expansions of Resources Mobilization Theory. *Am. Sociol. Rev.* **1984**, *49*, 583–600. [CrossRef]

62.   Baumeister, R.F.; Leary, M.R. The need to belong: Desire for interpersonal attachments as a fundamental human motivation. *Psychol. Bull.* **1995**, *117*, 497–529. [CrossRef] [PubMed]

63.   Boelens, R.; Claudin, V. Rooted rights systems in turbulent waters: The dynamics of collective fishing rights in La Albufera, Valencia, Spain. *Soc. Nat. Resour.* **2015**, *28*, 1059–1074. [CrossRef]

64.   Polletta, F.; Jasper, J.M. Collective identity and social movements. *Annu. Rev. Sociol.* **2009**, *27*, 283–305. [CrossRef]

65.   Tarrow, S. *Power in Movement: Social Movements and Contentious Politics*, 3rd ed.; Cambridge University Press: New York, NY, USA, 2011.

66.   Zwarteveen, M.Z.; Boelens, R. Defining, researching and struggling for water justice: Some conceptual building blocks for research and action. *Water Int.* **2014**, *39*, 143–158. [CrossRef]

67.   Mouffe, C. Artistic activism and agonistic spaces. *Art Res.* **2007**, *1*, 1–5.

68.   Schlosberg, D. Reconceiving environmental justice: Global movements and political theories. *Environ. Politics* **2004**, *13*, 517–540. [CrossRef]

69.   Loveman, M. High-Risk Collective Action: Defending Human Rights in Chile, Uruguay, and Argentina. *Am. J. Sociol.* **1998**, *104*, 477–525. [CrossRef]

70.   Bavinck, M. The role of informal fisher village councils (ur panchayat) in Naga-pattinam District and Karaikal, India. In *Strengthening Organizations and Collective Action in Fisheries: Towards the Formulation of a Capacity Development Programme*; Siar, S., Kalikoski, D., Eds.; FAO: Rome, Italy, 2016; Volume 41, pp. 383–404.

71.   Agrawal, A. Sustainable governance of common-pool resources: Context, methods and politics. *Annu. Rev. Anthropol.* **2003**, *32*, 243–263. [CrossRef]

72.   Agnew, J. The territorial trap: The geographical assumptions of international relations theory. *Rev. Int. Political Econ.* **1994**, *1*, 53–80. [CrossRef]

73.   Swyngedouw, E.; Boelens, R. " . . . And Not a Single Injustice Remains": Hydro–Territorial Colonization and Techno-Political Transformations in Spain. In *Water Justice*; Boelens, R., Perreault, T., Vos, J., Eds.; Cambridge University Press: Cambridge, MA, USA, 2018; pp. 115–133.

74.   Antonsich, M. Rethinking territory. *Prog. Hum. Geogr.* **2010**, *35*, 422–425. [CrossRef]

75.   Hommes, L.; Boelens, R.; Maat, H. Contested hydro-social territories and disputed water governance: Struggles and competing claims over the Ilisu Dam development in southeastern Turkey. *Geoforum* **2016**, *71*, 9–20. [CrossRef]

76.   Hommes, L.; Boelens, R. Urbanizing rural waters: Rural-urban water transfers and the reconfiguration of hydrosocial territories in Lima. *Political Geogr.* **2017**, *57*, 71–80. [CrossRef]

77. Hommes, L.; Boelens, R. From natural flow to 'working river': Hydropower development, modernity and socio-territorial transformations in Lima's Rímac watershed. *J. Hist. Geogr.* **2018**, 85–95. [CrossRef]

78. Baletti, B. Ordenamento territorial: Neo-developmentalism and the struggle for territory in the lower Brazilian amazon. *J. Peasant Stud.* **2012**, *39*, 573–598. [CrossRef]

79. Moreyra, A. Multiple Territories in Dispute: Water Policies, Participation and Mapauce Indigenous Rights in Patagonia, Argentina. Ph.D. Thesis, Wageningen University, Wageningen, The Netherlands, 2009.

80. Levitas, R.; Pantazis, C.; Fahmy, E.; Gordon, D.; Lloyd, E.; Parsios, D. The Multidimensional Analysis of Social Exclusion. 2007. Available online: http://dera.ioe.ac.uk/6853/1/multidimensional.pdf (accessed on 16 September 2016).

81. Agnew, J.; Oslender, U. Overlapping territorialities. Sovereignty in dispute: Empirical lessons from Latin America. In *Spaces of Contention: Spatialities and Social Movements*; Nicholls, W., Beaumont, J., Eds.; Ashgate: London, UK, 2013; pp. 121–140.

82. Sack, R.D. *Human Territoriality: Its Theories and History*, 1st ed.; Cambridge University Press: Cambridge, MA, USA, 1986.

83. Hammersley, M.; Atkinson, P. *Ethnography: Principles in Practice*, 1st ed.; Tavistock Publication: London, UK; New York, NY, USA, 1984; pp. 1–26.

84. Zaman, S. Native among the Natives: Physisian Antropologists doing hospital ethnography at home. *J. Contemp. Ethnogr.* **2008**, *37*, 135–154. [CrossRef]

85. O'Reilly, K. *Ethnographic Methods*, 1st ed.; Routledge: Abingdon, UK, 2005.

86. Tsuda, T. Ethnicity and the Anthropoligist: Negotiating Identities in the Field. *Anthropol. Q.* **1998**, *17*, 107–124. [CrossRef]

87. Ramanathan, K.; Abeygunawardena, P. *Hydropower Development in India: A Sector Assessment*; Asian Development Bank: Metro Manila, Philippines, 2007; Available online: http://bit.ly/2dioqDY (accessed on 10 November 2014).

88. Hoogendam, P.; Boelens, R. Dams and Damages. Dams and Damages. Conflicting epistemological frameworks and interests concerning "compensation" for the Misicuni project's socio-environmental impacts in Cochabamba, Bolivia. *Water* **2019**, *11*, 408, doi:10.3390/w11030408.

Article

Hydraulic Order and the Politics of the Governed: The Baba Dam in Coastal Ecuador

Juan Pablo Hidalgo-Bastidas [1,2,*] and Rutgerd Boelens [1,2,3]

[1]   Centre for Latin American Research and Documentation (CEDLA), University of Amsterdam,
        Roetersstraat 33, 1018 WB Amsterdam, The Netherlands; rutgerd.boelens@wur.nl
[2]   Faculty of Agricultural Sciences, Universidad Central del Ecuador, Ciudadela Universitaria,
        170129 Quito, Ecuador
[3]   Water Resources Management Group, Department of Environmental Sciences, Wageningen University,
        P.O. Box 47, 6700 AA Wageningen, The Netherlands
*   Correspondence: juanhidalgo_b@hotmail.com; Tel.: +593-02-2232402 (ext. 112)

Received: 2 July 2018; Accepted: 3 January 2019; Published: 26 February 2019

Abstract: Mega-dams are commonly designed, constructed, and implemented under governors' rule and technocrats' knowledge. Such hydraulic infrastructures are characteristically presented as if based on monolithic technical consensus and unidirectional engineering. However, those who are affected by these water interventions, and eventually governed by the changes brought by them, often dispute the forms of knowledge, norms, morals, and operation and use rules embedded in mega-hydraulic engineers' designs. Protests may also deeply influence the design and development of the technological artifacts. By using approaches related to the Social Construction of Technology and Partha Chatterjee's politics of the governed, this article shows (i) how protests against the Baba dam in coastal Ecuador greatly influenced the dam's designs, protecting communities' lands from being flooded; and (ii) how, at the same time, techno-political decision-makers deployed hydraulic design as a dividing rule, turning potentially affected communities against each other. We conclude that megadam designs are shaped by the power interplay among governors and governed, with the latter being internally differentiated. By critically analyzing the role of technology development—materializing changing 'political context and relationships'—we show how contested and adapted dam design may favor some stakeholders while simultaneously affecting others and weakening united dam-resistance movements.

Keywords: megadams; social construction of technology; politics of the governed; anti-dam resistance movements; technological design; contested knowledge; Ecuador

---

1. Introduction

"There is almost nothing, however fantastic, that (given competent organization) a team of engineers, scientists, and administrators cannot do today. Impossible things can be done. [...] When these men have imagination and faith, they can move mountains; out of their skills they can create a way of life new to this world"—David Lilienthal, director Tennessee Valley Authority [1] (p. 3).

Megadams are the material epitome and pride of expert, engineering knowledge. Proclaiming their origin as 'technical' has portrayed these works, their promoters and knowledge as if they were neutral, objective, apolitical elements of water management [2]. Such assumptions have consolidated these projects as part of a longstanding dominant paradigm which is universally unquestionable and technically necessary. Hence, scientists, engineers, and technocrats, following their own worldviews and knowledge systems, have designed, constructed and implemented hydraulic mega-infrastructure to attempt to correct nature's 'imbalances' while governing society [1,3]. Lilienthal's utopian dreams, for instance, when glorifying the Tennessee Valley Authority model that would modernize backward regions' infrastructure through electricity, flood control and multiple water uses, attract industry, and improve the economic and social lives of rural people, resound throughout the world [4–11] and also were key to Ecuador's mega-dam development analyzed in this paper [12,13].

However, in Ecuador as in many places worldwide, these type of projects, their promoters and knowledge frames have not escaped criticism [3,4,14–16]. Apart from academic critique, they have been fiercely contested by those societal sectors who end up paying the price of such undertakings' impacts [17–19]. Together, indigenous communities, peasant federations, environmental Non-Governmental Organizations (NGO), critical scholars and water professionals, urban leaders, among others, are influencing the very structures of knowledge and materials that comprise these mega-projects. On this basis, the objective and technical pedestal traditionally reserved for these projects is increasingly challenged, revealing how its foundations are profoundly social and political. Some scholarly work already shows, though often in general terms, how social struggles eventually 'succeed' in influencing mega dams' designs and knowledge (e.g., [20–22]). A number of examples show 'successful' anti-dam movements [22]. In India, we see, for example: (i) the protest against the Silent Valley Hydroelectric Project in the Kerala region, 1984; (ii) the protest movement against the Bedthi dam project in Karnataka in the early 1980s; (iii) the protest against the mega projects in Narmada river organized by among others the Save the Narmada movement. They resisted in the 1990s and managed to influence the designs of the project. However, the movement did not succeed in stopping the project altogether [20]. In other regions more examples can be found. For instance, "the most successful anti-dam campaign in the United States was the Grand Canyon campaign against the 525 feet-high Eco Park dam in the Green River. This dam was halted in 1963 after six years of construction as civil society groups opposed it" [21] (p. 70). This article seeks to contribute to this work through a detailed account of the anti-dam movement that fought against the Baba dam in coastal Ecuador. We aim, specifically, to understand how and to what extend such 'success' stories challenge dominant rule, and how and which actors are involved in what ways.

In related fields of water control, such as irrigation and drinking water supply studies, dissatisfaction with the poor performance of water control systems designed technocratically has urged inclusion of water users' knowledge in their design. Even though mainstream hydraulic schools and conventional irrigation engineering departments still continue to develop and promote high-tech water technologies in the top-down, old-fashioned ways (now framed as 'inter-disciplinary' since third, natural sciences and economic, disciplines (in particular 'new-institutional economics') have been 'added'); for more than three decades [23], the so-called social turn in irrigation has been making interesting efforts to not just integrate social and technical academic visions in irrigation technology design, but also *co-create* water technology in transdisciplinary ways [24–28]. Beyond influencing the means used to design irrigation systems, this has questioned and challenged mainstream technical engineering knowledge and its pretensions of objectivity, appropriateness, social efficiency, and societal relevance [27,28]. Such questioning has only incipiently expanded into other types of water technologies, even though in particular megadams continue to inform the prestige and pride of water engineering culture and to constitute the planet's most controversial water projects [4].

There are various efforts coming from Science and Technology Studies (STS) and, particularly, from Social Construction of Technology (SCOT) investigations, which have challenged the apolitical and purely technical/managerial conception of hydraulic dams and water technology [29–31]. This approach tell us that technology is not a 'thing' that is separate from social processes, but an essential part of them [31,32]. Departing from such contributions and considering that we are witnessing a new era of mega dams building [33–35], it is urgent to scrutinize them critically. We must understand dams by taking into consideration how rules, norms, discourses, designs, values, and their very material existence are negotiated and contested by 'non-technical' and vulnerable stakeholders.

In Ecuador, conflicts about megadams have not gone unnoticed [12,35]. Since the mid-1900s, racing toward development and modernity, the State has planned and built dozens of megadams on the country's main rivers. No doubt these efforts have brought 'development' to some people and geographies, however these mega projects also have been the cause of far-reaching socio-environmental conflicts and unleashing societal struggles. This article will examine resistance led by the inhabitants of rural parish Patricia Pilar, organized against design, construction and implementation of the Baba multi-purpose dam (Figure 1). Our aim is to understand how mega-projects can be also influenced by those who are/may be affected (governed) by the Baba dam, and how their socio-environmental demands shaped the way technology was materially designed. We argue that dam technology, beyond expressing and materializing expert knowledge and its ideals of progress, is the materialized track record of social struggles and of the interaction among diverse and divergent actors and knowledge systems that face off in contexts and under conditions of unequal power.

After this introduction, the second section gives details about the methodology for research and information analysis. The third section outlines a theoretical framework to understand technology as a two-way social construct and to analyze anti-dam social movements as grounded in the 'politics of the governed' [36]. The fourth section presents the case's empirical data, showing how the dam and the social movement against it unfolded, with societal influence on technical designs and the designs' effects on social resistance. The article ends with a discussion and final conclusions.

2. Methodology

The empirical research presented in this article was carried out in Patricia Pilar parish and its surrounding peasant and Afro Ecuadorian communities, from October 2015 until September 2017. Field work consisted of two visits to the research area. The first visit took place from October 2015 until April 2016. The second was in September 2017. This case study is based on historical and ethnographic research [37,38]. It is also based on an 'ethnography of technology' [37,38]. Water management involves several dimensions: technical, organizational, normative/socio-legal, cultural and socio-economic/political. While it is usual to research these areas separately, it remains a challenge to integrate all those dimensions in an interdisciplinary manner. Technography: the ethnography of technology provides a methodological approach that intends to integrate technological processes as part of human-technology interactions [38]. Particularly, technography allows us, in the case of the Baba dam, to integrate SCOT's conceptual approach with the empirical findings.

Participatory observation, semi-structured interviews, literature and secondary sources review (historical archives, newspaper articles, official reports) were the main data collection methods, including in total 36 in-depth interviews. Interviews included State and hydroelectric company officials, action-researchers, NGO representatives, peasant and Afro Ecuadorian leaders, and critical scholars. Our main selection criteria for choosing interviewees were based on the reconstruction of relevant (diagnostic and process-explanatory) life histories and crucial events. After an initial period of literature review and preparatory field visits, we selected our first contacts in Patricia Pilar, from whom we applied snowball sampling to reach other relevant actors. During field work, interviews were conducted in Spanish, most recorded and transcribed by the authors. The names of all interviewees are pseudonyms.

Interviews and other collected data were classified and analyzed according to two focal points: how the technocratic designs of Baba Dam evolved over time, and how the different actions and resistance events organized by Patricia Pilar's inhabitants and their allies evolved and informed the final designs of the dam. In order to analyze the collected data, we used mapping, qualitative chart building, comparative time-frame analysis and data triangulation.

3. Social Construction of Hydraulic Technology and Anti-Dam Social Movements from the Politics of the Governed

3.1. Megadams: Socially Constructed Technology and Its Hydraulic Order

SCOT (Social Construction of Technology) is a critical approach to technology that comes from the field of Science and Technology Studies (STS) [39]. From that basis it aims to challenge technological determinism [40,41]. The constructivist approach adopted by SCOT treats technology not as an universal truth, built upon scientific facts and provided with neutral, intrinsic properties, but it understands technology as a socio-technical system [42] which is *being constructed* [39] (p. 135). In this line, STS and SCOT scholars have largely shown how water technologies are the result of complex social processes (cf. [29,30,43,44]). Further research needs to be done to scrutinize how the most contested technological endeavors, such as megadams [14], are co-shaped by its 'non-technical' protagonists (e.g., dam-affected peoples). There is incipient literature, such as recent articles showing the interesting strategies of successful anti-dam movements in Thailand and Myanmar; but this does not explicitly consider how technology is influenced by social actors and how technology influences local context [22].

In his study on the social construction of dams and dikes, Wiebe Bijker tellingly explains how these technologies are 'thick with politics' and embed particular social and cultural patterns, and relationships [30]. He states that "studying artifacts—how they are socially constructed as well as how they shape society—yields crucial insights into the history and development of science and into the history and development of societies" [30] (p. 110). Bijker further suggests that "a focus on the 'things' of water management can help us to understand the cultural and democratic makeup of societies and at the same time is important for addressing questions about the further socio-technical development of those societies". Far from maintaining a deterministic view of technology, assuming that technology is the output of autonomous, linear, one-way, unavoidable development, SCOT attempts to understand it as the outcome of interacting social visions and political encounters, and in turn, with an effect also over the society in which it is produced and embedded [39,45]. "One of the central tenets of this approach is the claim that technological artifacts are open to sociological analysis, not just in their usage but especially with respect to their design's technical content" [41] (p. xiii).

Such approach does not just relate to (conscious and unconscious) engineers' and policy-makers' assumptions and decisions but equally apply to the influence of civil society groups on hydraulic development—whether this is recognized and allowed by government and expert institutions or not. Worldwide, struggles against dams are all different from each other. Their stakeholders are different, their strategies vary from locality to locality, their contexts and mobilization domains are distinct, their timing is unique, their calls to battle are manifold, and they even speak different languages. Nevertheless, these fighters have in common that their actions aim to stop or modify the construction and implementation of a technology. That is, these movements place the dam (and its direct territorial effects)—rather than any other aspect—at the center of their opposition actions. Hence, it all starts with the dam—then societal anti-dam movements expand their objections to related dam-network objects and subjects, such as the promoters of that technology (e.g., governments, builders, financiers, technocrats, and experts), the legitimating discourses (e.g., green energy, the well-being of the masses, development, progress, climate change, and democracy) and the knowledge frames that foster mega-dam development. Therefore, starting from this empirical fact—that the dam is a crucial element of these social struggles—what can the dam itself tell us about the conflict, about its own constitution and about society itself? In this sense, SCOT's perspective allows us to jointly understand how the Baba dam and Patricia Pilar's anti-dam movement developed along the time, and which consequences have occurred both in technology and society as consequence of such interrelated developments.

For this, as mentioned earlier, technography is required. Technography is defined as the ethnography of technology. It is an interdisciplinary, interrelating analysis of technology, nature and society [38]. In this particular case it helps to examine the concrete shaping, use, and impact of dam technologies in social situations, and how they (re)configure livelihoods, territories and create specific water access and control arrangements (see also [45,46]). This methodological approach together with a critical view on technology: considering technology as a *subject* of research enables a different epistemological way to identify and understand conflicts, in which social and political clashes between dam proponents and opponents literally become 'materialized'. This means that technology is loaded with language, values, norms, practices, and discourses [47].

Technology is socially constructed and politically negotiated or, as Pfaffenberger proposes [45] (p. 244): technology is "humanized nature", insisting "that [technology] is a fundamentally social phenomenon: it is a social construction of the nature around us and within us, and once achieved, it expresses an embedded social vision, and it engages us in what Marx would call a form of life". We, thus, argue that technology is molded by power relationships and societal visions and, once it materializes, transforms and affects society on the basis of those intrinsic social and political relationships that inform it [30,40,48,49]. On this basis, we suggest that technology is not a neutral element within megadam conflicts, but constitutes (at least partially) an explanatory element of them.

Based on this ontological position regarding technology, this article argues that a megadam, as the Baba dam, is by no means one-directional or solely influenced by the designers or a single group of societal stakeholders. In fact, dams—as a technology—are not built or implemented without being contested by other societal groups (e.g., rural people and their communities, indigenous peoples) who perceive them as a problem for different reasons: social, environmental, and economic [50]. Although hydraulic engineering models for dams are predefined or "closed" on engineers' desks, during their design, construction, and implementation phases, several "relevant social groups" [41] (p. 22) are able to influence these models' criteria. This means that the material product—technology = dam and ancillary works—is not the exclusive result of dominant power and its interests, but also of those who, from more vulnerable positions, contest the dam's implementation and presence. In this regard, technology becomes a hybrid, comprising different and usually antagonistic visions of what the relationships between society and nature are, or ought to be. The dam is a battlefield—of interests, values, meanings, norms, and discourses. This approach, therefore, enables us to understand how technology is constituted by society, and the effects and function that technology has on society.

3.2. Anti-Dam Social Movement: An Approach from the Politics of the Governed

"Several attempts have been made to approach the President of Ecuador, but they have all been fruitless; they continue with the mistaken decision to turn our province into a cesspool, to expel thousands of families from their habitat, and drive us into miserable poverty. Using force is our only chance for the President to realize that Ecuadorians also live here, citizens, with rights and duties, and that it is his obligation to listen to us" (Statement by an inhabitant of Patricia Pilar in resistance, (Diario Digital Ecuador Inmediato, 15 November 2005)).

In many parts of the so-called South, social movements against dams navigate their struggles within a political context marked by their country's colonial history. Ecuador is a case in point. As part of the contemporary post-colonial project to build the Republic, the Ecuadorian nation-state has planned and implemented hydraulic mega-infrastructure to develop and modernize the country. These government interventions have elicited different reactions from the local population group affected. Although, prior to 2002, there was social mobilization against these policies and their outcomes, they were isolated grievances, which were not even anti-dam per se, because (among other reasons) when their complaints were raised, the dams were already built and operating. Those struggles focused, therefore, on demanding the basic benefits that the State, as the provider of well-being, should ensure.

An example of this is the Daule-Peripa dam, the country's largest, built in the 1980s. Its implementation affected over ten thousand people and flooded nearly 30 thousand hectares of farms and forests. Up to the year 2017, hundreds of families had received no indemnity yet, but are living under conditions of involuntary isolation, with no access to basic services [12,14,22,51]. Since then, taking the bitter experience of Daule-Peripa's neighbors as a reference, several local populations have led anti-dam movements to keep megahydraulism out of their territories, while negotiating with the State and its representatives for a more full recognition of their 'citizenship'. "Megahydraulism" (see [3]) does not refer solely to the policy of building dams as material expressions, but also covers the system of knowledge, and the institutional, technocratic and financial processes that are legitimized under the dominant approach of 'good' governance of water and watershed management.

Key members of the mega dam regime [3], the State and its officials are usually fierce proponents enabling hydraulic mega-projects. Without their consent, promotion and political backing, and that of their government representatives (technocrats, politicians, technicians, bureaucrats) such projects would usually never happen. (See Patricio Silva for a discussion of the differences between technicians, technocrats, politicians, and bureaucrats [52]. For their empirical manifestation in Ecuadorian water and natural resource governance, see [53–56], and for their relationship with divergent Ecuadorian water user groups and the politics of national water governance, see [12,13,27,35,57–63]). In policy practice this entails under-estimating the negative socio-environmental impacts and over-valuing the benefits that such water development projects and mega-constructions will bring [21,56,64,65]. Therefore, as Chatterjee argues, this involvement means the modern State plays a crucial and discriminatory role in relation to this overarching discourse of equality and universal citizenship for "most of the world" [36]. That is, large masses of people who live within the boundaries of the State never see their citizens' rights materialize. Nevertheless, vulnerable populations' consciousness-raising about the unequal treatment they get from government officials has given rise to societal movements that are dissatisfied with State policies—in this case, against policies promoting mega-dams. In Ecuador the study of social anti-dam movements must address their relationships with the State. With this premise, it is particularly useful to examine the "politics of the governed" [36] to understand how these social movements deal with the State and its water policies. In his proposal, Chatterjee distinguishes between citizens and populations: the former mobilize in the theoretical (or formal) arena, whereas the latter belong to the political (or real) domain [66] (p. 6), [67]. In practice, this dualism means that those whom the State has not managed to include or totally consider as citizens (i.e., individuals fully enjoying their rights, with demands for equality, actively participating in decision-making by the nation-state, and backed by formal legal arrangements to deal with the State) have a different relationship with that State, in the real domain, on a political basis [66] (p. 8). Instead of acting from the civil society, the latter take action and mobilize from the field of political society—"an arena for negotiation and contestation", navigating between legal and para-legal issues, appealing to and/or (re)constructing the bonds of "moral solidarity" to make collective demands of the State and its institutions [68] (p. 150). "It is here that most political mobilization takes place and where the state has to find and reproduce its legitimacy as provider of well-being to its citizens" [69] (p. 22). Although political society's fundamental field of action is para-legal, this does not mean that the governed cannot use laws and civil-society institutions to enable their actions and demands to succeed. It is generally as a political society that the marginalized are able to reorient state benefits, policies, and programs in their favor: by applying, at the right time and in the tactical context "precise pressure on the right points of governmental mechanisms" [68] (p. 139).

So, while officials—since the dawn of the republic—have attempted to consolidate a modern nation-state under the abstract promise of citizen sovereignty and equal rights for all; according to Partha Chatterjee, millions of poor, marginalized people—located at the edges, between the formal and actual realities of that nation-state—"are devising new ways in which they can choose how they should be governed" [36] (p. 77). In other words, "people are learning, and forcing their governors to learn, how they would prefer to be governed" [36], (p. 78).

A successful politics of the governed, "viable, and able to obtain results, entails a considerable dose of mediation" [68] (p. 137). So, the success of a policy of the governed depends, first, on the capacity of particular individuals or groups to mobilize support and influence implementation of public policies in their own favor [68] (p. 132). Next, it depends on the capacity of leaders or mediators to generate societal cohesion by coating "the empirical form of a population group with the moral attributes of a community" [68] (p. 128). However, even if these requirements are met, success will be situational and temporary. Since political society moves (predominantly) in the political arena, if the political context in which they are acting shifts, it is quite possible that this will keep them from attaining their goals, or from attaining them completely. As we examine in this case of coastal Ecuador, focusing on the politics of the governed helps to understand how and to what extend those affected by large dams as Baba are able to claim co-decision in and on social-political and material-technological dam development, disputing with governors and engineers about 'the ways they would to be governed'.

4. The Baba Dam: Its Technological Development and the Social Struggle against It

Patricia Pilar is a parish (district) belonging to the canton of Buena Fé in the province of Los Rios. Since 1974, it began constituting as a town on the land sold by one of the large agro-industrial companies. Patricia Pilar got onto the country's formal political-administrative map on 19 September 1996, after being granted the status of a parish (which in contemporary Ecuador is similar to a secular, public administration district). It is located in the upper basin of the Guayas River, about 150 km southwest of Quito and some 20 km from the top end of the Daule-Peripa reservoir (Figure 1). Like many other towns along Ecuador's coastal highway, Patricia Pilar is a human settlement born in the heat of colonizing land that was inaccurately termed vacant (*tierras baldías*) during attempts at agrarian reform in the 1960s and 1970s [55,70,71]; and by the expansion and consolidation of large neighboring agri-business plantations (rubber, balsa wood, bananas, or oil palm). The high fertility of their soils and availability of water from the Baba River and its dozens of tributaries made this land's potential for agriculture very appealing.

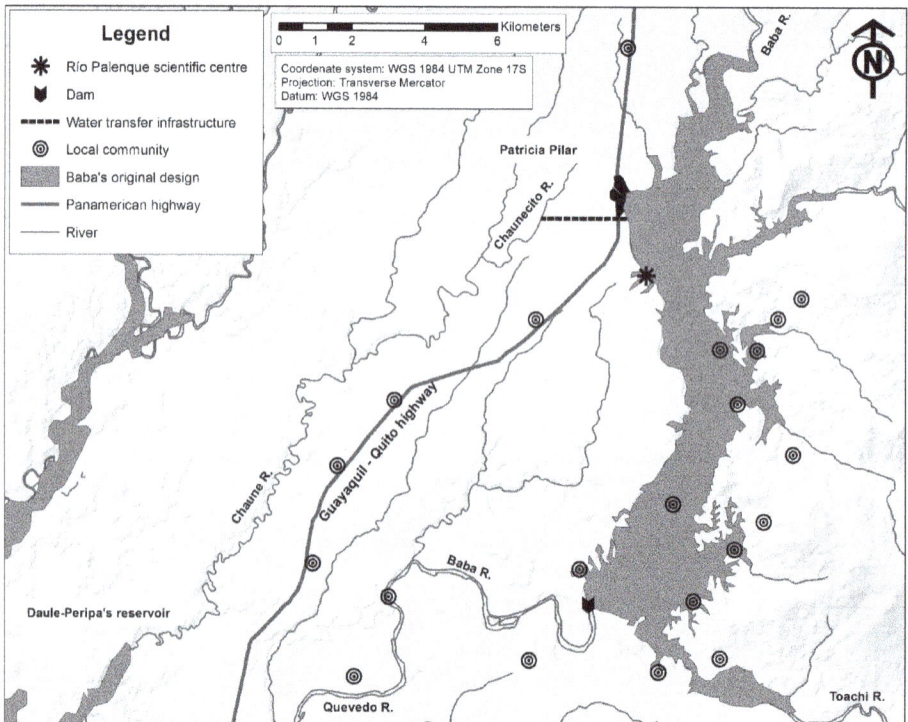

Figure 1. Baba dam's original design and main affected local communities. Source: HidroPacífico S.A. Prepared by: Juan Pablo Hidalgo-Bastidas.

The zone is a social mosaic of people from various origins, backgrounds and interests. Since the first half of the 20th century, the mix has included small farmers, Afro-descendant communities, capitalist agri-businesses, merchants, and rural workers. The zone has also received people displaced by other dams, such as those affected by the Daule-Peripa dam. Most local people came from the provinces of Pichincha, Manabí, Loja, and El Oro. This built up Patricia Pilar's social, productive and economic structure. Having people in this geographical location from so many diverse places would apparently mean a fragmented society: with no common recent past on which to build a sense of community. However, as we shall explore, situations such as the fight against the Baba Dam reveal that inhabitants are able to build quite strong social cohesion, with significant effects on their own well-being and on decision-making by government officials and technocrats.

4.1. The Original Design, and the Struggle in Patricia Pilar (1977–2005)

As a government's technical expert observed, "all these dams suffer from the same disease: they need to transfer water from other watersheds, because at some point they run dry". (Interview with a technician from the former CEDEGE, responsible for hydraulic mega-projects in the National Water Secretariat (SENAGUA), 15 January 2016). Therefore, when Patricia Pilar was being settled on the banks of the Baba River, plans for the multi-purpose dam project were already being made on desks of the Commission for Studies to Develop the Guayas River Basin (CEDEGE): to resolve the country's serious energy crisis and increase the inflow in the Daule-Peripa reservoir and hydroelectric plant, built with an overestimated design capacity [14,33]. The CEDEGE, from 1965 to 2008, was the governmental agency responsible for managing and building projects to manage water resources in the country's largest watershed, the Guayas River Basin. Designing of Baba project took nearly three decades and ran through several consultant firms. Nevertheless, they never consulted with the people living back then in the zone. After long years of studies, the final designs for the project proposed implementing a dam to hold 600 hm³ with one inter-basin transfer, near the town of Patricia Pilar on the Chaunecito River, with an area to be flooded of 3760 ha, a 54 MW hydropower plant, and a dam 55 m high (Figures 1 and 2). One important feature of the dam, as we will show later in the text, was the overflow spillover with a rectangular section (Figure 1).

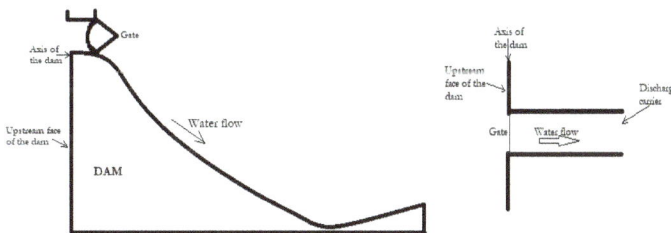

Figure 2. Cross section (**left**) and upper view (**right**) of the dam spillover. Elaboration: Juan Pablo Hidalgo-Bastidas.

What designers never expected (or probably under-estimated or ignored) at the time was that this design would affect a large number of communities living along the lower Baba and Toachi Rivers. As the parish priest of Patricia Pilar recalls, referring to a conversation with a CEDEGE's official:

"When they presented the project [before starting the construction phase], they [CEDEGE] said: 'it [the Baba project] was prepared before Patricia Pilar existed'. They meant there were only isolated villages of small farmers. At that time [when the project was being designed], the State did not hesitate to toss these farmers out" (Interview 28 November 2015).

As shown on Table 1, at the time of construction, the effects on the people there were sizable.

Table 1. Socio-environmental impacts and technical specifications of the original design.

Description	Figures
Impacts on displaced population	778 inhabitants
Homes affected by the reservoir	240 homes (including villages)
Infrastructure affected by the reservoir	6 schools/2 churches/6 bridges
Area flooded	3760 ha
Expropriations	4420 ha
Dam height	55 m
Water storage capacity	600 hm^3
Installed electrical generation power	54 MW
Reservoir water quality and public health risk	Reservoir regulated by gates (semi-stagnant water). Severe impact on health.

Source: [72,73].

With the studies ready, CEDEGE engineers applied in 1999 for the National Planning Secretariat to declare the Baba project as a national priority. They argued that it was part of the process of 'rationalizing' the use of water resources in the Guayas River Basin, an ideal introduced by the creators of CEDEGE in the mid-1950s (Official letter from CEDEGE Number 1000-E-0435, March 1999). This confirmed, as asserted by the government's technical expert, that the new project was an extension of CEDEGE's flagship: the Daule-Peripa dam complex. This priority was granted and ratified by the Secretariat-General of the Presidency of the Republic (Official letter from the Secretariat-General of the Republic of Ecuador Number ODEPLAN-99-605, 28 May 1999), which cast a mantle of legitimacy over the project as a necessary part of the country's development.

According to CEDEGE, the Baba project's benefits were multiple. Electrical generation was the first and main benefit expected from building the dam. Electricity generation would take place both in Baba and in downstream Daule-Peripa. The latter would benefit from the extra water volume derived from Baba project. The second benefit was flood control downriver, on the Quevedo River. Third was an increase by thousands of hectares under irrigation. And fourth, it would provide water supply for human consumption [74,75].

Although the project was approved for construction under the technical parameters of the original designs, they changed after prolonged contestation by local communities. As one of the leaders of the social mobilization explained: "the dam building project was interrupted and altered […] because the people rose up against it!" (Interview 16 October 2016). We describe this process with more details below.

Formation and Struggles of the 'Bi-Provincial Committee for Non-Construction of the Baba-Vinces Dam'

It was 2002. While CEDEGE pushed the project with increasing urgency, Patricia Pilar began hearing rumors that a dam might be built. Having no access to official information, some local leaders and politicians began mobilizing to find out, and warn their people. One of the first leaders involved was the then-President of the Parish Board, Carlos Méndez. He came to Patricia Pilar as a child, with his parents, from the province of Manabí. Over the years, his family purchased land and, with it, power, becoming one of the most well-heeled families in the zone. In the sub-national elections held prior to the conflict, he was elected President of Patricia Pilar parish, making him the local leader, especially for the urban part of Patricia Pilar.

In late 2002, some peasants looked up Méndez and told him, because of his position as the parish representative, what they had noticed about the future project. "These farmers began telling me about it […]. The problem was that technicians came to conduct their studies secretly, without telling anyone, or even misleading people" (Interview 26 October 2015), recalls the politician. After these complaints, Méndez called a meeting of representatives of the 17 communities around the town of Patricia Pilar, to discuss the issue openly for the first time. Méndez was the first leader in charge of organizing communities and finding out about the project. He also became the first mediator for local concerns, vis-a-vis other stakeholders and the central Government itself. Thus, he sent letters to CEDEGE, to other governmental and non-governmental institutions, to request information about the project and support incipient resistance.

Hand in hand with these actions, by late 2003, local peasants and women organizations contacted Lorena Zambrano. She was a young leader, living in a nearby town, Buena Fe. Her relevant experience was her work with Rural Social Security members in Los Ríos province and as Coordinator of an organization of rural and urban stakeholders in the province. With this organization, years ago, Zambrano had worked with several communities affected by the Daule-Peripa reservoir, seeking reparations from the Ecuadorian Government. Her work with local societal organizations had built up her contacts with several environmental NGOs, other political parties, and human rights organizations. This would also prove to be essential to reinforce the struggles beginning in Patricia Pilar, with her ability to work on multiple scales. Thanks to her contacts with other stakeholders and the people affected by Daule-Peripa, Zambrano, played a fundamental role in the early years of resistance in Patricia Pilar.

At the time, another leader was also outstanding, leading Afro-Ecuadorian communities: Patricio Hurtado. Unlike other leaders and most local peasants, he was born in the zone. He belonged to a family who had been living there for over 100 years. For this reason, he could claim that the Afro-Ecuadorian communities' land should be categorized and protected as ancestral. This claim was supported by the National Constitution issued in 2008. (See Article 60: "The ancestral people, indigenous people, afro-Ecuadorians and *montubio* people will be able to constitute territorial circumscriptions for the preservation of their culture. The groups that enjoy collective property over the land will be recognized as a form of ancestral territorial organization"). Before the conflict he worked on a medium-sized farm (26 ha) owned by his mother, where they grew crops on the banks of the Baba River. Although his own property would not be affected directly by the original dam design, he decided to back Patricia Pilar's resistance and that of other Afro-descendant rural communities there.

By early 2004, as CEDEGE prepared to tender for the project's construction stage, leaders from Patricia Pilar had organized 31 communities. Throughout that year, Zambrano and Méndez did major political work. They were both trusted and backed by communities and had contacts with NGOs, political parties, societal organizations, and labor unions. While Méndez warned political stakeholders external to the parish, Zambrano and other local leaders such as Hurtado organized and participated in meetings with each community to encourage local people to protest against the dam's construction. They worked intensely, especially to build, merge and solidify a community identity among people from different origins or, in Chatterjee's terms, a "moral solidarity". As Zambrano recalls:

> "We prepared for several months, with meetings in each community [. . .]. Organizing was not easy, and required long discussions. This organizing is a question of talking things over, like forming a family where everyone can understand each other and make commitments, because otherwise you'll get nowhere!" (Interview 26 November 2015).

Meetings to organize were held in two settings: large assemblies every Sunday in the urban area of Patricia Pilar, and smaller ones during the week in rural communities. General assemblies were held in the central park of the parish, some 500 persons attended regularly. Smaller meetings would gather people mostly from local communities only. One of the peasants confirms this: "Meetings were held in Patricia Pilar, but the main constituency was here in the communities" (Interview 28 October 2015). This statement also shows the feeling that would emerge concretely in the future: the resistance was rooted in rural communities set to be affected by the original project and, to a lesser degree, in the town (urban part) of Patricia Pilar.

Several elements began uniting the local villagers to mobilize. The main ones were the fear of suffering similar impacts to what their neighbors suffered from the Daule-Peripa dam: loss of their land, livelihoods, neighborhood relationships, and being isolated by the massive overgrowth of aquatic weeds preventing navigation. Memory of the past and roots in the present reinforced the arguments that consolidated collective interest in Patricia Pilar's resistance. As printed in El Comercio daily newspaper: "Bad experience with Daule-Peripa warns about the Baba dam" (Diario El Comercio, 17 May 2004). This memory was amply used by leaders to motivate protestors. They organized exchange visits with testimonials and life stories from those affected. The results were impactful: the Daule-Peripa experience became the banner Patricia Pilar's people marched under, used to motive protests against building the new dam.

This social effervescence and moral solidarity enabled the 31 Afro-Ecuadorian and other rural communities to organize under the name of 'Bi-provincial Committee for Non-Construction of the Baba-Vinces Dam'. The Committee became a de facto organization, with the sole and main purpose of preventing construction of the Baba dam. It was born and continued throughout the conflict and struggles over the Baba project.

In April 2004, while CEDEGE vigorously pursued their project, the Committee grew even stronger and gained public attention after the first incident between their opposition and the dam technicians. As a local newspaper put it: "Inhabitants prevented soil studies. Approximately 200 residents of the parish of Patricia Pilar went to the site yesterday morning where the Baba Dam is to be built, to prevent the work from continuing" (Diario La Hora, 7 April 2004). That action, according to the newly-created Committee and its leaders, was a crucial step to show residents that the dam posed a real, imminent threat.

From then on, the case attracted local players plus provincial, national and international stakeholders. Another organization was formed regionally: the Coordinating Agency to Defend Life and Nature in the Guayas River Basin (COORDENAGUA). Unlike the Committee, comprising rural folk from the 31 communities, the Coordinating Agency had other members outside the zone directly influenced by the dam, almost all provincial organizations (agricultural centers, the association of artisans of Los Ríos, professional drivers, etc.). The Coordinating Agency was led by Julio Moreno, a teacher from a local university, self-identified as a scholar-activist (who had contacts with several NGOs engaging him with international conferences in Guatemala and Argentina about anti-dam movements). He felt that the Coordinating Agency was "... the most visible spokesperson entity. The Committee by contrast was the gatekeeper, to prevent any attempts to placate the people. We [the Coordinating Agency] held press conferences, participated in debates and in public events. That way our resistance had different levels, to achieve visibility and spread our issue around: local, territorial, regional and Latin American visibility" (Interview 16 October 2015, Julio Moreno). This created a platform that did elicit support from other stakeholders but not the total support of local residents, possibly because the Coordinating Agency included discourses extending beyond local, immediate concerns. For instance, Moreno stated in an interview:

> "Baba's issues are not isolated, but part of the transformation of the Guayas River Basin. And Baba is not disconnected from transformation of large-scale capital. Baba has to do with the whole project to develop oil palm, teak, and banana plantations. Baba involves the problems with Daule Peripa. It has to do with the issue of agro-fuels. And with the Manta-Manaos multimodal route" (Interview 16 October 2015).

Such broader discourses somehow disconnected the new organization from local villagers' claims and interests. Local villagers' concerns mainly were about direct socio-environmental impacts (e.g., the loss of land, livelihood changes, etc.). The latter concerns, as we will show, were crucial for subsequent re-design of the Baba mega-project. First, local inhabitants were afraid about the area of land to be flooded, which would inundate local livelihoods, isolate thousands of persons and leave large areas of productive land unusable. It would also isolate the protected area inter-community ecological reserve. This area is situated around the Río Palenque Research Station, catalogued as the last relict of primary tropical forest anywhere on the coast of Ecuador. In 1970, the center was established as such by the University of Miami. It was declared a protected area by the national Government. Second, the stagnant water would be a public health hazard, leading to outbreaks of diseases transmitted by mosquitoes. Third, they were concerned about how uncontrolled buildup of water lily in the stagnant water might affect them. Fourth, they complained that project design did not include adequate, timely community participation. Fifth, the irrigation would not benefit small-scale farmers or rural people, but only large-scale agro-industrial businesses. And finally, local protestors claimed that the dam's projected flood control would not be effective.

In late 2004, based on these concerns, local communities asked CEDEGE officials for more detailed and accurate information about the Baba project; however, these claims were unattended. Neither access to official information nor an open negotiation with the Government was achieved. On the contrary, CEDEGE issued a tender for a private strategic partner to provide financial and technical support to build the project. As well as strongly promoting the project, these disparities deepened the gap between proponents and opponents.

In this context, in mid-2005, an event that intensified protests but also weakened part of the Committee's communities, was the assassination of the Afro leader, Patricio Hurtado. This happened just a few days after he was appointed President of the Committee and gave an energetic speech against building the dam. Despite rumors that the rural leader's disappearance was led by the Baba project promoters, this murder's motives have remained unclear until today. One the one hand, this tragic incident left the Afro communities leader-less and their voice faded in the Committee and mobilizations. On the other hand, protests gained public attention. Besides a growing media attention, local authorities such as presidents of other parishes, mayors, members of Congress and even the Prefect of Los Ríos province showed support for mobilizing against the dam.

In that year CEDEGE chose, as its strategic partner, the Hydropower Consortium of the Coast (CHL). Meanwhile, the Committee and the COORDENAGUA, taking advantage of such social effervescence and media attention, organized several actions to stop dam construction. The most significant action, in the last months of 2005, was blocking one of Ecuador's (and Latin America's) main highways: the Pan-American Highway. That is, the track leading from the national capital, Quito, to the country's largest city, Guayaquil (Figure 1). In this sense, 2005 was decisive for the dam conflict. When local communities' complaints went unheard, the Committee and the COORDENAGUA organized three big highway blockades. The most forceful one was in November 2005. The Committee's former secretary recalls:

> "The highway was full of people, everywhere. To figure out the strategy, NGOs and other organizations came from Quito. We were totally prepared. The strike was a success, out of this world!" (Interview 13 October 2015).

This was the climax of the fight. The strike materialized the collective morality that glues together local communities, popular leaders and urban activists. They closed the highway for three days. Even secondary roads were blockaded. Each community organized to get everyone to the Pan-American Highway. Nearly 3000 people took part in the strike. The Government knew that, since this is such an important highway, traffic could not be shut down for long. So, on day three, the police came in violently, by air and by land, to scatter demonstrators. (The increase of acts of State and private army violence against dam opponents is well-described; see e.g., [76]). However, this event had already gotten plenty of attention in the press and international news through the NGOs involved.

After this strike, years of complaints finally reached the ears and plans of government's project proponents. Before the end of the year, CEDEGE decided to change the Baba dam's original designs, taking into account protestors' arguments. The President of CEDEGE at the time announced that "the layout of the Baba dam will change substantially, to affect as few rural people as possible in the province of Los Ríos, who protested publicly last week about this" (Diario El Mercurio, 24 November 2005).

4.2. The Alternative Design, and the Resistance's Division (2006–Present)

The alternative design suggested by CEDEGE's new strategic partner, CHL, differed from the original in various technical infrastructural aspects and its geographical location as shown in Table 2.

Table 2. Socio-environmental impacts and technical specifications of the alternative design.

Description	Figures
Impacts on displaced population	191 inhabitants
Homes affected by the reservoir	41 homes (including villages)
Infrastructure affected by the reservoir	1 school
Area flooded	1099 ha
Dam height	20 m
Water storage capacity	110 hm^3
Installed electrical generation power	42 MW
Reservoir water quality and public health risk	Constant spillover dam without gates. Minimal health impact.

Source: [72–74].

As shown on Figure 2, the dam site was changed to 15 km south of Patricia Pilar. In consequence, the dam would not flood most of the local communities. The dam height was significantly reduced by 35 m, and its hydraulic design changed, from gate regulation with a normal spillway, to a gateless design with a duckbill spillover (Figure 3). Unlike the original design (Figure 2), the alternative design would enable constant flow of water.

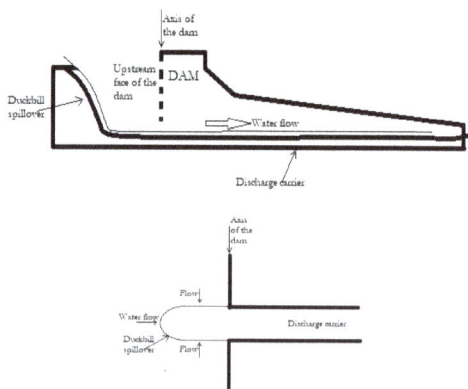

Figure 3. Cross section (**above**) and upper view (**below**) of the duckbill spillover. Elaboration: Juan Pablo Hidalgo-Bastidas.

The larger perimeter of the duckbill spillover (Figure 3) in comparison to the overflow spillover (Figure 1) allowed (partially) for the reduction of the height of the dam without significantly affecting the dam technical purposes (e.g., hydroelectricity generation). The change on the design of the spillover also would flood less than one-third of the originally planned area, with a reservoir of running water, to significantly reduce the massive proliferation of aquatic weeds and maintain relatively good water quality. These changes, clearly, incorporated local communities' main socio-environmental concerns into the very materiality of the infrastructure.

As shown in Figure 4, with the alternative design most of the flooded land belonged to large haciendas and agro-industrial companies. Figure 5 makes clear that most of the local communities were not affected by the new project. Only two percent, belonging to small farmers, fundamentally to Afro-descendant communities was affected (see Figure 5).

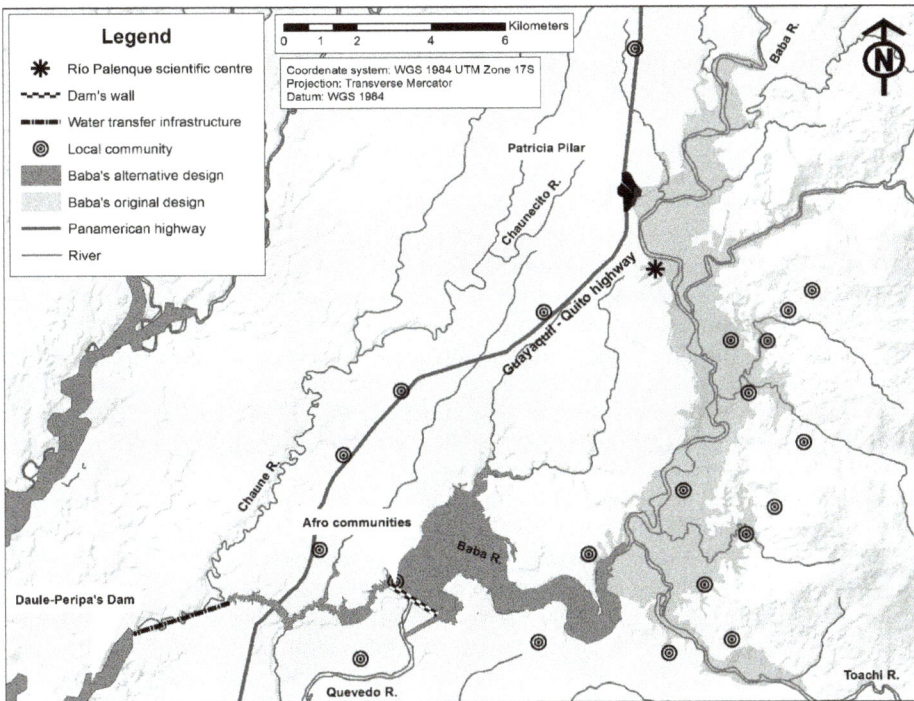

Figure 4. Baba dam's alternative design and affected Afro-Ecuadorian communities. Source: HidroPacífico S.A. Prepared by: the authors.

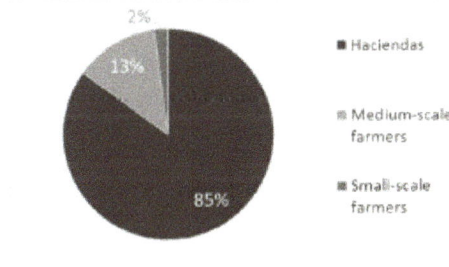

Figure 5. Types of properties affected by the alternative Baba project design [75].

Although the protest was born to stop dam construction, the proposed changes left most protesting peasant communities un-flooded (Figure 4). This made most local communities lose their interest in continuing their struggle: after CEDEGE announced the changes made in the dam design, community support for resistance diminished significantly. University teacher Moreno, leading the Coordinating Agency, said:

"When they took the dam further downriver, it was strategic for them, because we no longer had any way to mobilize people from Patricia Pilar; [. . .] so, it was difficult for the leadership to reconstruct a new discourse backing the struggle" (Interview 16 October 2015).

These changes gave a sense of legitimacy to the governmental authorities to commit to building the project. In January 2006, the project was awarded to Odebrecht S.A. (part of CHL) and in May 2006 the contract was signed to capitalize shares in these private companies. At this event, it was announced that "once the dam is ready, the consortium will operate the new hydropower station for 35 years, during which time it can use the electricity generated for its own companies and sell any surplus to the State. After that period, the project will return to the State" (Diario Expreso, 10 May 2006). As well as revealing the urgency of implementing the project, such statements showed the true scope and use of the project, leaving the initial government agencies' rationales—i.e., urgent public energy provision and flood control, in combination with irrigation and eventually drinking water provision–unjustified.

There is no doubt that the anti-dam movement exerted a great political pressure over CEDEGE and the government to adopt an alternative project design. As a high-rank CEDEGE official asserted: "Do you ask me if the social protest influenced our decision to change the designs? Yes, of course! They and their coherent demands importantly redirected the design. [...] The project was now more expensive, but it significantly diminished the negative socio-environmental impacts" (Interview 27 November 2015). However, such a far-reaching decision to implement an alternative dam design is not based on popular demands only. Besides social and political pressure we argue that it was also informed by rent-seeking relations between government officials and the construction company that was contracted for the job. Although the rent-seeking aspects were not explicitly revealed during field work, an economic audit requested by the Ministry of Energy and Mines in May 2007 disclosed serious financial and economic inconsistencies in the Baba project negotiation process between the government and Odebrecht. ("Estudio Económico y Financiero del Proyecto Multipropósito Baba de Hidronación." Elaborated by Carlos Cortéz in May 2007). In recent years, Odebrecht's Latin American corruption scandals have shocked the world because of their depth and broadly networked practices. (See an article published by the Ecuadorian digital platform for research journalism Plan V, on 22 December 2016: 'The dirty hand of Odebrecht', http://www.planv.com.ec/investigacion/investigacion/ecuador-y-la-mano-sucia-odebrecht. Also see the court proceedings of the United States District Court Eastern District of New York number F.# 2016R00709, pages 7, 16, 19 and 20).

Meanwhile, the impact of changing the dam's technological design was multiple and ambivalent. Given most communities' lack of interest in continuing to fight, as they no longer felt threatened, the sense of moral solidarity was lost. In this context, no more large assemblies were held, and there was, for instance, no more interest in blocking the highway. Strategies for resistance changed, from the streets, to the desks of national NGOs and international organizations, which began taking cases to court and urging international agencies to stop supporting the project. Nevertheless, these actions had no favorable results. Further, it was rumored that leaders had wangled individual benefits out of the struggle, which discouraged local communities' members. This eroded local leaders' credibility, and ultimately both organizations dissolved, first the Committee and then the COORDENAGUA.

By changing their designs, project proponents gained access to some communities and leaders, already partially de-mobilized. They publicly discussed the project and negotiated positions. These negotiations included land for peasants, especially with the Afro communities who were to be the only ones directly affected by the alternative design. At this stage, the strategies commonly used by promoters of such large projects pried at the cracks and further weakened the social organization. In July 2006, with no approved environmental license, the construction contract was signed for the Odebrecht S.A. Company to build the dam. After several years of construction and delays, the project was proudly inaugurated in June 2013 by then-Ecuador's President Rafael Correa.

Afro Communities: The Most Vulnerable Faction in the Resistance

Although Afro communities, through their leader, were well-represented at the outset, after the leader was murdered and the alternative design was adopted, they became the most vulnerable members of the weakened social movement. Farmers from other parts of Patricia Pilar, who had marched alongside the Afro communities when protests began, now accused them of being sell-outs, with most of the blame for the dam project's ultimate construction. However, analyzing the situation in which the Afro-descendants were left after the design change, these communities' best option for success was to negotiate effectively, in time, for land from the building company. They no longer had any popular support to resist and, further, the company's technicians threatened that: "if you don't negotiate soon, you will end up with nothing". In that context, the Afro-descendant communities negotiated their lands. One rural woman remembers that situation:

"We heard that, in Patricia Pilar, they [leaders] had already received money, so we were not going to be left behind, as the only village against the dam, like fools. So, we negotiated as well" (Interview October 2015).

Still, they ended up being the most affected population group, losing their fundamental networks and means of subsistence. Although most did receive economic compensation for their land and crops lost, these would not meet by far their loss of collectivity and community-rooted subsistence. Moreover, along with the land, they lost one of their main livelihoods: fishing. And having a huge reservoir a few meters from their homes was no guarantee that many families would have access to water: neither drinking water, nor irrigation water. At this time, large banana plantations and haciendas are the only ones extracting water from the reservoir, because of their location and access to pumping technology. These communities are caught between the questionable benefits offered by the community relations department of the company operating the power plant, and the negative impacts that it brought them.

5. Discussion and Conclusions: Hydraulic Order and the Politics of the Governed

This article has portrayed the process of organization and contestation led by the people of Patricia Pilar and neighboring communities against construction of the Baba multi-purpose dam. This has shown how such mega-projects are material expressions not exclusively of expert criteria, but actually hybrid manifestations that can be structurally influenced by more vulnerable stakeholders who are threatened to become affected. The Baba-dam water planning and implementation process shows that hydraulic designs transcend the technical realm where they are commonly located and discussed. On the contrary, we have shown that they evince contested knowledge, social constructs, in which different types of stakeholders—technocrats, politicians, rural communities, marginalized people, etc.—become political agents with the capacity and claiming the right to participate as such in decision-making and change the so called 'technical' processes.

Local claims, understood as part of the politics of the governed, had a significant effect on development of the Baba project. The pressure brought by social mobilization and resistance managed to insert their interests into the overall project design.

So, their hydro-social notions and knowledge were materialized in iron, cement, and new hydraulic flow patterns. The dam was relocated below the town of Patricia Pilar and below most of their rural communities. The height of the dam was reduced by 35 m, which then reduced the area to be expropriated and flooded by over two thousand hectares. Further, the hydraulic design, implemented with a duckbill spillway, made it possible to lower the water level and enable constant water flow in the reservoir. This change reduced potential negative health issues and the proliferation of aquatic weeds on the surface of the reservoir that would disrupt local water transport routes and obstruct the building of new (or conversion into alternative) water-based livelihoods.

In addition to the social cohesion achieved, and their capacity to upscale their struggle, part of the social movement's gains can be attributed to the geographical location of the protests. Closing one of the main highways from the Coast to the Highlands was fundamental to enhance the Committee's and Coordinating Agency's negotiation and contestation capacities. Their political-geographical location was crucial to be able to bring pressure on government officials.

Nevertheless, this social success reflected in the hydraulic infrastructure is relative, since it displaced its impacts onto the most vulnerable members of the social movement. What happened to the Afro-descendant communities after the mobilizations in Patricia Pilar manifests how every design assumption and choice in dam technology has divergent social and political consequences for the actors involved, re-organizing them as stakeholders and stakelosers. It shows the importance of including, when analyzing conflicts over mega-dams, not only the history of proponents and opponents, but also their relationship with the genealogy and contents of the technological artifacts themselves. The latter analysis brings to the fore how dam development imaginaries and hydraulic design and building knowledge are political and disputed. It asks for scrutinizing the power relationships informing the design and construction of dam technology, and the effects the latter have over society.

By including both the design process and the design contents of technology in the analysis, the article contributes to the understanding of the motivations and disincentives co-determining the success or failure of a social mobilization. Chatterjee affirms that, when the political context changes, the capacity of political society to achieve success also changes. This article shows that, in conflicts about hydraulic mega-projects such as Baba, this 'political context' is not just a matter of disembodied societal relationships or power structures: political context is co-embedded in technology and artifact design. Technology as (partly and momentarily) 'hardened knowledge, morals, norms, skills, and social relationships' [26,45,77] significantly affects the development and capacity for advocacy of mobilizers—both in terms of artifacts' contents and the process of designing, constructing and implementing these artifacts. In Baba, the modified dam design benefitted particular stakeholders (e.g., rent-seeking actors) and undermined collective moral solidarity as well as mediators' capacity to reassemble the resistance's discursive framework; framing the social mobilization became a major challenge.

Revealing the social and political foundations of technology design, development, and implementation emphasizes people's political capacities. This offers options for recognition, participation and empowerment for the governed—those who are characteristically left out as knowledgeable agents in formal hydraulic design processes.

In sum, anti-dam struggles as in Baba show how the socio-political context and dominant relationships steer technology's designing norms and practices; but also, vice versa, how the technical-hydraulic contents of such technology, in their turn, significantly shape the very socio-political context and impact on locally prevailing socio-economic and cultural-political relationships.

Author Contributions: For this article, research preparation, literature research, and fieldwork investigation were done by J.P.H.-B.; J.P.H.-B. and R.B. have written the article.

Funding: This research was funded by Secretaría de Educación Superior, Ciencia, Tecnología e Innovación (SENESCYT-Ecuador).

Acknowledgments: We thank the editors and anonymous reviewers for their insightful comments, and Patricia Pilar's people, peasant leaders and government officials for their support to this research. Special thanks to the social movement that struggled against the Baba dam and those families who endure its negative impacts.

Conflicts of Interest: The authors declare no conflict of interest. The founding sponsors had no role in this study's design; in collecting, analyzing, or interpreting data; in writing the manuscript, or in the decision to publish the findings.

References

1. Lilienthal, D.E. *TVA-Democracy on the March*; Harper & Brothers Publishers: New York, NY, USA; London, UK, 1944.
2. ICOLD, International Commission on Large Dams. International Commission on Large Dams. 2017. Available online: http://www.icold-cigb.net/ (accessed on 14 December 2017).
3. Boelens, R.; Shah, E.; Bruins, B. Contested Knowledges: Large Dams and Mega-Hydraulic Development. *Water* **2019**, *11*, 416. [CrossRef]
4. McCully, P. *Silenced Rivers: The Ecology and Politics of Large Dams*; Zed Books: London, UK, 2001.
5. Kaika, M. Dams as symbols of modernization: The urbanization of nature between geographical imagination and materiality. *Ann. Assoc. Am. Geogr.* **2006**, *96*, 276–301. [CrossRef]
6. Nixon, R. Unimagined communities: Developmental refugees, megadams and monumental modernity. *New Form.* **2010**, *69*, 62–80. [CrossRef]

7. Boelens, R.; Uiterweer, N. Hydraulic heroes: The ironies of utopian hydraulism and its politics of autonomy in the Guadalhorce Valley, Spain. *J. Hist. Geogr.* **2013**, *41*, 44–58. [CrossRef]
8. Duarte-Abadía, B.; Boelens, R.; Roa-Avendaño, T. Hydropower, encroachment and the re-patterning of hydrosocial territory: The case of Hidrosogamoso in Colombia. *Hum. Organ.* **2015**, *74*, 243–254. [CrossRef]
9. Hommes, L.; Boelens, R.; Maat, H. Contested hydrosocial territories and disputed water governance: Struggles and competing claims over the Ilisu Dam development in southeastern Turkey. *Geoforum* **2016**, *71*, 9–20. [CrossRef]
10. Molle, F.; Mollinga, P.P.; Wester, P. Hydraulic Bureaucracies and the Hydraulic Mission: Flows of Water, Flows of Power. *Water Altern.* **2009**, *2*, 328–349.
11. Ekbladh, D. "Mr. TVA": Grass-roots development, David Lilienthal, and the rise and fall of the Tennessee Valley Authority as a symbol for US Overseas Development, 1933–1973. *Dipl. Hist.* **2002**, *26*, 335–374. [CrossRef]
12. Hidalgo, J.P.; Boelens, R.; Isch, E. Hydro-territorial configuration and confrontation. The Daule-Peripa Multipurpose Hydraulic Scheme in coastal Ecuador. *Lat. Am. Res. Rev.* **2018**, *53*, 517. [CrossRef]
13. Hidalgo-Bastidas, J.P.; Boelens, R. The Political Construction and Fixing of Water Overabundance. Rural-urban Flood Risk Politics in Coastal Ecuador. *Water Int.* **2019**, online. [CrossRef]
14. Sneddon, C.; Fox, C. Struggles over dams as struggles for justice: The World Commission on Dams (WCD) and anti-dam campaigns in Thailand and Mozambique. *Soc. Nat. Resour.* **2008**, *21*, 625–640. [CrossRef]
15. Baird, I.G. The Don Sahong Dam. *Crit. Asian Stud.* **2011**, *43*, 211–235. [CrossRef]
16. Ansar, A.; Flyvbjerg, B.; Budzier, A.; Lunn, D. Should we build more large dams? The actual costs of hydropower megaproject development. *Energy Policy* **2014**, *69*, 43–56. [CrossRef]
17. Bose, P.S. Critics and experts, activists and academics: Intellectuals in the fight for social and ecological justice in the Narmada Valley, India. In *Popular Intellectuals and Social Movements. Framing Protest in Asia, Africa, and Latin America*; Baud, M., Rutten, R., Eds.; Cambridge University Press: Cambridge, UK, 2004; Volume 12, pp. 133–157.
18. Gomez, A.; Wagner, L.; Torres, B.; Martin, F.; Rojas, F. Social resistance against hydraulic megaprojects in Latin America. *Eur. Rev. Lat. Am. Caribb. Stud. Eur. Estud. Latinoam. Caribe* **2014**, 75–96.
19. Rothman, F.D. A comparative study of dam-resistance campaigns and environmental policy in Brazil. *J. Environ. Dev.* **2001**, *10*, 317–344. [CrossRef]
20. Swain, A. Democratic consolidation? Environmental movements in India. *Asian Surv.* **1997**, *37*, 818–832. [CrossRef]
21. Nayak, A.K. Big dams and protests in India: A study of Hirakud dam. *Econ. Polit. Wkly.* **2010**, *XLV*, 69–73.
22. Kirchherr, J. Strategies of successful anti-dam movements: Evidence from Myanmar and Thailand. *Soc. Nat. Resour.* **2018**, *31*, 166–182. [CrossRef]
23. Coward, E.W. *Irrigation and Agricultural Development in Asia: Perspectives from the Social Sciences*; Cornell University Press: Ithaca, NY, USA; London, UK, 1980.
24. Uphoff, N.T. *Getting the Process Right: Improving Irrigation Water Management with Farmer Organization and Participation*; Cornell University Press: Ithaca, NY, USA; London, UK, 1986.
25. Chambers, R.; Pacey, A.; Thrupp, L.A. *Farmer First: Farmer Innovation and Agricultural Research*; Intermediate Technology Publications: London, UK, 1989.
26. Mollinga, P.; Mooij, J. *Cracking the Code: Towards a Conceptualization of the Social Content of Technical Artefacts*; Occasional Paper No. 18; Milton Keynes Open University, Technology Policy Group: Milton Keynes, UK, 1989.
27. Boelens, R. *Water, Power and Identity: The Cultural Politics of Water in the Andes*; Earthscan and Routledge: Washington, DC, USA; London, UK, 2015.
28. Veldwisch, G.J.; Bolding, A.; Wester, P. Sand in the engine: The travails of an irrigated rice scheme in Bwanje Valley, Malawi. *J. Dev. Stud.* **2009**, *45*, 197–226. [CrossRef]
29. Law, J. Technology and heterogeneous engineering: The case of Portuguese expansion. *Soc. Constr. Technol. Syst. New Dir. Sociol. Hist. Technol.* **1987**, *1*, 1–134.
30. Bijker, W.E. Dikes and dams, thick with politics. *Isis* **2007**, *98*, 109–123. [CrossRef]
31. Winner, L. Technologies as forms of life. In *Ethics and Emerging Technologies*; Palgrave Macmillan: London, UK, 2013; p. 48.

32. Law, J. (Ed.) *A Sociology of Monsters: Essays on Power, Technology and Domination*; Routledge: London, UK, 1991; Volume 171.

33. Einbinder, N. *Dams, Displacement and Development: Perspectives from Río Negro, Guatemala*; Springer: Cham, Switzerland, 2017.

34. Nüsser, M.; Baghel, R. The emergence of technological hydroscapes in the Anthropocene: Socio-hydrology and development paradigms of large dams. In *Handbook on Geographies of Technology*; Warf, B., Ed.; Edward Elgar: Camberley Surrey, UK, 2017.

35. Warner, J.F.; Hoogesteger, J.; Hidalgo, J.P. Old Wine in New Bottles: The Adaptive Capacity of the Hydraulic Mission in Ecuador. *Water Altern.* **2017**, *10*, 322–340.

36. Chatterjee, P. *The Politics of the Governed: Reflections on Popular Politics in Most of the World*; Columbia University Press: New York, NY, USA, 2004.

37. Kien, G. Technography = technology+ ethnography. *Qual. Inq.* **2008**, *14*, 1101–1109. [CrossRef]

38. Jansen, K.; Vellema, S. What is technography? *NJAS Wagening J. Life Sci.* **2011**, *57*, 169–177. [CrossRef]

39. Bijker, W.E. Technology, Social Construction of. In *International Encyclopedia of the Social & Behavioral Sciences*; Elsevier Ltd.: New York, NY, USA, 2015; Volume 24.

40. Pfaffenberger, B. Social anthropology of technology. *Annu. Rev. Anthropol.* **1992**, *21*, 491–516. [CrossRef]

41. Bijker, W.E.; Hughes, T.P.; Pinch, T. *The Social Construction of Technological Systems: New Directions in the Sociology and History of Technology*; MIT Press: Cambridge, MA, USA, 2012.

42. Hughes, T.P. *Networks of Power: Electrification in Western Society, 1880–1930/Thomas P. Hughes*; Johns Hopkins University Press: Baltimore, MD, USA, 1983.

43. Pfaffenberger, B. The harsh facts of hydraulics: Technology and society in Sri Lanka's colonization schemes. *Technol. Cult.* **1990**, *31*, 361–397. [CrossRef]

44. Shah, E. Social Design: Tank Irrigation Technology and Agrarian Transformation in Karnataka, South India. Ph.D. Thesis, Wageningen Universiteit, Wageningen, The Netherlands, June 2003.

45. Pfaffenberger, B. Fetishised objects and humanised nature: Towards an anthropology of technology. *Man* **1988**, *23*, 236–252. [CrossRef]

46. Verbeek, P.P. The Struggle for Technology: Towards a Realistic Political Theory of Technology. *Found. Sci.* **2017**, *22*, 301–304. [CrossRef]

47. Sanchis-Ibor, C.; Boelens, R.; García-Mollá, M. Collective irrigation reloaded. Re-collection and re-moralization of water management after privatization in Spain. *Geoforum* **2017**, *87*, 38–47. [CrossRef]

48. Winner, L. Do artifacts have politics? *Daedalus* **1980**, *109*, 121–136.

49. Meehan, K. Disciplining de facto development: Water theft and hydrosocial order in Tijuana. *Environ. Plan. Soc. Space* **2013**, *31*, 319–336. [CrossRef]

50. Swyngedouw, E.; Boelens, R. ... And Not a Single Injustice Remains: Hydro-Territorial Colonization and Techno-Political Transformations in Spain. In *Water Justice*; Boelens, R., Perreault, T., Vos, J., Eds.; Cambridge University Press: Cambridge, UK, 2018; pp. 115–134.

51. CAIC. *Informe Final de la Auditoría Integral de la Deuda Ecuatoriana*; Ministerio de Finanzas: Quito, Ecuador, 2008.

52. Silva, P. *En el Nombre de la Razón: Tecnócratas y Política en Chile*; Ediciones Universidad Diego Portales: Santiago, Chile, 2010.

53. Cremers, L.; Ooijevaar, M.; Boelens, R. Institutional reform in the Andean irrigation sector: Enabling policies for strengthening local rights and water management. *Nat. Resour. Forum* **2005**, *29*, 37–50. [CrossRef]

54. Terán, J.F. *La Sequedad del Ajuste: Implicaciones de la Gobernanza Global del Agua Para la Seguridad Humana en Ecuador*; Centro Andino de Estudios Internacionales, Universidad Andina Simón Bolívar: Quito, Ecuador, 2005.

55. Brassel, F.; Herrera, S.; Laforge, M. *Reforma Agraria en el Ecuador? Viejos Temas Nuevos Argumentos*; SIPAE: Quito, Ecuador, 2008.

56. Boelens, R.; Hoogesteger, J.; Baud, M. Water reform governmentality in Ecuador: Neoliberalism, centralization, and the restraining of polycentric authority and community rule-making. *Geoforum* **2015**, *64*, 281–291. [CrossRef]

57. Gaybor, A. Acumulacion en el campo y despojo de agua en el Ecuador. In *Justicia Hídrica: Acumulación, Conflicto y Acción Social*; Boelens, R., Cremers, L., Zwarteveen, M., Eds.; Instituto de Estudios Peruanos: Lima, Peru, 2011; pp. 195–207.

58. Boelens, R. Water rights arenas in the Andes: Upscaling networks to Strengthen Local Water Control. *Water Altern.* **2008**, *1*, 48–65.

59. Hidalgo, J.P.; Boelens, R.; Vos, J. De-colonizing water. Dispossession, water insecurity, and Indigenous claims for resources, authority, and territory. *Water Hist.* **2017**, *9*, 67–85. [CrossRef]

60. Hoogesteger, J.; Verzijl, A. Grassroots scalar politics: Insights from peasant water struggles in the Ecuadorian and Peruvian Andes. *Geoforum* **2015**, *62*, 13–23. [CrossRef]

61. Hoogesteger, J.; Boelens, R.; Baud, M. Territorial pluralism: Water users' multi-scalar struggles against state ordering in Ecuador's highlands. *Water Int.* **2016**, *41*, 91–106. [CrossRef]

62. Mena-Vásconez, P.; Boelens, R.; Vos, J. Food or flowers? Contested transformations of community food security and water use priorities under new legal and market regimes in Ecuador's highlands. *J. Rural Stud.* **2016**, *44*, 227–238. [CrossRef]

63. Valladares, C.; Boelens, R. Extractivism and the rights of nature: Governmentality, "convenient communities" and epistemic pacts in Ecuador. *Environ. Politics* **2017**, *26*, 1–20. [CrossRef]

64. Kopas, J.; Puentes, A. *Grandes Represas en América: Peor el Remedio que la Enfermedad*; AIDA: San Francisco, CA, USA; Bogotá, Colombia, 2009. Available online: https://aida-americas.org/es/grandes-represas-en-am-rica-peor-el-remedio-que-la-enfermedad-0 (accessed on 12 January 2018).

65. Fearnside, P.M. Environmental and social impacts of hydroelectric dams in Brazilian Amazonia: Implications for the aluminum industry. *World Dev.* **2016**, *77*, 48–65. [CrossRef]

66. Chatterjee, P. *La Nación en Tiempos Heterogéneos*; Cuaderno de Trabajo Nro. 1; PUCP-CISEPA: Lima, Perú, 2007.

67. Thomas, M.S. *The Challenge of Legal Pluralism: Local Dispute Settlement and the Indian-State Relationship in Ecuador*; Routledge: London, UK; New York, NY, USA, 2016.

68. Chatterjee, P. *La Nación en Tiempo Heterogéneo y otros Estudios Subalternos*; Instituto de Estudios Peruanos: Lima, Perú, 2007.

69. Baud, M. Indigenous Politics and the State: The Andean Highlands in the Nineteenth and Twentieth Centuries. *Soc. Anal.* **2007**, *51*, 19–42. [CrossRef]

70. Barsky, O.; Furche, E.; Mizrahi, R.C. *Políticas Agrarias, Colonización y Desarrollo Rural en Ecuador: Reflexiones Sobre el Proyecto de Desarrollo Rural Integral Quininde-Malimpia-Nueva Jerusalem*; Centro de Planificación y Estudios Sociales: Quito, Ecuador, 1982.

71. Ditto, J.S. *Leyes y sangre en el agro*; Universidad de Guayaquil: Guayaquil, Ecuador, 1986.

72. Caminosca. *Diseños Definitivos de la Central Hidroléctrica del Proyecto Quevedo-Vinces (Presa Baba). Informe Final Provisional. Informe General*; CEDEGE: Guayaquil, Ecuador, 2004.

73. De, Q. *Eficacitas, U.T.E. TDRs del Estudio de Impacto Ambiental Definitivo Proyecto Multipropósito Baba*; Consorcio Hidroenergético del Litoral (CHL): Guayaquil, Ecuador, 2006.

74. CEDEGE. *Proyecto de Propósito Múltiple Quevedo—Vinces. Primera Etapa: Presa Baba y Trasvase hacia Daule—Peripa*; CEDEGE: Guayaquil, Ecuador, 1999.

75. IDB. *Multipurpose Baba Project—Environmental and Social Management Report*; Inter-American Development Bank: Washington, DC, USA, 2007.

76. Del Bene, D.; Scheidel, A.; Temper, L. More dams, more violence? A global analysis on resistances and repression around conflictive dams through co-produced knowledge. *Sustain. Sci.* **2018**, *13*, 617–633. [CrossRef]

77. Boelens, R.; Vos, J. Legal pluralism, hydraulic property creation and sustainability: The materialized nature of water rights in user-managed systems. *Curr. Opin. Environ. Sustain.* **2014**, *11*, 55–62. [CrossRef]

water

MDPI

Article

Mobilizing Water Actors and Bodies of Knowledge. The Multi-Scalar Movement against the Río Grande Dam in Málaga, Spain

Bibiana Duarte Abadía [1,*], Rutgerd Boelens [1,2,3,4] and Lucas du Pré [2]

1   Centre for Latin American Research and Documentation (CEDLA), University of Amsterdam, 1018 WB Amsterdam, The Netherlands; rutgerd.boelens@wur.nl
2   Department of Environmental Sciences, Wageningen University, 6708 PB Wageningen, The Netherlands; lucasdupre23@gmail.com
3   Faculty of Agricultural Sciences, Universidad Central del Ecuador, Ciudadela Universitaria, Quito 170129, Ecuador
4   Department of Social Sciences, Pontificia Universidad Católica del Perú, Av. Universitaria 1801, San Miguel, Lima 15088, Peru
*   Correspondence: bibiana.duarte@gmail.com or B.A.DuarteAbadia@cedla.nl; Tel.: +31-685-615-869

Received: 4 July 2018; Accepted: 3 January 2019; Published: 26 February 2019

check for updates

Abstract: Just as in other parts of Spain, the Guadalhorce Valley, Málaga, has a long history of policies based on 'hydraulic utopianism' (regenerationist and Franco-ist), bent on 'reorganizing' political, geographic, and human nature. Residents of the neighboring sub-basin, the Río Grande valley, have seen how these policies, designed to transfer rural water to modern urban centers, have turned the Guadalhorce hydrosocial territory into a 'hydraulic dystopia'. In this article, we examine how Río Grande valley residents mobilized to maintain control over the development and use of their resources, livelihoods, and knowledge systems, when modernist-urbanist policies planned to take their water from a major dam on the Río Grande. Interviewing actors at different scales we examined how this anti-dam movement organized massively in a creative, multi-actor, and multi-scale network. Our results also show that this unified, successful fight against the 'common enemy', the mega-hydraulic construction, has become more complex, as threats crop up not only from the 'city over there' but also from 'internal' hydro-territorial transformations. These sprout from policies to modernize traditional irrigation systems, supposedly to 'save water', but critical voices assume that it is all about passing on the 'surplus' to Málaga city, or using that water to expand agribusiness. We conclude that the challenge lies in critically integrating multiple forms of knowledge, stakeholders, and scales to both defend collective water management and creatively construct anti-hegemonic alternatives.

Keywords: hydrosocial territory; knowledge encounters; hydraulic utopia; modernity; commensuration; anti-dam movement; Málaga; Spain

1. Introduction

Spain's history of political and geographic re-organization is intimately linked to the 'hydraulic utopia' urged by the Regenerationist movement, and paradoxically, implemented during the Franco period [1–4]. The aim was to technocratically control rivers by establishing interconnected dams, to geographically distribute water more fairly, as hydro-solidarity between the country's rainy and dry zones, to enhance agricultural production, and to integrate rural economies into the industrial and energy sectors. However, in Andalucía, as in other regions, creating hydraulic utopia has significantly contributed to the environmental degradation of rivers, abandoned croplands, the urbanization of nature, the bureaucratization of traditional autonomous irrigation systems, the flooding and displacement of towns, and the marginalization of farmer knowledge systems regarding local management of water sources [2,3].

The legacy of these historical actions and the emergence of new, modernistic plans threatening to re-organize water flows, territories, and human population groups now meet with a growing response. Civil society alliances have resurrected folk vernacular knowledge, practices, and forms of social organization, to propose alternative ways to manage and use their rivers' water. One such alliance is the 'Coordinating Body to Defend the Río Grande', belonging to the Guadalhorce basin in Málaga Province. Through this case, we will illustrate how social mobilization has shown the capacity to construct social networks with diverse stakeholders, mobilize much of civil society, and challenge dominant modernistic discourses backing dam building and hydrosocial transformation of the sub-basin, also see References [3,5–7]. This gave rise to the Río Grande movement (hereafter the 'Coordinating Body') seeking to exercise hydro-ecological democracy and impact policy-making, in order to set the Río Grande's water free.

Despite the social movement's success in stopping the hydraulic mega-project on the Río Grande in 2001 and 2006, dominant political and economic powers continue to insist on damming the Río Grande, a project revived lately by tourism growth in the city of Málaga [7]. In this arena, stakeholders with diverging interests wield their different strategies, resources, and discourses to get the upper hand. Repercussions include contrasting, legitimizing or de-legitimizing different knowledge systems or repertoires. This article analyzes the different historical phases through which the different knowledge systems face off, mobilizing discourses and counter-discourses, and hefting their power to influence how the Río Grande hydrosocial territory is configured.

The article is structured as follows: The second section shows the methodology, the third one conceptualizes how encounters among the different life-worlds co-constituted confrontations among diverse knowledge systems [8], which also directly impact the constitution of 'hydrosocial territories' [9]. The fourth part illustrates and analyzes how, in the course of time, a powerful knowledge system has imposed itself over the others in the Guadalhorce basin. We particularly analyze experts' (engineering) knowledge versus grassroots knowledge, their respective alliances and powers that are constituted to debunk locally existing truths and thereby transform realities. We will also examine how, as a response, alternative knowledge systems revalue the sub-watershed's history, ecology, and local socio-economic livelihoods.

In the last part, we explain and conclude how the unified struggle to prevent the Río Grande dam's construction becomes more complex and fragmented by the subtle penetration of powerful interests tied to less visible threats, in particular the policies of modernizing and upgrading technology in traditional irrigation systems, which—like building the dam—equally aim to transfer rural water to urban centers. While some farmers and communities therefore oppose this modernization of their irrigation systems, others see modernization and investment in their irrigation technologies as a guarantee to conserve their water rights and defend them when confronted with the powerful water interests of the city of Malaga.

2. Methodology

Applying a political ecology approach to water control, the article discusses the findings from various periods of fieldwork (2015–2017). We conducted 68 semi-structured interviews with different groups of actors in the Río Grande sub-watershed. These were selected through snowball sampling and grouped according to their place of action. The first group involved farmers and their irrigators' organization using the seven community-operated canals (*acequias*), comprising the Río Grande irrigation system. With this group, including small and middle-class farmers and agribusiness enterprises, we conducted 30 interviews.

The second group refers to the social leaders who formed the 'Coordinating Body to Defend the Río Grande', a local societal movement representing the inhabitants of Coín Municipality, who opposed the dam's construction. With this group we conducted 16 interviews. In the third group, we did 10 interviews with scholars who supported the societal movement. Additionally, the last group (12 interviews) consists of officials of the water-supply company (EMASA Municipal Water Company of Málaga) and entities responsible for coordinating water issues, under the Environment Council for the Province of Málaga. Most officials were in favor of the dam construction. The interview questions, structure and settings were adapted according to the group of actors interviewed. Having this diversity of actors, living and acting in different contexts, our research—involving both academic and action-research—includes a balanced sample of the divergent knowledge frames and values that confront in the Río Grande watershed (see also [10]). The interviews were bolstered by field observations and secondary literature review, which includes newspaper articles, official reports, and historical files. The data were triangulated. During the fieldwork the interviews were recorded, transcribed, and coded, and 14 field reports were elaborated to examine the collected data, reflect on research strategies, enrich the upcoming interviews and compare actors' narratives.

3. Political Ecology of Water: Transforming Hydrosocial Territories and Struggles over Knowledge

In this paper we use a political ecology focus on 'water', scrutinizing how political forces and power relations shape human knowledge of and interventions in the water world, and how this affects water allocation, management and practices. It examines how dominant modes of water knowledge, water intervention, and water control aim to determine ways of governing nature and society, and simultaneously, define a particular hydrosocial order that organizes access and distribution of water and power among different actors [11]. Our political ecology of water focus therefore concentrates on the (re)constitution of 'hydrosocial territories' [9] as dynamic, disputed spatial, political, and geographic configurations, in which persons, technology, financial resources, legal and economic arrangements, cultural institutions and practices constantly interact to order, distribute, and control water flows and establish ways to govern them. Boelens et al. conceptualize a 'hydrosocial territory' as: "the contested imaginary and socio-environmental materialization of a spatially bound multi-scalar network in which humans, water flows, ecological relations, hydraulic infrastructure, financial means, legal-administrative arrangements and cultural institutions and practices are interactively defined, aligned and mobilized through epistemological belief systems, political hierarchies and naturalizing discourses" [9] (p. 2). Therefore, in addition to aligning diverse social and natural elements, and defining water governance models, competing hydrosocial territories deeply express divergent notions of territoriality, featuring the stories of fights over water appropriation and dominion, contrasting values, meanings, norms, knowledge, identity, and authority.

These contestations to establish 'hydrosocial territory' define processes of inclusion and exclusion, marginalization and development, and distribution of benefits and burdens [12–16]. Significantly, these spatial configurations are informed and aligned by divergent knowledge systems [13,16–18]. Generally, re-organizing hydrosocial territories has responded to the interests of the most economically and politically powerful stakeholders [4,19–22]. However, during this disputed re-configuration, marginalized and anti-hegemonic groups commonly struggle to defend their water, and often mobilize politically and strategically through alliances that combine local, national, and global scales [23–27]. In this way, the hydrosocial territory concept highlights the existence of other, alternative worldviews and vernacular knowledge and how these may co-exist and overlap with, and/or contest, hegemonic forces, structures, and discourses that affect socio-natural reconfiguration. (Related concepts such as 'waterscape' focus particularly on the hegemonic forces that produce water techno-scientific knowledge, discourses and socionature; for a discussion, see Swyngedouw and Boelens [14] (p. 129,130); also see [9,13,16,18,28]).

Spain is an important, illuminating case. The country has historically moved toward modernity by implementing water mega-projects and reorganizing nature politically and geographically, attempting to impose a homogenizing, totalizing order [1–4,29]. Universalizing objectivist reason and imposing technical and scientific knowledge have been central factors [2,14,30]. This drive to tame and control natural order through technical-positivist science and large-scale engineering—a universalizing phenomenon and force that is also present in other parts of the world—has been portrayed as indispensable for cities to develop and grow, and conceptualized as 'urbanizing nature' [31] (p. 276) (for comparison with similar experiences in other regions, see for instance [12,13,16,22,28,31–41]). For this, a range of modernist 'commensuration' mechanisms are put to work that oversimplify, standardize, and sideline the diversities and complexities intrinsic in life, to build a common metric that will make territories, livelihoods, values, meanings, and knowledge comparisons, which are manageable and governable [42] (see also the Bolivian Misicuni case, in this Special Issue, on defining 'rightful' compensation for dam damages [37]). As a political utopian project, it is rooted, among others, in philosophical thinking that ranges from Francis Bacon [43] to Jeremy Bentham [44,45]—seeking 'the greatest happiness for the majority' by means of calculability, technology, and socio-technical control [46,47].

As we have mentioned above, this utopian national hydraulic project surged in Spain in the late 19th century, when the long-standing transnational empire lost its last colonies. Along with the resulting economic crisis, this materialized the idea that the only way to make up for the country's chaos and backwardness was to colonize itself: By building hydraulic projects, to interconnect the country's bodies of water, mobilizing them to achieve economic development, and build the nation's unity [48–51]. (This process of entwined state-building and nation-building through techno-political water intervention and hydro-territorial reconfiguration is well described in the Menga/Swyngedouw edited volume "Water Technology and the Nation-State"—space, territory, and society are materially, socially, and politically constructed at various, intertwined scales [51] (p. 7)). Building dams demonstrated progress, independence, and modernity, the ability to govern the people and their territories through engineering breakthroughs to control and master the Iberian peninsula's water [52]. Spain's utopian plan of channeling, controlling and damming all water flows to satisfy electrical demand, domestic consumption and cities' industrial production, has been designed with the idea to govern both water and people at once [47]. Political strategies to materialize this 'hydraulic utopia' [1–3,14,29,47] illustrate the mechanisms of standardized modernist domestication of socio-natural geography, making mega-dam construction the epitome of uniformizing control, rationalistic organization, modernist progress, and nationalism [14,30,31,35,53].

In this way, technological developments have accompanied the course of Spain's hydraulic modernity, transforming, appropriating, and instrumentalizing nature and rivers, while orienting patterns of power, generating new social structures, and hydro-territorial orders. Establishing and expanding mega-dams and their massive interconnection would determine new relations among urban, rural and natural life. Spain's history over the last century shows how progressive 'regenerationist' imaginaries regarding political-hydraulic modernization has driven social-ecological transformation, in which imposing reason led the way to irrational acts, violent practices, and mechanisms of exclusion that denied other forms of knowledge and alternative cultural practices [1,3]. What history tells us about modernity is that its policies and politics of invisibility erase diversity and 'inconvenient' social and natural communities [36]; see also [21,35,54–58]. See, for example, the enclosure of the commons in Senegal valley by the completion of Manantali dam in 1987; the powerful bureaucracies building large dams in the Soviet Union; the British colonialist Empire in the early 20th century damming the Ganges and Nile rivers to control exportations of cotton, sugar cane and opium; or the right to expropriate lands that Tennessee Valley Authority created in 1933 for dam building in US, among others [59].

However, contemporary modernity, more than excluding and erasing, also includes controlling anything different, and attempts to involve water-user communities in hydro-territorial projections and the rationalities of governing groups, in order to mold and reinforce the dominant hydro-territorial order [60]. As we will also see in the Río Grande case, this new face of modernity pursues efficient schemes for efficient water use and saving, but in practice it is worsening the unequal distribution of water as well as unsustainable extraction, affecting, and appropriating rural people's natural livelihoods and commons [29,61–64]. This way, progress through modernity and hydro-territorial appropriation by powerful groups stands on and expands at the expense of local realities: Those domains that still maintain some independence from market forces and universalistic and governmental laws.

In the Guadalhorce region, as elsewhere in Spain, these domains with vernacular values, rooted in diverse cultural practices, have not surrendered. In many places, they have been able to coexist with modernistic transformations, maintaining and fostering values and knowledge that often remain concealed, but are intrinsic within their diverse cultural reproduction strategies and the permanence of othernesses—of that which is distinctly different normatively, ontologically, and epistemologically [65]. This vernacular knowledge emerges through day-to-day practices, as they evolve over time, wisdom—'farmers' knowledge'—is constantly recreated through experience. It is based on rural peoples' capacities to coordinate and adapt to a broad scenario of political and agro-economic uncertainties, generating techno-ecological opportunities that enable them to get the hoped-for results to make their living [66]. This is creative and dynamic, built constantly by selectively incorporating new ideas and prior experiences, based on cultural beliefs and historical and modern sentiments [67].

Foucault [68] conceptualizes such vernacular knowledge, often concealed, as 'subjugated knowledges', referring to them as the bodies of knowledge that have been disqualified for not having been elaborated within scientifically and formally accepted conceptual frameworks. Such knowledge is ignored, deemed irrelevant, and therefore remains underground; ranked as hierarchically inferior by those who hold the power to validate and judge the way they want the majorities to see and understand reality. Therefore, a knowledge hierarchy is implicit in society's interlinking with aspects of power, authority, and legitimacy [8]. Bodies of knowledge do not have absolute validity or truth. Foucault would argue that knowledge, truth, and power are co-defined in a triangular relationship (1980)—it is the nexus of knowledge and power that creates, legitimizes, and lends 'truth' to a certain order of knowledge. It has direct implications to uphold and impose a certain hydraulic policy and the consequent political and geographic reorganization of nature. Here, Long [8] adds that relations between power and knowledge emerge from social interactions, where knowledge involves clashes between stakeholders who want to control and dominate others in their plans to create society and territory. On the battlefield among bodies of knowledge, they try to get their particular frameworks of meanings accepted, to position their ways of viewing life. The hierarchy of knowledge then results from interactions, dialogues and contestations over values and meanings; the legitimacy or invisibility of knowledge hinges on power relationships, establishing forms of authority, normative frameworks, discursive guidelines, and orientation in allocating, controlling and distributing resources.

In this battle of epistemological domains, as we show in the following sections, local groups and their culture and place-born knowledge respond actively to the imposition of knowledge to alter water distribution and governance patterns; they contest the reorganization of their hydrosocial territory. Consequently, political and economic power relations directly impact how and what forms of knowledge will prevail in a given territory. As stakeholders, their ideas, knowledge, and values engage and confront each other, they perpetuate the possibilities of de-constructing dominant discourses and consequently transform their realities. According to Bebbington et al. [23], success in questioning dominant knowledge systems and discourses lies in forming an alliance of multiple stakeholders at multiple scales—local, regional, national, and international. Multi-scale and multi-stakeholder movements, hence, are more effective to actively respond to the imposition of knowledge that alters water distribution and governance patterns. This shapes the political arenas where values, rules, rights, techniques, practices, and knowledge are decided, determining which hydrosocial order will be dominant.

The end of this article will analyze how most rural and grassroots contestations focus not on rejecting modernity, but on reformulating and including it as part of their own projects and proposals, to uphold their livelihoods and defend their water rights and territory (see [8,11,67,69]). Thus, modernity is and becomes 'multiple', incorporating the different interfaces and reflecting the conflicts and frictions among epistemological systems and discourses—see the elaborate analysis in the introduction paper to this Special Issue, [36]. This also appears directly in technological water development—subject to being contested, appropriated, modified, and altered by the claims and interests of the directly affected population groups [70–72].

4. The Utopia of Hydraulic Modernity in the Guadalhorce Watershed: A Dystopian Mirror for Residents of the Río Grande Micro-Watershed

Understanding the knowledge struggle involved in building the Río Grande dam requires illustrating and historically analyzing the political and economic factors impacting the hydro-territorial transformation of the wider Guadalhorce basin, which the Río Grande flows into (see Figure 1). We will analyze how building mega dams and imposing modernistic technical and scientific thinking have driven transformation of the Guadalhorce Valley.

As was mentioned, the thinking of the intellectual-political 'Regeneracionismo' movement and its leader, Joaquín Costa (1875–1911), are fundamental to start with. Costa laid the foundation for Spain's water policy in the late 19th century. His utopian desire centered on designing an idealized society by engineering its water, to transform both humans and nature. As explained above, at that time, Spain was in an economic and political post-colonial crisis, casting about for a water utopia to emerge: Regenerationist intellectuals proclaimed the need to start colonizing internally to regenerate the country. Among Costa's proposals, this self-colonization was geared toward recovering the Arabic hydraulic heritage, empowering rural people, eliminating bourgeoisie, and building a new national identity by technically and politically modernizing water [1,14,29,47,73].

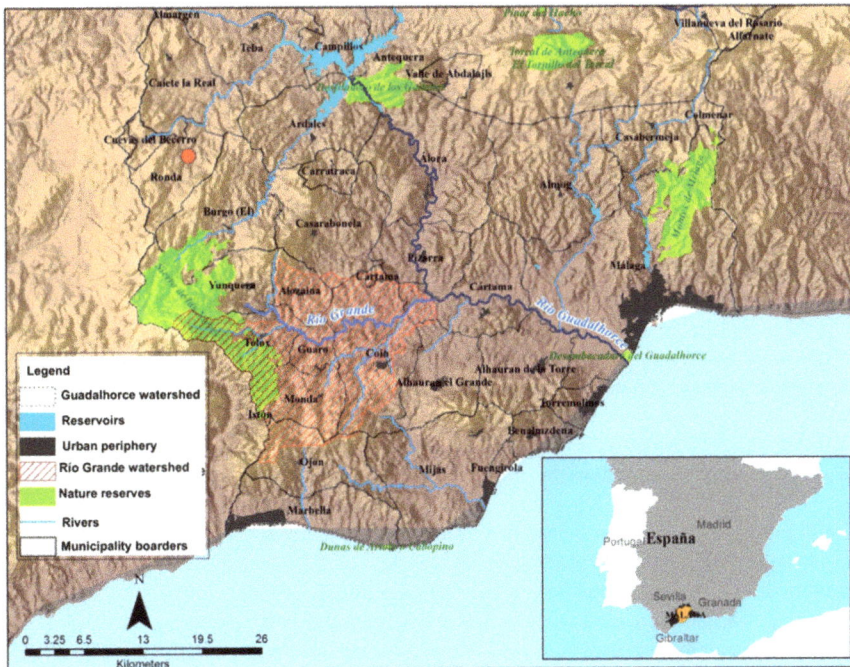

Figure 1. Guadalhorce and Río Grande Basin Location. Source: Andalucía cartographical base, 2015. Prepared by: Pacheco, 2017 [74].

As Achterhuis [46] explains, political and historical utopias always come to mind, and once 'realized', become dystopias. Spain's hydraulic utopia, too, when as all utopias do, it defined the need for a total break with the core of the old society, to build a new one. This break has justified violent interventions in and against nature and people cf. [29,47,61].

Following regenerationist ideas, in Spain, the construction of hydraulic projects was driven by modernizing that impoverished society, regenerating culture, land and the whole political-economic system. Constituting irrigation policy, by expanding dams and irrigation systems nationwide, would resolve the country's water scarcity: Its water inequities would be corrected in terms of geographical distribution, and this would also reinforce electrical generation, especially when cities began growing and agro-industrial development took off. Setting up Spain's hydraulic modernity would entail reinforcing the state's role and centralizing political power in the state, as the representative of the whole nation's general interest. For this purpose, water was nationalized. To manage effectively, Costa proposed to create 'the new man' through engineering sciences, so ideological and political alliances between regenerationists and engineers led new 20th century hydro-territorial transformations. The purpose was both to restore the country's geography and to build a new order [1,2,29].

Even when regenerationist projects failed because of the inherent political and economic conflicts between 'old' and 'new', the Franquista mission made sure to implement Costa's hydraulic dream [4]. Socio-spatial power relationships were interwoven between the military, the church, the national industrial bourgeoisie, large landowners, and the state to—among other goals—transform watersheds, which were seen as the integrating backbones of water development and territorial management. Thus, new hydrographic confederations, established in the early 20th century, would connect and integrate plans for the engineering corps' water projects. This would mean fully tapping and controlling rivers' water throughout their course; a totalizing system in which confederations were conceived as a river basin governance organizations uniting all people as a single family—laying the foundations to integrate and colonize the nation's territory after Spain's civil war. The engineers began materializing the regenerationists' water policy thinking, so each dam they built became a symbol of modernity, a concrete step forward, and above all, a homage rebuilding the weakened national identity. Engineers pursued a technical, social, and patriotic mission [29]. A key rhetorical idea was to enhance and unify the vernacular wisdom of farmers, to improve their living conditions through constant, controlled deliveries of water [2]. In reality, though, engineers' knowledge and scientific progress were dominant [30] (p. 7) and it imposed technical-scientific thinking upon rural peoples' local knowledge.

To promote hydraulic policy and thinking in Guadalhorce Valley, engineer Rafael Benjumea, a believer in Joaquín Costa's hydraulic policy, attempted to fix the problems of Málaga society using water mega-projects. One of the first projects was the 'Conde del Guadalhorce' Dam on one of this river's main tributaries. Building this dam was the first step to develop the 'Coordinated Guadalhorce Plan': It centered on capturing, storing, and regulating all the basin's water sources to foster integrated development of agriculture, the electric power industry, drinking water for Málaga city, and flood control. This plan was implemented once dictator Franco came into power (1939–1975). To close up the basin, upgrade the water regime technology, and keep a single drop from reaching the sea [14], two other major dams were built in the basin, on the Guadalhorce and Guadalteba Rivers, from 1966 to 1973. With 328 hm^3 capacity, along with the 86 hm^3 already in the 'Conde del Guadalhorce' Dam, this would provide enough water to irrigate an area of 20.000 ha. However, the priority was to supply water to Málaga, followed by generating electrical energy [2,3] (for an overview of the Guadalhorce dams and hydraulic infrastructure along the river, see Figure 2).

Building these dams entailed dramatically expropriating the affected farms and land. This was enabled by the Forcible Expropriation Law. Franco's state displaced people 'for a price it judged fair'. This Law included forming 'pueblos de colonización' ('colonization towns') and a program of agricultural indoctrination and Franquista discipline. One was the town of Peñarrubia, located at the headwaters of the Guadalhorce River. It was flooded and its inhabitants were displaced to these new towns. Uprooting and de-localizing people was a Franquista strategy to neutralize territories and make them manageable. By pulling up roots, exterminating culture, it was possible to mold a new society [3].

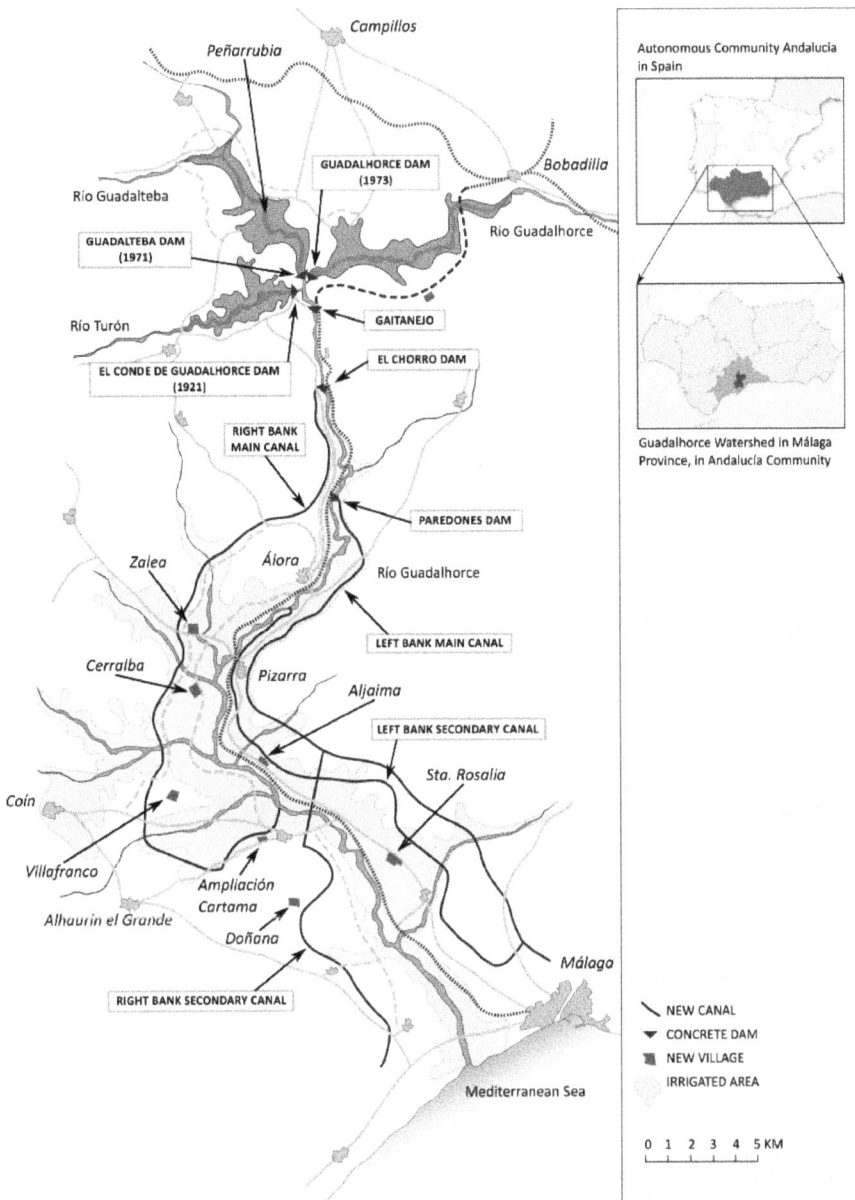

Figure 2. Guadalhorce dams and hydraulic infrastructure along the river. Source: Boelens and Post Uiterweer [2].

These 'colonization towns', located in zones without irrigation, became the buffer zones for migration by rural people whose lands were seized for water projects in Guadalhorce Valley. This hid away the memories, which stayed alive in hiding, of taking away people's towns and zones to flood them. Stories by people from Peñarrubia, for example, stress how rootlessness is an indelible feeling, closely linked with a refusal to erase the past. For this reason, they continue holding a reunion every year to commemorate the love they felt for their town [3]. The story of Guadalhorce Valley shows how building the Utopia became a violent, horrible dystopia.

At the same time, the history of the Guadalhorce Valley was the historical mirror, enabling neighbors to continue fighting against the zone's water Utopia, which continued to replicate, now building the Cerro Blanco dam in their sub-watershed, the Río Grande. This work planned to dam the only river still flowing freely to the Guadalhorce, in the mid-basin. The purpose of this dam was to provide water for the tourism industry, concentrated in the city of Málaga. The following sections describe the project's different phases and the diverse societal contestations emerging to prevent its construction.

5. The Río Grande's Conquest

5.1. Planning the Cerro Blanco Dam—Phase 1 (2001)

When the dictatorship ended and Spain entered the age of democratization, the *Política Hidráulica* technocratic model continued to reign over water governance. Since the mid-1980s, it was planned to dam the Río Grande with the Cerro Blanco Dam, already part of the Hydrological Plan for the Southern Basin (PHCS) from 1995. The dam was designed to supply water to Western Málaga city [5,75]. It was also justified to regulate Río Grande's considerable high flows during winter.

The dam on the Río Grande would be constructed in the municipalities of Guaro and Coín (see Figure 3), its reservoir capacity of 32 hm^3 would flooding 240 ha of land, foreseeing a regulation capacity of 25 hm^3/year [76]. In 2001, ACUASUR (Sociedad Estatal Aguas de las Cuencas del Sur), a company created by the government, announced 'The conquest of the Río Grande'. This project was also stipulated in the National Hydrological Plan (2001–2004).

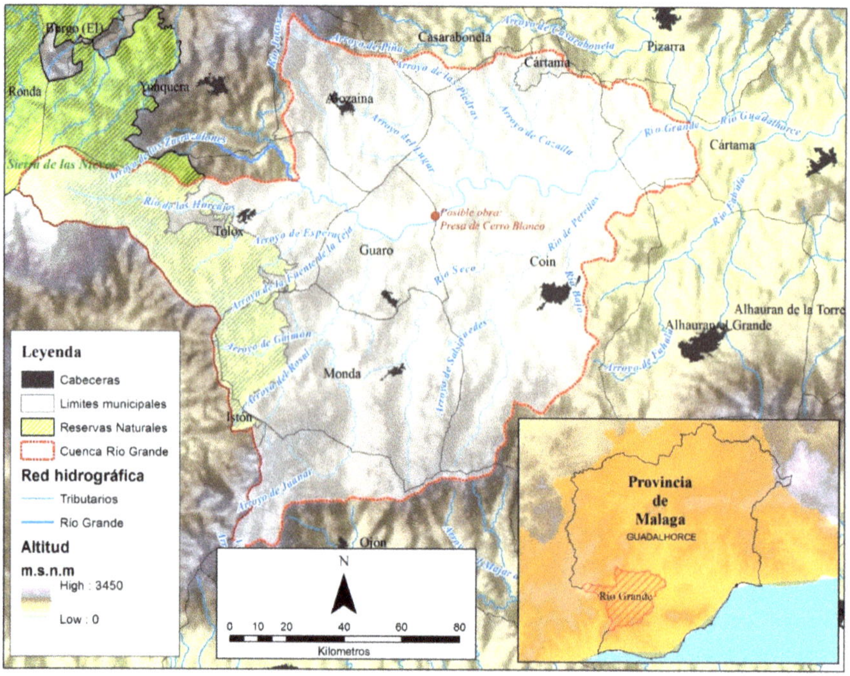

Figure 3. Río Grande basin and Dam location. Source: Andalucía cartographical base, 2015. Prepared by: Pacheco, 2017 [45].

This drove the creation of the 'Cerro Blanco Anti-dam Platform', led by farmers from Guaro, in 2001, supported by environmentalists and activists from Coín, grouped in the 'Jara Association'. The members of local offshoots of 'Ecologists in Action' (a federation of over 300 ecologist groups throughout Spain, unified in 1998) also took part, plus the academic sector, who were active members of the national 'Nueva Cultura del Agua' (NCA) movement. (The New Water Culture is a societal movement that emerged in opposition to the inter-basin transfer of the Ebro River and then proposed alternative thinking to change Spain's water management paradigms—appearing in the mid-1990s). Mobilizing the 'Cerro Blanco Anti-dam Platform' was ideologically accompanied and interconnected with the diverse demonstrations opposed to building major hydraulic works contemplated in the National Hydrological Plans of 1992 and 2001. On the national level, the platform joined with the demonstrations against water transfers from the Ebro River valley to the Mediterranean—Platform to Defend the Ebro—and with the Coordinating Body for People Affected by Large Dams and Inter-Watershed Transfers (COAGRET) [56,75,77]. "We also support demonstrations in Catalonia or other parts of the country ... People in Catalonia were surprised greatly to see us take part, because they thought that, since Andalucía would receive more water with the National Hydrological Plan, we should agree with it" (interview, local leader in Río Grande, April 2017).

These movements refused to continue bearing the consequences of deterioration in their aquatic ecosystems and social displacements derived from Franquismo-style national hydraulic policy. Their contestations aimed to protect the few 'living rivers' remaining in Spain that were threatened by the construction of 120 new dams and inter-basin transfer projects. The Cerro Blanco anti-dam platform successfully mobilized to defend the natural environment, with all its biodiversity, orchards and the Arabic-rooted traditional irrigation culture that remains alive in the Río Grande Valley. Guaro farmers' rootedness in their fields and traditional acequias, especially by the elder population, has been passed down to younger generations, and is perceived in local official entities, which supported their protests and were able to present their positions in regional political agencies and curb this dam's construction in 2001 [6,7] (interviews, Río Grande inhabitants, September 2016–September, 2017).

In this phase, we can see how vernacular values and knowledge from the valley's smallholder farmers emerge and revive, contesting hydro-techno-scientific paradigms that were imposed by modernistic hydraulic designs and utilitarian dogmas seeking 'the greatest benefit for the majority'. The Cerro Blanco anti-dam movement expresses that alternative notions of territoriality are still alive, and are very vivid and vital. They powerfully sustain mobilization against a very visible threat, and yield the support of other anti-dam national movements that fight for alternative a world and water-views.

5.2. Direct Conduction of the Río Grande's Water, Building a Diversion Weir rather than a Regular, Higher Dam: Phase 2 (2006)

Although dam construction was halted in 2001, in June 2006, a different, but similar, project was approved by the Ministry of Environment. This time, rather than announcing construction of a large dam, they referred to a diversion weir ('*azud*'), a smaller dam to reroute water by fixing a permanent water level (it would be 7 m high, to hold 8.4 hm^3 of water). This way, rather than flooding 171 hectares, water would be directly transferred to Málaga (see Figure 4). A 38-km conduction canal would be built, with a pipe of 1.60 m in diameter, with the capacity to transport 20 hm^3/year, at a flow rate of 4 m^3/s, from the river's high-water flow during winter. Members of the NCA felt the project meant conducting the whole river, because the derivation pipe's diameter could reach full operating capacity of 126 hm^3/year, more than the river's total flow, calculated to an average of 80 hm^3/year [76].

In fact, the former director of the Andalusian Water Agency (AAA, Agencia Andaluza del Agua) stated in 2010 [5], to reassure the Southern Basin engineers and technicians, that building the diversion dam was only a first step toward later building of the full dam. Likewise, the project officer in EMASA reiterated this: "I drafted the project ... what I did was not the dam, just the conduction, with the diversion dam to carry the water through a pipeline, leaving one area dry, and then be able to make the dam" (interview, project manager EMASA, March 2017).

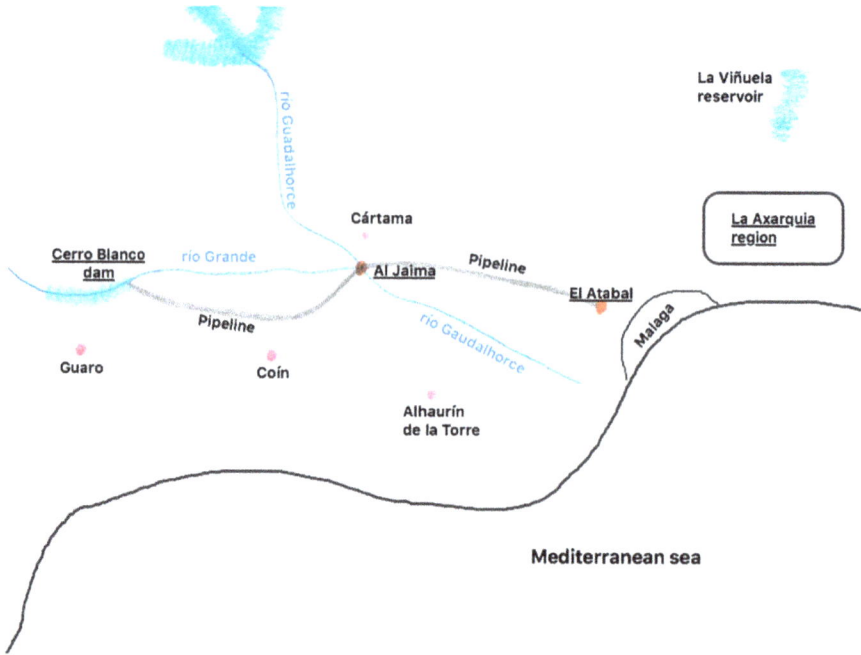

Figure 4. Hydraulic construction to transfer Río Grande water to Málaga. By Lucas du Pré [7].

That same year when the project was approved, 2006, people owning land in Río Grande valley started getting expropriation notices. About 25,000 persons would be affected by the project's implementation, and more than 2500 ha of irrigated croplands would be lost. Coín and Guaro were all living in fear and uncertainty, so they first turned to the local ecologist groups, 'The Jara Cultural Association'. They joined forces also with the Cerro Blanco Anti-dam Platform, and they all formed, in September 2006, the 'Coordinating Body to Defend the Río Grande'. They were also joined by the towns of Pizarra, Cártama, and Alora. " ... We united the people to save it, with the analogy, on our movement's logo, of a heart and the river flowing" (interview, Jara Association president in Río Grande, 22 June 2015).

The first strategy by the Coordinating Body was to connect with other local collectives, activists, and academic entities belonging to the NCA. While forming this coalition, the Coordinating Body began an information campaign for residents of Río Grande valley and neighboring citizens. They analyzed the project documentation with intellectual support from the NCA, to then disclose and diffuse it, promoting alternatives. This was their second strategy to reinforce their control over their water and defend their rights to the river: to master their overall knowledge about the project, associating it with urban-rural relations regarding water supply and demand. Accordingly, they organized the data, conducting studies and drafting legal arguments to show that the dam was not viable but there were alternatives to supply Málaga with water (such as watering gardens in summer with treated water and setting up water purification plants to reuse water). "We gave them data on the dam's environmental risks, biophysical data that would make it impossible to build. The protest had everyone well-informed. We moved around, receiving help from scholars and from people in Madrid. It was a very successful movement. The people of Málaga were not the ones asking for that water. It was other interests.

The politicians started to run out of argument" (interview, Jara Association president in Río Grande, 22 June 2015).

The Coordinating Body's third strategy consisted of maintaining political independence during their mobilization. They did not receive support from any political party, or allow parties to speak for them during the mobilization, which earned them citizens' rapport and credibility. During their mobilization, the Coordinating Body realized that the dam building was being pushed by political interests seeking European Union grant money to build projects. "The large building companies wanted to get grants, because it is not profitable to build a dam. Not even the state makes it profitable, much less the farmers" (interview, local leader in Río Grande, 22 June 2015). The Coordinating Body's independent position and increasing mobilization in 2006 and 2007 got politicians to shift their position, especially since this coincided with the municipal elections in 2007. Candidates had to commit publicly to reject the project to avoid losing votes.

The Coordinating Body's fourth strategy, with financial backing from the NCA, was to conduct a study to value the ecological status of the Río Grande's water, to demonstrate its great biodiversity, and the purity of this sub-watershed's water. The Coordinating Body was working to get this zone declared as a European Union recognized '*Lugar de Importancia Comunitaria*' (LIC) to legally forestall any new attempts to build the dam (for more details about LIC see [78]). Even so, in 2009 the Ministry of Environment asked them to pay for an independent study, endorsed by a university to lend it more weight. The Coordinating Body turned to the University of Málaga, but could not afford the price it quoted, so the Jara Association was forced to abandon the legal approach to protecting the Río Grande. "Just now, one of the few things they are afraid of is LICs, because they know Europe values them. So LICs are a good safeguard to prevent the dam" (interview, scholar Málaga Universiy, 24 April 2017).

Due to all the pressure mobilized, in May 2007 the construction of the diversion weir on the Río Grande was discarded, and in December an alternative project was announced to replace it. This 21-km conduction would connect the Aljaima diversion dam (on the Guadalhorce River, just below where the Río Grande joins it) to the desalinization plant in Málaga city (this project was completed in August 2012). This alternative had been proposed from the outset by the Coordinating Body, and studied by the Andalusian Water Agency director, Joan Coraminas, a founder of the NCA. The Jara Foundation put it this way: "We offered alternatives, for Málaga to take its water from the downstream part of the river and not the upstream part; there had already been a diversion dam there for a long time . . . its biodiversity." He explained—"That project would cause no impact on the river's course, while achieving the same aims as the large dam" (interview, former Director AAA. 23 March 2017).

In this second phase, expert engineers in alliance with the government continued insisting on transferring water from Río Grande to Málaga. As a response, the social movement showed the capacity to create a diverse social network formed by farmers, scholars, state employees, local inhabitants, youngsters, civil society organizations and platforms, and national/regional/local activists to contest the new project. This multi-stakeholder movement was able to connect different knowledge systems successfully disputing the socio-ecological feasibility of constructing a water offtake dam upstream in the Río Grande. They offered an alternative solution.

In the following section, however, we will describe the new challenges that the social movement has to face when new actors appeared, with new techno-modernist and 'green', 'inclusive' discourses that, however, keep seeking to transfer the rural waters.

5.3. The Paradoxes of Defending the Río Grande: Modernizing the Traditional Irrigation System and Silently Transferring Water (2009–2017)

The slogan '*defending the Río Grande as a living river*' generated a unified, strong mobilization to prevent construction of canals and diversion dams to transfer rural water from Coín to Málaga city. Paradoxically, this idealist slogan became the discourse that other ecologist groups, living in other zones, would use to demand changes in the traditional practices of irrigators' communities using the river, also see Reference [7].

During the summers of 2009–2010, the Río Grande community of irrigators was sued by the Environmental Prosecutor's Office. The suit was brought by the 'Association of Fish Conservation and Aquatic Systems of the South' (ACPES). They argued that building traditional diversion dams that irrigators used to catch the river's water and get it into their ditches was killing fish, harming the balance of natural systems. ACPES felt the summer drought on the Río Grande was caused by multiple river water extractions, many 'illegal' (that is, informal), and therefore the most visible ones had to be brought under control. Consequently, the river guard began ongoing surveillance during the summer in 2009–2010 to prevent irrigators' communities from deploying the practices they had always used. Irrigators felt such procedures were a legal instrument being used against them to call the attention of the Andalusian Council and de-legitimize their customary water rights.

In addition to this pressure, also motivated by the EU Water Framework Directive that aims for water saving, protection, and river restoration, recent Spanish water policy considers water as a scarce resource [7] and has introduced market laws in which water management must obey economic rules of efficient use [41,57]. Among others, this entailed regulating irrigators' communities, persuading them to officially register and renew their concessions, according to actual water availability and subject to the new demands. Not all Río Grande irrigators' organizations abided by this legislative procedure and consequently some lost their legal recognition and water rights. Simultaneously, besides aiming to minimize their existing volumetric water rights, the Department of Water Authorizations and Concessions of the Andalusian Council has restricted the granting of new concessions for irrigation in the Río Grande zone, because the water table has been judged to be lowering and in poor condition. Moreover, the Department of Hydrological Planning and the Andalusian Irrigation Plan (1986) introduced the need to adopt 'water saving' measures as strategy to mitigate the impacts of climate change and support sustainable rural development (see: "Plan Nacional de Regadíos, horizonte 2008" (Real Decreto 329/2002) and "Plan de Choque de Regadíos" (Decreto287/2006), [79]).

The main regulatory strategy to prevent water scarcity and optimize water redistribution to the other societal sectors is to promote 'irrigation modernization', especially replacing 'riego a manta' (blanket watering, traditional surface irrigation) by drip techniques [80]. "They will have to stop surface irrigation, because it squanders water! The Administration will help them with pressurized pipe systems or other modernizing, by granting subsidies" (interview, former Environment Delegate, Andalusian Council, 28 April 2017). Modernization is justified as a way to overcome 'the shortage of water in the river. In 2010–2011, the first dialogues began with and within the community of irrigators, to start modernizing their irrigation. This technology changeover began with Andalusian Council support, offering grant funding. However, a key issue was that, to receive grants, the irrigators' community would have to change its concession rights, reducing the volume habitually allocated. In fact, they dropped from 7500 m^3/ha/year to 5500 m^3/ha/year. In the end, however, bureaucratic procedures made did not get economic benefits, but their water rights were cut, anyway. Nevertheless, a group of Río Grande micro-entrepreneurs and agribusiness producers have subsequently taken the lead in modernization. They have a personal interest in technological modernization, because year-round water access will benefit their agri-business companies. The irrigation system with traditional 'acequia' canals works only in summer, but the modernized system would work all year round. With an expert in modern irrigation engineering, they organized and persuaded other members to cover the costs of modernization.

6. Current Hydrosocial Territorial Transformations

6.1. Contested Knowledges and Internal Frictions

Contemporary modernization has created new legal and epistemological authorities who have now begun to wield technical, legal and financial control, setting up new water rights arrangements for irrigators. One involves the rights to access river water. The community of irrigators has stopped building their own little diversion dams and now get their water by digging down to the water table, near the river's course. They pump water from these wells directly to the piping laid over the former ditches. Measures devices allocate each farmer a flow rate, according to the land they own. "Now we are forbidden to touch the river, and we have to dig a well to get water. The ecologists, environmental authorities, don't want us to touch the river because of its biodiversity . . . so they have changed our water concession" (interview, smallholder from Valenciana ditch, 6 April 2017).

These changes in rights to access the river also transform the hydro-ecological self-regulation used to operate traditional irrigation systems, and make new arrangements to distribute and allocate water to each irrigator. Modernization, according to Río Grande residents' local knowledge, by changing groundwater flows, will take away water from other local sources: Wells, aljibe cisterns, and aquifers. The reason is that water has always been conducted through ditches, to flood their fields, which simultaneously recharges groundwater: Feeding aquifers, wells, or aljibe cisterns for other fields downstream, and finally returning percolated and excess water to the river. Understanding this, many residents and farmers do not view surface irrigation as a 'squandering' of water. They feel that the government, experts, and agribusiness farms are actually promoting a strategic discourse to marginalize traditional systems and thereby justify drip irrigation. "Whenever we flood our fields, a few hours later all our wells are totally full. These wells are more than 20 meters deep. People call me to say their wells are dry, and after a few hours of irrigation, they have filled right up" (interview, water distributor of Guaro ditch, 10 March 2017).

Many irrigators also hesitate to join in the modernization, because this undermines their local water self-governance. Once ditches are replaced by pipes, water distribution is no longer a matter of shared work or collective decision-making. It becomes an individualized affair, controlled by a few who use technological artifacts to take over water distribution. Moreover, many feel that modernization is not a profitable alternative, because of low produce prices, versus the high start-up costs and energy costs to extract water. Most farms are small, cultivated for self-supply. This means that most cannot afford to modernize and will lose their water rights in the future. Further, some farmers fear that modernization, in the medium term, could facilitate transfer of presumably 'shared' control of the new water infrastructure with the Andalusian Council, to elites and formal rulers exclusively. Agribusiness farmers, who are leading modernization, would get control over water management. "I don't like this modernization, because we lose our ditches. They are making some installations . . . It is a political plan. The Council will never be stopped. They have decided to implement the project, and then take it over" (interview, inhabitant Río Grande, 24 March 2017).

However, many irrigators who did decide to join in the modernization (approximately 30% of the total community) said it was their only alternative to avoid forfeiting their water rights to the river, to keep farming, keep their cropland and leave it to their children in better conditions. Entering the modernization would guarantee them their right to community water, because it would show that they were adapting to environmental standards oriented toward economizing water and protecting the ecosystem, giving them a legal, administrative guarantee and therefore water security to confront the new intention to transfer Río Grande water to other places such as Málaga. Their traditional irrigation practices, building a small dam to catch water from the river and do 'surface irrigation', would no longer be accused by environmental authorities, ecologists, or fishers' associations. "I have a private well . . . but I want to be in the community to be able to pressure for that water right. Because if the well dries up . . . Who will listen to my problems? I want a common project, where we can exert more pressure when we need to complain. I want to conserve our common rights, for a sector that is respecting diversity" (interview, farmer Río Grande, 4 March 2017). "There are still many of us who want to grow some food. . . . Modernizing was the only way to continue irrigating" (interview, farmer Río Grande, 3 March 2017).

According to critical voices, modernization aims to free up water flows to transfer it to places that concentrate demand and economic power. This is another way to transfer local water to urban and business centers. By saving water and preventing the supposed 'loss of water' in the irrigation system, they will assure that water is available for the tourist industry and urban demand concentrated on the

Mediterranean coast. In the words of a Universitat Politècnica de València scholar: "In Andalucía, they keep promoting that type of infrastructure to make maximum use of available resources. The intention to improve is not for the environment, or to recover the river flow or continue applying the water framework directive . . . but in response to the old model's interests, modernistic capitalism, and put as much land as possible into production" (interview, 24 March 2017).

In fact, irrigation modernization continues to uphold the same supply management model that has developed while building large water projects. It has been turned into a subtle social strategy, reconciling the notion of 'maximum water saving' under two social positions: Protecting ecological flows, and allocating water to places where population is concentrated. However, their effects are contradictory: extraction of ground water is multiplying, escaping from collective and even official control. Further, we see in the case of the Río Grande's communities how irrigation modernization divides these communities from within, undermining local water self-governance and resulting in a new hierarchy of knowledge and values where expert know-how and standards dominate, along with their allied political and economic power groups. The next section clarifies how these consequences influence the fight against the 'common enemy', the hydraulic mega-project on the Río Grande—a fight that still carries on.

6.2. New Announcement of the Dam (2016–2017)

In late 2016, the heavy rains flooded Málaga, the city was on red alert. Consequently, officials on the Andalusian Council announced, in different newspapers, the need to revisit construction of the Cerro Blanco dam on the Río Grande.

Beyond controlling and regulating water flows to prevent flooding, there are multiple other needs, interests, and pressures. The mayors of neighboring municipalities have urbanization projects on hold because they cannot guarantee their water supply. Further, the agri-businesses in Axarquia, a province of Málaga to the East, intensively growing tropical crops, express the urgent need for this dam, because they have over-pumped their own aquifers; if the dam is not built, they warn of serious repercussions in the province's economy. They want to freely use water from another dam near them, the Viñuela dam, which at this moment, against their interests, is mostly to supply Málaga.

EMASA, the water supply company in Málaga, is another sector pressuring the Andalusian Council to implement the project. They want to access the clean, pure water from the Río Grande mid-range. This would also 'correct' the technological deficiencies (e.g., saline water) of Franquismo dams on the Guadalhorce River, which at this moment are the key water suppliers for the city. Finally, it will represent a savings in energy costs for the company; the current water diverted to Málaga must be pumped, whereas water from the Río Grande would run down by gravity.

These alarms reactivated the mobilization of leading members of the 'Coordinating Body to Defend the Río Grande'. Their messages rejecting this type of projects were broadcast on local television and direct appointments with the mayor of Coín. Their demands this time focused mainly on defending the Río Grande's territory and croplands—these farms provided livelihoods for a number of families who lost their jobs during the country's economic crisis from 2008–2014.

The Coordinating Body demands, before thinking about damming the Río Grande, that they invest in technologies to clean and recycle waste-water for each urban area in the Río Grande sub-watershed—money already granted by the EU but not yet implemented due to political negligence. Further, the social movement declared that the late 2016 flooding was not for lack of a dam, but because of irregularities in urban planning; the places flooded were wetlands and old river courses, which used to belong to the river. In addition, the Coordinating Body argues that the dam argument is false, and anyway, surplus water, flooding urban zones near Málaga, was not from the Río Grande sub-watershed, but from the Guadalhorce River itself, which is exactly the river that was totally dammed by the hydraulic dystopia and its enlightened modernistic experts' know-how.

Farmers who have joined in irrigation modernization, especially the leaders who have promoted it, avoid discussing at meetings the possibility of building the dam. This would intensify conflicts among irrigators, and they fear that many will drop out or withdraw from the modernization. For the time being, what they have agreed is that, if they join the fight against the dam, they will do so individually, but not on behalf of the irrigators' community, because they fear that the Andalusian Council will again retaliate against them, especially affecting all those families who are investing their own money to defend their water rights.

Clearly, the move toward modernizing irrigation is generating divisions within the community of farmers and debilitating the unified front to protect the Río Grande; especially breaking up collective, community water management. As drip irrigation expands, the threat is no longer only external, but also internal; now everyone is attempting to access underground water, but without any actual control over available water. This is how one of the Coordinating Body leaders expressed this: "This is a threat that is not so easy to see, and everyone who has water next to their field is filching from the river. The problem is that this is no longer a threat against us all ... when each is consuming water uncontrolledly, then we are the threat" (interview, local leader of JARA, 20 June 2016).

The Coordinating Body is aware that the dam project was discarded from the official policies and plans. However, the fight is not over yet. A local leader describes the situation: "There is nothing official, but the threat is there. We have won some battles, but we have not won the war yet" (interview, 4 May 2017).

7. Conclusions

This article has analyzed how Spain's late 19th-century hydraulic-utopian modernity project, increasingly dominated by positivist scientific-technological knowledge, has deeply colored the last century's efforts to tame and unify rivers, territories, and people, configuring new power relationships that would cram vernacular political-normative and agro-cultural diversities into a single hydraulic-administrative mega system that aligns norms, resources, practices, discourses, and human behaviors. Illustrative is the case of the Guadalhorce Valley, where regenerationist hydro-territorial utopia and disciplinary Franquist dystopia neatly entwine, breaking down traditional irrigation management practices to integrate them into a single, totalizing project. New hydrosocial connections transferred water from the Guadalhorce Valley to Málaga city, weakening and rearranging rural livelihoods, drowning headwater communities and containing populations in colonization towns. Hydraulic modernity for the Guadalhorce basin has expanded at the expense of local, vernacular realities, feeding uncontrolled tourism, and urban growth in Málaga. Consequently, water demand pressures intensify, provoking shortages, paradoxically giving rise to building new modernist water projects.

After the political regime change, in response to this crisis, alternative bodies of knowledge have emerged and/or been revitalized, now joining together to reject technocratic management of rivers, which—far from resolving the water management crisis—worsen it by monopolizing resources, truth claims and power. When construction of the Río Grande dam in the Guadalhorce's sub-basin was announced in 2001, social contestations emerged, especially from smallholder and elder residents, expressing their rootedness in their fields and traditional irrigation practices, and mobilizing vernacular knowledge and customary organizational norms and forms. They built strength by joining other mobilizations, and were supported by networks of scholars and intellectuals from different universities in Spain. Their counter-studies showed how scientific knowledge may be mobilized, both to defend powerful sectors' interests and to co-develop, hybridize and grow stronger with rural experiences and the knowledge of anti-dam movements. This way, contestations to dominant knowledge can position new stakeholders to defend their livelihoods and construct socio-environmentally fairer alternatives.

Along with the protagonism by rural people, the article has shown the key importance of associating through multi-stakeholder, multi-scalar networks. The Jara Association joined in the struggle, supporting defense of the Río Grande, making ecological knowledge of the river more visible and highlighting the importance of keeping the river alive and free of dams. Their mobilization connected heterogeneous stakeholders, facilitating extension toward diverse scales and strategically integrating diverse forms of knowledge: Grassroots and scientific know-how. Strategically studying and questioning the technical-scientific knowledge supporting dam construction has materialized their contestations in legal allegations, which—along with communicational strategies and trans-local cooperation with other social networks—managed to stop dam construction. In this way, we have shown how the dynamic knowledge contestations, embedded in changing institutional-political, socio-economic and techno-material networks and coalitions, have constantly reconfigured the hydrosocial territory of Río Grande valley.

We have demonstrated how the movement has had (and continues having) a catalytic effect to unite multiple stakeholders and bodies of knowledge, connected on different scales, when the threat is external and visible (i.e., the mega hydraulic project that extracts water for the benefit of external urban centers elsewhere). However, this becomes more complex when this endeavor is subtly disguised as an internal community project, 'bottom-up' and 'participatory' (technical-modernistic development of irrigators' communities for the 'common benefit'). Paradoxically, the ideology underpinning the social struggle in Río Grande—protecting the river, alive and free—became the strategy that official entities and environmentalist groups used to restrict and alter key societal sectors' and irrigators communities' water rights to the Río Grande. Contestations against this phenomenon divide the community and complicate the unified battle against transferring rural water to the city. Some farmers internalize the utopian discourses presented by proposals to modernize irrigation and transform their traditional practices, unconsciously or consciously, to maintain their linkages with the land. Other farmers resist modernization, using their customary rights in order not to lose local, collective water self-management.

In this context, from a political ecology approach, this article has reflected on the challenges that societal movements as in Río Grande's struggle face. Beyond opposing dam construction to defend the river, we argue that their challenges and efforts must center on standing up for water usage rights and collective water management, and thereby generate collective co-construction of knowledge, norms and practices to defend the river. Strategic, critical-conscious, publicly discussed integration of multiple types of knowledge, multiple stakeholders, and multiple scales fosters the autonomous construction of a deep-rooted hydrosocial territory, to ensure the survival and permanence of their legacy at the same time as ongoing renewal of their cultural practices.

Defending the Río Grande and its whole environment from the construction of any large hydraulic project transferring its water means getting free of that hydraulic utopia that drowned towns and channeled the Guadalhorce River. The fight for the Río Grande poses the challenge of rethinking water management in terms of less universalistic and homogenizing concepts, to yield more diversified collective water management, where economic sectors' performance matches their territorial capacities and does not compromise other rural zones' social well-being.

Author Contributions: For this article, B.D.A. (PhD research) and L.d.P. (MSc research) did literature research, research preparation, and fieldwork investigation; B.D.A., R.B. and L.d.P. have organized, conceptualized and written the article.

Funding: This research received no external funding. Bibiana Duarte-Abadía has a Colciencias PhD research fellowship.

Conflicts of Interest: The authors declare no conflict of interest.

References

1. Ortí, A. Política hidráulica y cuestión social: Orígenes, etapas y significados del regeneracionismo hidráulico de Joaquín Costa. *Agr. y Soc.* **1984**, *32*, 11–107.
2. Boelens, R.; Post Uiterweer, N.C. Hydraulic Heroes: The Ironies of Utopian Hydraulism and its Politics of Autonomy in the Guadalhorce Valley, Spain. *J. Hist. Geogr.* **2013**, *44*, 44–58. [CrossRef]
3. Duarte-Abadía, B.; Boelens, R. Colonizing Rural Waters: The Politics of Hydro-Territorial Transformation in the Guadalhorce Valley. *Water Int.* **2019**. [CrossRef]
4. Swyngedouw, E. Technonatural revolutions: The scalar politics of Franco's hydrosocial dream for Spain, 1939–1975. *Trans. Inst. Br. Geogr.* **2007**, *32*, 9–28. [CrossRef]

5. Jiménez Sánchez, M.; Poma, A. Lógicas en conflicto. Conocimiento experto y política en la movilización social en defensa de río Grande (Málaga). *Arxius de Sociología* **2011**, *25*, 59–70.
6. Poma, A.; Gravante, T. Analyzing Resistance from below: A Proposal of Analysis Based on Three Struggles against Dams in Spain and Mexico. *Cap. Nat. Soc.* **2015**, *26*, 59–76. [CrossRef]
7. Du Pré, L. Río Grande's Troubled Water: The Struggles between Rural water Users and External Actor Alliances over the Materialization of the Rural Hydrosocial Territory. Master's Thesis, Wageningen University, Wageningen, The Netherlands, August 2017.
8. Long, N. *Development Sociology: Actor Perspectives*; Routledge: London, UK, 2001; pp. 9–92.
9. Boelens, R.; Hoogesteger, J.; Swyngedouw, E.; Vos, J.; Wester, P. Hydrosocial Territories: A Political Ecology Perspective. *Water Int.* **2016**, *41*, 1–14. [CrossRef]
10. Zikos, D.; Thiel, A. Action research's potential to foster institutional change for urban water management. *Water* **2013**, *5*, 356–378. [CrossRef]
11. Boelens, R. *Water, Power and Identity. The Cultural Politics of Water in the Andes*, 1st ed.; Routledge: London, UK; New York, NY, USA, 2015; p. 359. ISBN 978-0-415-71918-6.
12. Duarte-Abadía, B.; Boelens, R.; Roa-Avendaño, T. Hydropower, encroachment and the repatterning of hydrosocial territory: The case of Hidrosogamoso in Colombia. *Hum. Organ.* **2015**, *74*, 243–254. [CrossRef]
13. Hommes, L.; Boelens, R.; Maat, H. Contested hydro-social territories and disputed water governance: Struggles and competing claims over the Ilisu Dam development in southeastern Turkey. *Geoforum* **2016**, *1*, 9–20. [CrossRef]
14. Swyngedouw, E.; Boelens, R. And not a single injustice remains—Hydro-territorial colonization and techno-political transformation in Spain. In *Water Justice*; Boelens, R., Perreault, T., Vos, J., Eds.; Cambridge University Press: Cambridge, MA, USA, 2018; pp. 115–134.
15. Zinzani, A. Development initiatives and transboundary water politics in the Talas waterscape (Kyrgyzstan-Kazakhstan): Towards the Conflicting Borderlands Hydrosocial Cycle. In *Water, Technology and the Nation-State*; Menga, F., Swyngedouw, S., Eds.; Routledge: London, UK, 2018; pp. 147–166.
16. Marks, D. Assembling the 2011 Thailand floods: Protecting farmers and inundating high-value industrial estates in a fragmented hydro-social territory. *Political Geogr.* **2019**, *68*, 66–76. [CrossRef]
17. Hoogesteger, J.; Boelens, R.; Baud, M. Territorial pluralism: Water users' multi-scalar struggles against state ordering in Ecuador's highlands. *Water Int.* **2016**, *41*, 91–106. [CrossRef]
18. French, A. Webs and Flows: Socionatural Networks and the Matter of Nature at Peru's Lake Parón. *Ann. Am. Ass. Geogr.* **2019**, *109*, 142–160. [CrossRef]
19. Baletti, B. Ordenamento Territorial: Neo-developmentalism and the struggle for territory in the lower Brazilian Amazon. *J. Peasant Stud.* **2012**, *39*, 573–598. [CrossRef]
20. Linton, J.; Budds, J. The hydro-social cycle: Defining and mobilizing a relational-dialectical approach to water. *Geoforum* **2014**, *57*, 170–180. [CrossRef]
21. Rodríguez-de-Francisco, J.C.; Boelens, R. PES hydrosocial territories: De-territorialization and re-patterning of water control arenas in the Andean highlands. *Water Int.* **2016**, *41*, 140–156. [CrossRef]
22. Del Bene, D.; Scheidel, A.; Temper, L. More dams, more violence? A global analysis on resistances and repression around conflictive dams through co-produced knowledge. *Sustain. Sci.* **2018**. [CrossRef]
23. Bebbington, A.; Bebbington, D.H.; Bury, J. Federating and defending: Water, territory and extraction in the Andes. In *Out of the Mainstream: Water Rights, Politics and Identity*; Boelens, R., Getches, D., Guevara-Gill, A., Eds.; Earthscan: London, UK; Washington, DC, USA, 2010; pp. 307–327.
24. Boelens, R. Water Rights Arenas in the Andes. Upscaling the Defense Networks to Localize Water Control. *Water Altern.* **2008**, *1*, 48–65.
25. Hoogesteger, J.; Verzijl, A. Grassroots scalar politics: Insights from peasant water struggles in the Ecuadorian and Peruvian Andes. *Geoforum* **2015**, *62*, 13–23. [CrossRef]
26. Meehan, K. Disciplining de facto development: Water theft and hydrosocial order in Tijuana. *Environ. Plan. D Soc. Space* **2013**, *31*, 319–336. [CrossRef]
27. Swyngedouw, E. The political economy and political ecology of the hydrosocial cycle. *J. Contemp. Water Res. Educ.* **2009**, *142*, 56–60. [CrossRef]
28. Hommes, L.; Boelens, R. Urbanizing rural waters: Rural-urban water transfers and the reconfiguration of hydrosocial territories in Lima. *Political Geogr.* **2017**, *57*, 71–80. [CrossRef]

29. Swyngedouw, E. *Liquid Power: Contested Hydro-Modernities in 20th Century Spain*; MIT Press: Cambridge, MA, USA, 2015; p. 320.
30. Recercat. Historia, Política y Ciencia: El Papel de los Expertos en el Debate Sobre el Agua en España. Available online: http://www.recercat.net/bitstream/2072/4783/1/Recerca+historia+aigua+Espanya.pdf (accessed on 14 September 2015).
31. Kaika, M. Dams as symbols of modernization: The urbanization of nature between geographical imagination and materiality. *Ann. Assoc. Am. Geogr.* **2006**, *96*, 276–301. [CrossRef]
32. Illich, I. *H₂O and the Waters of Forgetfulness*; University of California: Marion Boyars, CA, USA, 1985; p. 92.
33. Kaika, M. Cities of flows. In *Modernity, Nature and the City*; Routledge: London, UK, 2005; p. 200.
34. Heynen, N.; Swyngedouw, E. Urban political ecology, justice and the politics of scale. *Antipode* **2003**, *34*, 898–918.
35. Nixon, R. Unimagined communities: Developmental refugees, megadams and monumental modernity. *New Form.* **2010**, *69*, 62–80. [CrossRef]
36. Boelens, R.; Shah, E.; Bruins, B. Contested Knowledges: Large Dams and Mega-Hydraulic Development. *Water* **2019**, *11*, 416. [CrossRef]
37. Hoogendam, P.; Boelens, R. Dams and Damages. Conflicting epistemological frameworks and interests concerning "compensation" for the Misicuni project's socio-environmental impacts in Cochabamba, Bolivia. *Water* **2019**, *11*, 408. [CrossRef]
38. Hoffmann, C. From small streams to pipe dreams—The hydro-engineering of the Cyprus conflict. *Mediterr. Politics* **2018**, *23*, 265–285. [CrossRef]
39. Zikos, D.; Sorman, A.H.; Lau, M. Beyond water security: Asecuritisation and identity in Cyprus. *Int. Environ. Agreem. Politics Law Econom.* **2015**, *15*, 309–326. [CrossRef]
40. Hidalgo-Bastidas, J.P.; Boelens, R.; Isch, E. Hydroterritorial Configuration and Confrontation: The Daule-Peripa Multipurpose Hydraulic Scheme in Coastal Ecuador. *Latin Am. Res. Rev.* **2018**, *53*, 517–534. [CrossRef]
41. Fox, C.; Sneddon, C. Political Borders, Epistemological Boundaries, and Contested Knowledges: Constructing Dams and Narratives in the Mekong River Basin. *Water* **2019**, *11*, 413. [CrossRef]
42. Espeland, W.N.; Stevens, M.L. Commensuration as a social process. *Annu. Rev. Sociol.* **1998**, *24*, 313–343. [CrossRef]
43. Bacon, F. *The New Atlantis*; Kessinger Publishing Co: Montana, MT, USA, 2009; Published in 1626 and 1627.
44. Bentham, J. *The Principles of Morals and Legislation*; Prometheus Books: Amherst, NY, USA, 1988. First published in 1781.
45. Bentham, J. Panopticon; or the Inspection-house. In *The Panopticon Writings*; Bozovic, M., Ed.; Verso: London, UK, 1995; pp. 29–95, Published in 1787-1791.
46. Achterhuis, H. *De erfenis van de utopie [Utopia's Heritage]*; Ambo: Amsterdam, The Netherlands, 1998; p. 444.
47. Boelens, R. *Rivers of Scarcity. Utopian water Regimes and Flows Against the Current*; Wageningen University: Wageningen, The Netherlands, 2017.
48. Costa, J. *Política hidráulica: Misión social de los riegos en España*; Biblioteca J. Costa: Madrid, Spain, 1911; p. 353.
49. Estevan, A.E. *Herencias y Problemas de la Política Hidráulica Española*; Bakeaz: Bilbao, España, 2008; p. 163. ISBN 978-84-88949-95-0.
50. Fernández Clemente, E. *De la Utopía de Joaquín Costa a la Intervención del Estado: Un Siglo de Obras Hidráulicas en España*; Universidad de Zaragoza: Zaragoza, Spain, 2000; p. 65. ISBN 84-930255-6-9.
51. Menga, F.; Swyngedouw, S. *Water, Technology and the Nation-State*; Routledge: London, UK, 2018.
52. Gajic, T. Fronteras líquidas: Agua y bio-política de la territorialidad en España. *Arizona J. Hispan. Cultur. Stud.* **2007**, *11*, 25–41. [CrossRef]
53. Swyngedouw, E. Modernity and Hybridity: Nature, Regeneracionismo, and the Production of the Spanish Waterscape, 1890–1930. *Ann. Assoc. Am. Geogr.* **1999**, *89*, 443–465. [CrossRef]
54. Bauman, Z. *Liquid Times. Living in an Age of Uncertainty*; Polity Press: Cambridge, UK, 2007; p. 111. ISBN 978-07456-3986-4.
55. Valladares, C.; Boelens, R. Extractivism and the rights of nature: Governmentality, "convenient communities" and epistemic pacts in Ecuador. *Environ. Politics* **2017**, *26*, 1015–1034. [CrossRef]
56. Smith, N. *Uneven Development: Nature, Capital and the Production of the SPACE*; Blackwell: Oxford, UK, 1984; Volume 3.

57. Bromley, D.W. The 2016 Veblen-Commons Award Recipient: Daniel W. Bromley: Institutional Economics. *J. Econom. Issues* **2016**, *50*, 309–325.

58. Ibele, B.; Sandri, S.; Zikos, D. Endogenous versus exogenous rules in water management: An experimental cross-country comparison. *Mediter. Politics* **2017**, *22*, 504–536. [CrossRef]

59. McCully, P. Silenced Rivers. In *The Ecology and Politics of Large Dams*; Zed books: London, UK; New York, NY, USA, 2001; p. 359.

60. Boelens, R. Cultural Politics and the Hydrosocial Cycle: Water, Power and Identity in the Andean Highlands. *Geoforum* **2014**, *57*, 234–247. [CrossRef]

61. Achterhuis, H.; Boelens, R.; Zwarteveen, M. Water property relations and modern policy regimes: Neoliberal utopia and the disempowerment of collective action. In *Out of the Mainstream: Water Rights, Politics and Identity*; Boelens, R., Getches, D., Guevara, A., Eds.; Earthscan: London, UK, 2010; pp. 27–56. ISBN 184971455X.

62. Boelens, R.; Vos, J. The danger of naturalizing water policy concepts. Water productivity and efficiency discourses from field irrigation to virtual water trade. *J. Agric. Water Manag.* **2012**, *108*, 16–26. [CrossRef]

63. Espeland, W. The struggle for water. In *Politics, Rationality, and identity in the American Southwest*; University of Chicago Press: Chicago, IN, USA; London, UK, 1998; p. 281. ISBN 9780226217932.

64. Vos, J.; Boelens, R. Sustainability Standards and the Water Question. *Dev. Chang.* **2014**, *45*, 205–230. [CrossRef]

65. Illich, I. Vernacular values. *Philosiphica* **1980**, *26*, 2–32.

66. Stuiver, M.; Leeuwis, C.; van der Ploeg, J.D. The power of experience: farmers' knowledge and sustainable innovations in agriculture. In *Seeds of Transitions*; Wiskerke, H., van der Ploeg, J.D., Eds.; Royal Van Gorcum: Assen, The Netherlands, 2004; pp. 93–118.

67. Arce, A.; Long, N. The dynamics of knowledge interfaces between Mexican agricultural bureaucrats and peasants: A case study from Jalisco. *Boletín de Estudios Latinoamericanos y del Caribe* **1987**, *43*, 5–30.

68. Foucault, M. Power/knowledge. Selected interviews and other writings 1972–1978. In *Power/Knowledge: Selected Interviews and Other Writings 1972–1978*; Gordon, C., Ed.; Pantheon Books: New York, NY, USA, 1980; p. 282.

69. Bebbington, A. Movements, modernizations, and markets. In *Liberation Ecologies: Environment, Development, Social Movements*, 1st ed.; Peet, R., Watts, M., Eds.; Routledge: London, UK; New York, NY, USA, 1996; pp. 86–109. ISBN 0-203-03292-6.

70. Pfaffenberger, B. Technological dramas. *Sci. Technol. Hum. Values* **1992**, *17*, 282–312. [CrossRef]

71. Sanchis-Ibor, C.; Boelens, R.; García-Mollá, M. Collective irrigation reloaded. Re-collection and re-moralization of water management after privatization in Spain. *Geoforum* **2017**, *87*, 38–47. [CrossRef]

72. Winner, L. Upon opening the black box and finding it empty: Social constructivism and the philosophy of technology. *Sci. Technol. Hum. Values* **1993**, *18*, 362–378. [CrossRef]

73. Maurice, J.; Serrano, C. J. Costa: Crisis de la Restauración y Populismo (1875–1911); Siglo XXI Editores: Madrid, Spain, 1977; p. 245. ISBN 978-84-323-0258-9.

74. Pacheco, C. Patterns Changes of Land Use/Cover over Time in Río Grande, River Watershed—Spain: Intensification of the Irrigated Agriculture; Research Project; Institute for Biodiversity and Ecosystem Dynamics and CEDLA, University of Amsterdam: Amsterdam, The Netherlands, August 2017.

75. Gómez Moreno, M.L. El Genal apresado: ¿agua y planificación: Desarrollo sostenible o crecimiento ilimitado? 1st ed.; Bakeaz: Bilbao, Spain, 1998; p. 288. ISBN 9788488949332.

76. Puche, F. Río Grande: Cuaderno de Trabajo por la Nueva Cultura del Agua; Mesa de Amig@s de los Ríos: Málaga, Spain, 2003; ISBN 84-930029-4-1.

77. Gorostiza, S.; March, H.; Saurí, D. Piercing the Pyrenees, Connecting Catalonia to Europe: The ascendancy and dismissal of the Rhône Water Transfer Project (1994–2016). In *Water, Technology and the Nation-State*; Menga, F., Swyngedouw, S., Eds.; Routledge: London, UK, 2018; pp. 34–48.

78. Gobierno de España, Ministerio Para La Transición Ecológica. Available online: http://www.mapama. gob.es/es/biodiversidad/temas/espacios-protegidos/red-natura-2000/rn_pres_tipos_lugares_LIC.aspx (accessed on 20 January 2019).

79. Sampedro, D.; y Del Moral, L. Tres décadas de política de aguas. *Cuadernos Geográficos* **2014**, *53*, 36–67. Available online: http://hdl.handle.net/11441/43624 (accessed on 15 January 2017).

80. Genovés, J.C.; García Moyà, M.; Sanchis Ibor, C.; Vega Carrero, V.; Avellà Reus, L. Case Studies Synthesis, Spain—Institutional Framework for Local Irrigation Management in Spain: The Case of Upper Genil and Low Jucar Valleys. Available online: http://www.isiimm.agropolis.org/OSIRIS/report/Isiimm-SynthesisSpain_eng.pdf (accessed on 22 October 2018).

water

MDPI

Editorial

Reflections: Contested Epistemologies on Large Dams and Mega-Hydraulic Development

Esha Shah [1,*], Rutgerd Boelens [1,2,3,4] and Bert Bruins [1]

[1] Water Resources Management Group, Department of Environmental Sciences, Wageningen University, P.O. Box 47, 6700 AA Wageningen, The Netherlands; rutgerd.boelens@wur.nl (R.B.); bert.bruins@wur.nl (B.B.)

[2] Centre for Latin American Research and Documentation (CEDLA), University of Amsterdam, Roetersstraat 33, 1018 WB Amsterdam, The Netherlands

[3] Faculty of Agricultural Sciences, Universidad Central del Ecuador, Ciudadela Universitaria, Quito 170129, Ecuador

[4] Department of Social Sciences, Catholic University Peru, Avenida Universitaria 1801, Lima 32, Peru

[*] Correspondence: esha.shah@wur.nl; Tel.: +31-317-483904

Received: 18 February 2019; Accepted: 23 February 2019; Published: 26 February 2019

check for updates

The contributions to the Special Issue on *Contested Knowledges: Water Conflicts on Large Dams and Mega-Hydraulic Development* have looked at the politics of contested knowledge as manifested in the conceptualization, design, development, implementation and governance of large dams and mega-hydraulic infrastructure projects in various parts of the world. The contributing authors have amply demonstrated that the mega-hydraulic developments all over the world involve profound socio-technical, ecological and territorial transformations. The contributions have also abundantly shown how multiple knowledge claims are constructed using different grounds for claiming the truth about water design, development and implementation, and how both dominant and 'local', 'vernacular', or 'indigenous' knowledge frameworks underlying (or disputing) hydraulic projects and water control regimes, are not neutral nor 'independent', but culturally and politically laden and historically produced—and often, co-created. In this concluding chapter we aim to give an overview and also briefly discuss and summarize the main findings of the contributions addressing the core question: Which knowledge regimes and claims on mega-hydraulic projects are encountered, and how are they shaped, validated, negotiated and contested in concrete contexts? For that, the authors have focused also on the issue of whose knowledge counts and whose knowledge is downplayed in water development conflict situations, and how different epistemic communities and cultural-political identities (including class, ethnic, gender or professional forms of identification) have shaped the practices of design, planning and construction of dams and mega-hydraulic projects. They also scrutinized how these epistemic communities interactively shape norms, rules, beliefs and values about water problems and solutions, including notions of justice, citizenship and progress that subsequently are to become embedded in material artefacts.

The introductory article has laid out the theoretical and conceptual groundwork for examining the following issues, for instance: The notions of the dark legend of ungovernance; hydromodernity and modernizing paradigms; the depersonalization by objectifying and universalist water governance models and how they construct 'otherness' and manufacture ignorance; the issue of governmentality, power, epistemological contestations and subjugated water knowledges; the questions of constructing 'risk', commensurating values and (mis)calculating societal values; the contested reconfiguration of hydrosocial territories; the problem of reifying local and indigenous water ontologies and epistemologies; the multiple 'modes of power and response'; and multi-scalar mobilizations and the co-production of alternative knowledges. For conceptual elaborations we refer to this editorial paper [1].

Below we further discuss the findings of the contributions to this special volume in more detail. A number of articles pointed out the adverse hydrological and ecological impact of dam building and how these threaten the livelihood of especially marginal communities in and even beyond the river basin and watershed. Large dams profoundly transform local and regional hydrosocial territories, impounding water from the surrounding watersheds, their rivers, springs and aquifers, and often expropriating water resources that were previously, and are currently, used by subsistence communities, indigenous peoples, local fishermen and peasant families to satisfy their food security and livelihood needs. Frequently, such customary uses are brushed under the carpet to allow corporate profits. In this respect mega-hydraulic projects and large-scale river diversion schemes are frequently seen as representing the interests of powerful stakeholders from outside the project area, such as mega-cities and industries. To this respect, several authors have discussed how this increased understanding of the adverse consequences has not only had minimal impact on the state officials' decision-making and hydropower developers' design and construction practices, but also how these actors have increasingly turned to engineering expertise and technological assessments to further justify dam construction. Coleen Fox and Christopher Sneddon [2] indeed ask the question: Why does engineering/technological knowledge retain so much legitimacy and authority in the face of mounting knowledge about ecological change? In addition, how is engineering and technical knowledge elevated by powerful political and economic interests to serve a particular development agenda, despite the challenges that ecologists, scholars, and locally affected communities pose to these forms of epistemological knowledge?

These pertinent questions are further explored in the volume. Karen Bakker and Richard Hendriks [3] analyze the contested knowledge regimes in the regulatory review process of the Site C Project on the Peace River in north-eastern British Columbia, Canada, and argue that the regulatory approval of such projects involve what is termed as "pervasive appraisal optimism"—which entails under-estimating risk by relying on overly positive assessment of future gains and benefits while, at the same time, under-estimating and/or externalizing environmental and social-cultural costs [3]. In fact, academic research on the adverse environmental and social impact of the Site C Project was not even referenced in the government's public announcement, and the Project was pushed ahead on the ground that it was too far advanced to halt. Bakker and Hendriks refer to the Site C Project as an example of what we have discussed at length in the introduction paper as "manufactured ignorance" [1].

Barbara Lynch [4] and Amelie Huber's [5] contributions to this volume also show how systematic production of ignorance (Lynch), on the one hand, and willfully ignoring risks (Huber), on the other, were integral parts of the making of mega-hydropower knowledge paradigms. Lynch critiques heterodox and pioneering development economist Albert Hirschman's argument that the "accidental" and "benevolent ignorance"—what he called "hiding hand", a feature of the development of project processes especially in Latin America, Africa, Asia and southern Europe—makes it possible to conceal the difficulties and uncertainties inherent in such planning processes, and that such ignorance is benevolent in fostering "creativity". Hirschman argued that if the planners would know all the obstacles to a project's successful implementation they would not undertake such projects, but hiding uncertainties and difficulties would make them respond with creative solutions. Lynch discusses two projects, the San Lorenzo Dam in northern Peru (which was Hirschman's original case study) and the controversial Guatemalan Chixoy Dam, and argues that the hidden costs and suffering were not inadvertent, but came about as a result of a systematic production of ignorance. In the San Lorenzo case the planning staff deliberately ignored the potential impact of water diversion on peasants and herders, which resulted in a devastating social and cultural impact of displacement caused by the Project. In the case of Chixoy, although the state-sanctioned military violence against Maya communities in the Project area was well documented, these foreseeable impacts were also ignored by the World Bank and international contractors during the planning process. This manufactured ignorance was deliberately produced, as Lynch compellingly argues, for the planning actors to be absolved of any responsibility for the way the dam building would contribute to what was later seen as genocidal behavior by the Guatemalan state.

Based on the empirical research in the new hydropower hotspots in the Eastern Himalayan region of Northeast India, Amelie Huber [5] similarly argues in her article that a blind eye to environmental risks, which she calls strategic ignorance or manufactured production of risk, facilitates unequal distribution of benefits accelerating the process of social marginalization. She shows how experts and hydropower professionals manufacture scientific uncertainty to depoliticize and conceal the subject of risk in dam conflict. Huber further explores how influencing the production of knowledge about risk can create an opportunity to contest hazardous hydropower projects. She discusses the protracted conflict over the Lower Sabansiri Hydroelectric Project in Northeast India, and how it turned into a highly publicized controversy because civil society organizations were able to draw upon alternative knowledge sources to challenge the mainstream knowledge claims with powerful counter-claims. She shows that the successful challenge of this citizen-science alliance to institutionalized ignorance amounts to democratization of knowledge production.

Returning to the question that Fox and Sneddon ask, we think that engineering/technological narratives of mega-hydraulic projects have such legitimacy and authority because they are seductively coproduced with narratives of progress and development (see [1,2,6–8]). Tuula Teräväinen [8] analyzes multiple and contradictory expectations, socio-technical imaginaries and related knowledge regimes in the recently launched megaproject Coca Codo Sinclair in Ecuador, and shows how these imaginaries are performative in terms of creating actions, defining roles and responsibilities, and shaping political agenda. Teräväinen shows how the imaginaries of the Coca Codo Sinclair Project were deployed as a showpiece of national competence and pride and how they further nurtured the expectations that the hydropower project would ensure substantial economic benefits accompanied by enhanced energy security and self-sufficiency, climate friendliness, and local well-being—imaginaries that became seductively appealing. These dominant imaginaries, however, meet with counter-imaginaries of failed political promises, misleading information and secretive policymaking practices. In the similar vein, in another contribution to the special volume, Jeroen Warner and colleagues [9] use Lacanian psychoanalysis to describe the Grand Inga Hydroelectric Project on the River Congo as a grand fantasy rather than a reality. Going beyond the "pro" and "contra" arguments that lay behind the competing imaginaries for the Hydroelectric Project, Warner and colleagues argue that the Grand Inga as a fantasy instils agency and legitimacy to various groups working both for and against it. The idea of Grand Inga thus becomes an object; the desire for its existence or absence is "enjoyed" (a rough translation of the Lacanian concept of *jouissance*) equally by both sides proposing and opposing the dam. Warner and colleagues find the answer to the question why the idea of Grand Inga is so seductive in deeper psychoanalytical drives that produce "enjoyment" for imagined development projects among actors.

A number of papers also discuss at length the contestations to the techno-engineering knowledge regimes instigated by state, donor and private companies. Fox and Sneddon point at the key boundaries between the engineering and technological knowledge produced by consultants, state officials, hydropower companies and the knowledge produced by ecologists, engaged scientists, and local and affected communities. They discuss three contested dams in the Mekong basin in various stages of construction—the nearly complete Xayaburi Dam, the under construction Dan Sahond Dam, and the planned Pak Beng Dam—to show how contestations over the dams' shifting epistemological boundaries in meaningful ways create new spaces for knowledge production and transfer [2]. These new spaces, however, do not only imply new forms of relating to risks and uncertainties and hence new forms of knowledge production, but also new ways to relate to social context. To this respect, Bakker and Hendriks [3] show how in the regulatory decision-making process multiple contestations over knowledge production arise between opponents and proponents and how these contestations involve differing social values.

The contributions to the special volume have also shown how these contestations involve multi-scalar, multi-actor networks. Paul Hoogendam and Rutgerd Boelens [10] discuss the case of the Misicuni Multipurpose Hydraulic Project in Bolivia to show how the political processes and demands for fair compensation for the affected communities involved divergent knowledge systems.

The confrontations on these knowledge frameworks were embedded in wider struggles over territorial control and natural resource governance while they were simultaneously characterized by highly unequal economic, political, and discursive power relationships. This unequal epistemological arena defined what counted as compensation, and how it should be counted. The affected indigenous communities remained invisible in the compensation process while their demands were disregarded. Despite the relatively progressive nature of the Misicuni Project, described as an example of "vernacular modernism" by Bolivia's pro-rural-community and popular government, the decisions regarding compensation were taken top-down, valuing expert understanding only. During the whole process of negotiations, the affected communities were forced to prove negative consequences and accept the suggested "appropriate framework for compensation". Hoogendam and Boelens show that the issue of "compensation" is politically contested and fiercely fought, in particular because the issue of "commensuration of incommensurables" is at the core of the epistemological and material conflict. This raises many fundamental questions: Who has the authority and legitimacy to define the standards? What is the "common metric" to "measure" the value and meaning of social, material and cultural assets and socio-environmental relationships, and how is this decided? Clearly, different from the World Commission on Dams (WCD) principles, the harsh reality of affected communities shows that the issue of compensation is not a matter of shared and objective decision-making but a hard wrought contestation over meaning, values and worldviews.

The local epistemological alliances that challenge dominant knowledge regimes are, however, not without contradictions. Rinchu Dukpa, Deepa Joshi and Rutgerd Boelens [11] examine how in India's Eastern Himalayan state of Sikkim, indigenous local communities have successfully contested all proposed hydropower projects and sustained anti-dam opposition in their home region. Based on a detailed ethnographic exploration of such oppositional movement, they argue that the traditional system of self-governance—"vernacular statecraft"—known as Dzumsa, prevalent among indigenous Bhutia communities, played a central role. This form of self-organization mobilized people's attachment to their place and the corresponding notions of territoriality, in order to forge "agonistic unity" against large dams. The system of Dzumsa is often eulogized as an egalitarian and democratic institution, but Dukpa and colleagues argue that in its structure and operation the system is rather hierarchical, masculine and exclusionary. The authors here make a novel point stating that the successful resistance to dominant forms of knowledge regimes contributing to democratization of knowledge traditions may come from highly undemocratic forms of collectivity. A similar issue is raised by Juan Pablo Hidalgo-Bastidas and Rutgerd Boelens [7] when discussing the processes of organization and contestation led by the people of Patricia Pilar and neighboring communities against the construction of multipurpose Baba Dam in coastal Ecuador. These protracted local protests had significant impact on the development and designs of the Baba projects—the dam site was changed and height was reduced which, in consequence, reduced the submerged area significantly, thus benefitting the local communities. However, such a far-reaching decision to implement an alternative dam design was not based only on the local protests, but was also informed by the rent-seeking relations between government officials and the construction company. In addition, the impact of changing the dam's technological designs was multiple and ambivalent, the authors argue, because the negative impact was now displaced on the most vulnerable members of the social movement—the Afro-descendant community. Both these articles raise questions regarding the character and eventual impact of the protest movements against large dams.

The contributions to the special volume also engaged with the questions: How is citizen, vernacular or lay-expert knowledge deployed to produce alternatives for mega-hydraulic projects and/or strengthen anti-dam opposition? How is such oppositional action organized? But also, how do anti-dam alliances confront internal contradictions? Bibiana Duarte-Abadia, Rutgerd Boelens, and Lucas Du Pré [6] analyze how in late 19th early 20th century Spain "hydraulic utopianism," dominated by positivist scientific-technological knowledge, framed the way rivers, territories and people were to be controlled, and how this was key to reconfiguring most of Spanish river basins, as in the case of the Guadalhorce basin in Malaga. They explain how this triggered resistance in the neighboring Rio Grande valley, whose river was equally threatened to become dammed, dominated and diverted by a powerful coalition of engineering experts, politicians, bureaucrats and capitalist firms. Duarte-Abadía and colleagues explore how alternative bodies of knowledge have emerged and revitalized, now joining together to reject the technocratic management of the hydrosocial territory. These social contestations first emerged from smallholder peasants and local residents whose ideas about the river regime were rooted in longstanding agricultural and irrigation practices, and who mobilized vernacular grassroots knowledge and customary organizational forms and norms to defend their livelihoods. Their efforts were later joined by engaged scholars, ecologists, and NGOs from other parts of Spain who mobilized alternative forms of scientific know-how. This multi-actor, multi-scalar alliance between the vernacular and scientific forms of knowledge combined with legal action and trans-local communication, successfully stopped dam construction. However, while the struggle against the common threat of mega-hydraulic intervention in their territory enabled a strong opposition alliance that united diverse bodies of knowledge, the new policies to modernize traditional irrigation systems and to 'save water' subtly constitute a fundamental threat to these grassroots struggles and their knowledge.

In the end, we want to highlight that the contributions to the special volume have shown a diversity of knowledge contestation and co-production strategies (see [1]). Here we especially want to present following inferences on how power is deployed, manifested and contested in mega-hydraulic development projects. Firstly, we think that there has been a clash of expertise in which the dominant 'visible' power deployed through formal rules and hierarchical expert institutions has been characteristically challenged by the forms of counter-expertise emerging from the same systemic context that aim to delegitimize the hegemonic knowledge claims. Secondly, dominant 'hidden' power that is manifested through overstating positive results and underplaying negative socio-environmental impact—for instance, in terms of "manufactured ignorance" or "pervasive appraisal optimism"—is typically contested by the marginalized actors who have been empowered by means of producing alternative knowledge regimes based on grassroots epistemologies and ontologies brought to the arena of contestation. We find considerable evidences in the contributions to this special volume that the dominant 'manipulative' power that controls the production of mega-hydraulic knowledge by constructing favorable narratives is increasingly being contested by means of oppositional strategies that have raised questions about not just the mega-hydraulic projects but have also altered the understanding on wider processes of water knowledge production itself. Thirdly, we also think that the dominant 'normalizing' (Foucauldian disciplining) power that links knowledge, power and truth to unconsciously shape the legitimacy of mega-hydraulic order through processes of subjectification is increasingly responded by oppositional and advocacy alliances that have not only questioned the normality of mega-hydraulic development, but also have extended their critique to the related discourses of modernity, progress, and development.

In sum, to highlight, all contributions to this special volume have shown how the production of knowledge of mega-hydraulic structures and their impacts on local water cultures and societies constitute fierce contestations over interests, values and worldviews of diverse actors and divergent hydro-territorial objectives and projects. From these contestations hybridization of knowledge takes place, grounded in multiple social realities. These contributions have challenged the myth that knowledge on dam-development is a rational buildup of facts, or a coherently ordered water governance reality. They have further shown how, as part of epistemological contestations, multi-actor, multi-scalar alliances are formed that seek to challenge dominant forms of power, mega-hydraulic-territorial objects, hydro-political institutions and their claims to truth, and how they in turn have shaped new hydro-territorial subjects and realities. This way, alternative knowledge is co-produced, critically entwining positivist engineering, activist, grassroots and other knowledge systems, providing an important platform for those who suffer from the impacts of large dams and mega-hydraulic development. These challenges to the mega-hydraulic rationality have re-politicized large dam regimes, offering new inspiration for the democratization of the sector.

References

1. Boelens, R.; Shah, E.; Bruins, B. Contested Knowledges: Large Dams and Mega-Hydraulic Development. *Water* **2019**, *11*, 416. [CrossRef]
2. Fox, C.; Sneddon, C. Political Borders, Epistemological Boundaries, and Contested Knowledges: Constructing Dams and Narratives in the Mekong River Basin. *Water* **2019**, *11*, 413. [CrossRef]
3. Bakker, K.; Hendriks, R. Contested Knowledges in Hydroelectric Project Assessment: The Case of Canada's Site C Project. *Water* **2019**, *11*, 406. [CrossRef]
4. Lynch, B. What Hirschman's hiding hand hid in San Lorenzo and Chixoy. *Water* **2019**, *11*, 415. [CrossRef]
5. Huber, A. Hydropower in the Himalayan Hazardscape: Strategic Ignorance and the Production of Unequal Risk. *Water* **2019**, *11*, 414. [CrossRef]
6. Duarte-Abadía, B.; Boelens, R.; Du, P.L. Mobilizing water actors and bodies of knowledge. The multi-scalar movement against the Río Grande Dam in Málaga, Spain. *Water* **2019**, *11*, 410. [CrossRef]
7. Hidalgo-Bastidas, J.P.; Boelens, R. Hydraulic order and the politics of the governed: The Baba Dam in coastal Ecuador. *Water* **2019**, *11*, 409. [CrossRef]
8. Teräväinen, T. Negotiating water and technology—Competing expectations and confronting knowledges in the case of the Coca Codo Sinclair in Ecuador. *Water* **2019**, *11*, 411. [CrossRef]
9. Warner, J.; Jomantas, S.; Jones, E.; Ansari, M.S.; de Vries, L. The Fantasy of the Grand Inga Hydroelectric Project on the River Congo. *Water* **2019**, *11*, 407. [CrossRef]
10. Hoogendam, P.; Boelens, R. Dams and Damages. Conflicting epistemological frameworks and interests concerning "compensation" for the Misicuni project's socio-environmental impacts in Cochabamba, Bolivia. *Water* **2019**, *11*, 408. [CrossRef]
11. Dukpa, R.; Joshi, D.; Boelens, R. Contesting Hydropower dams in the Eastern Himalaya: The Cultural Politics of Identity, Territory and Self-Governance Institutions in Sikkim, India. *Water* **2019**, *11*, 412. [CrossRef]

MDPI
St. Alban-Anlage 66
4052 Basel
Switzerland
Tel. +41 61 683 77 34
Fax +41 61 302 89 18
www.mdpi.com

Water Editorial Office
E-mail: water@mdpi.com
www.mdpi.com/journal/water

www.ingramcontent.com/pod-product-compliance
Lightning Source LLC
Chambersburg PA
CBHW051313020426
42333CB00028B/3323